D1124291

Process Optimization

PROCESS OPTIMIZATION

WITH APPLICATIONS IN METALLURGY
AND CHEMICAL ENGINEERING

W. HARMON RAY and JULIAN SZEKELY

Department of Chemical Engineering and
Center for Process Metallurgy
State University of New York at Buffalo

A WILEY-INTERSCIENCE PUBLICATION

JOHN WILEY & SONS, New York · London · Sydney · Toronto

Library of Congress Cataloging in Publication Data:

Ray, Willis Harmon, 1940–
Process optimization.

"A Wiley-Interscience publication."
Includes bibliographical references.
1. Process control. 2. Mathematical optimization.
3. Metallurgy. 4. Chemical processes. I. Szekely, Julian, 1934– joint author. II. Title.

TS156.8.R39 658.5′3 73-936

ISBN 0-471-71070-9

Printed in the United States of America

10 9 8 7 6 5 4 3 2 1

Preface

During the last decade considerable advances have been made in the development of mathematical and computational techniques for process optimization. The application of these techniques to many chemical and petroleum engineering operations has met with considerable success, and process optimization is becoming a standard tool in chemical and petroleum engineering practice. However, in the current textbook literature the emphasis appears to be on analysis and theoretical developments, and rather less attention is paid to the actual practical application of these methodologies to real, complex systems.

It is perhaps a truism that the success of any optimization procedure must depend critically on the accuracy of the process models used in the computational scheme. For this reason, we believe that there is a real need, from both the practical and the pedagogical viewpoints, to present the methodology of optimization so that it can form a ready interface with the disciplines devoted to the modeling of processes. The presentation of the material in this context has been one of our principal motivations in writing this textbook.

In contrast to the widespread application of optimization in chemical and petroleum engineering, little work has been done on the optimization of metallurgical operations; in fact, the accomplishments in this area are largely restricted to the routine application of linear programming to inventory control, scheduling, and the like.

This state of affairs results from the fact that metallurgical engineers are not familiar with the concepts and techniques of process optimization.

Although many large metals processing corporations do have operations research groups, the real problems are often not communicated to them in a form suitable for process optimization studies. Thus there seems to be a real need to provide the "process man" with an understanding of the concepts, techniques, and jargon of optimization so that this crucial communication link is bridged. If these lines of communication can be established, there is little doubt regarding the vast potential for process improvement through optimization in the metallurgical industry.

The purpose of this monograph is to introduce the reader to the theory and practice of process optimization. In the selection of the material, particular emphasis has been placed on the practical application of the various techniques to the optimization of "real" rather than "synthetic" systems. In order to emphasize this point, many substantial worked examples are introduced in the latter part of the book (from Chapter 4 onward), and the final chapter is devoted in its entirety to the discussion of practical problems. In addition, frequent references are made to standard optimization programs which are available in the computer libraries.

This book is intended for chemical, petroleum, or metallurgical engineers, and could form a text suitable for a one semester senior or first year graduate level course. Alternatively, the material may also be helpful to practicing engineers faced with optimization problems.

From a fundamental viewpoint, the techniques for process optimization must be the same, irrespective of the field of application. However, practical considerations may dictate quite different approaches, depending on one's understanding of the process, the inherent complexity of the situation, and finally in the sophistication of the technical personnel. In these respects, chemical and metallurgical operations may differ appreciably, and proper cognizance has been given to this fact within the text, wherever possible.

The major difference between chemical and metallurgical processing operations is that in general, the former are very much better understood, and the availability or ready development of suitable process models may be taken for granted in the majority of cases. In contrast, we are only beginning to attain a quantitative understanding of many metallurgical operations. Therefore, in metallurgical systems, practical considerations may force us either to settle for less sophisticated process models or to exert a considerable amount of model-building effort before any optimization proper may commence.

This simultaneous coverage of chemical and metallurgical systems is thought to provide a major pedagogical advantage in stressing the generality of these optimization techniques while emphasizing the practical problems of process description that have to be taken into consideration and which could vary from system to system.

In concluding these remarks it is a pleasure to acknowledge the help and encouragement we received from many colleagues in the preparation of this manuscript. In particular we would like to thank Professors Robert W. Bartlett, John F. Elliot, Norman Parlee, and H. W. Smith who encouraged us in this undertaking; we also thank Professor J. G. Vermeychuk, Dr. J. Downing, Mr. Gordon Lilley and Mr. Robert W. Bouman, who reviewed the manuscript. Professor Vermeychuk's detailed and penetrating comments were particularly helpful. We owe a special debt to Professor N. R. Amundson for his timely and helpful suggestions.

We would also like to express our appreciation to the many students and practicing engineers who tested this material in the classroom over the past six years. In particular, we acknowledge the efforts of Dr. Y. K. Chuang, Mr. S-D Fang, Mr. M. A. Ajinkya, and Dr. C. Patel who worked out many of the examples and homework problems.

Heartfelt thanks are due to Mrs. Lucille Delmar for her patient and very competent typing (and retyping) of the manuscript and to Nell Ray who typed and proofread several versions of the text. Finally, we wish to thank our families for their tolerance and understanding during the time this manuscript was prepared.

W. HARMON RAY
JULIAN SZEKELY

Buffalo, New York
January 1973

Contents

1 Introduction

The objective of this text is to introduce the reader to the theory and practice of optimization. The emphasis is on problem solving, and for this reason special attention is given to the discussion of efficient computational techniques and to the principles on which these techniques are based. To minimize the mathematical background required, we present the theory in a somewhat nonrigorous manner but with sufficient detail to justify the computational techniques to interested readers. Although the text is addressed primarily to metallurgical and chemical engineers and most of the examples are drawn from these areas, the basic material is appropriate to other engineering fields.

As necessary background, it is expected that the reader is familar with matrix algebra and has some experience with differential equations. Some of the basic concepts of matrix algebra, which are frequently used in the text, are summarized in Appendix A.

1.1 WHAT IS OPTIMIZATION?

Broadly speaking, optimization means to select the *best course of action from the available alternatives*. We note that this is a rather broad definition, with which no one would quarrel, although considerable practical difficulties may arise when we attempt to define our terms "best" and "available alternatives." In particular, the definition of the "*available* alternatives" could pose problems, since the wording implies that our choice may not be entirely free or, in other words, that we may be *constrained* in our choice.

Before plunging into these details, let us recall a very familiar optimization problem, the establishment of the optimum thickness of insulation. Let us consider a processing unit, a furnace or the reboiler of a distillation column, which contains materials at a temperature higher than that of the environment. This temperature difference will cause heat losses from the system, and our objective is to reduce these losses in an optimal fashion.

Clearly, the heat losses can be reduced by putting thermal insulators onto the walls of the unit; the thicker the insulating layer, the lower the heat losses, but at the same time there will be an increase *in the capital charges* which must be placed against the cost of insulation. In the absence of insulation the cost of the heat losses may be high, whereas if we were to apply a very thick insulating layer, the capital charges would predominate.

Intuition suggests that there exists an intermediate value of the thickness of the insultation, where the sum (cost of heat losses + capital charges on insulation) shows a minimum. This minimum in the cost would correspond to the optimal thickness of insulation.

This familiar problem is readily solved by plotting both the cost of heat losses and the capital charges against the thickness of the insulation, as shown in Fig. 1.1. We observe that for our example this optimum thickness of insulation (minimizing the total cost) was about $\frac{1}{2}$ ft.

We note that in the form stated, this was an *unconstrained* optimization problem, because we were at liberty to choose any value for the thickness of the insulation.

In practice, constraints may be introduced by specifying a given maximum outside temperature, which may force us to use a thicker layer of insulation than the optimum shown in Fig. 1.1; alternatively, restrictions of space may force us to specify an insulating layer that is thinner than the optimum.

Let us now proceed with a slightly more involved example, and assume that in the operation of an autoclave in a hydrometallurgical process or some other chemical operation we wish to maximize profits F as well as to minimize the amount of water pollution A. Then the objectives, which must be put into mathematical form in order to use quantitative optimization techniques, can be expressed as an *objective function* or performance index, $I(F, A)$. This objective function could take the following form

$$I(F, A) = F - \lambda A \tag{1.1.1}$$

where λ is a penalty for water pollution, which depends on many intangibles

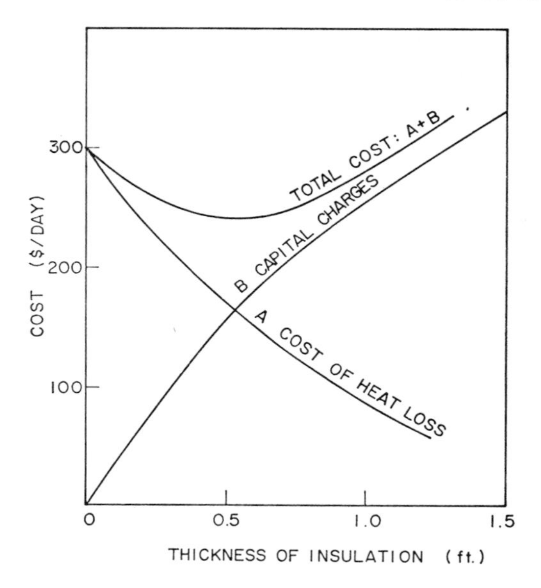

Fig. 1.1 Plot of A: cost of heat loss, B: capital charges, and total cost versus the thickness of insulation.

like government pressure, the degree of civic responsibility of the company, and so on.

If λ were a government tax on pollutant discharge, then the function $I(F, A)$ would designate the exact value of the net profit. However, if λ is not such a precise economic parameter, then $I(F, A)$ would have more nebulous meaning. It follows that $I(F, A)$ defines what is meant by "best" and thus must be chosen with care. The selection of the profit function is an art in itself and often can only be done by trial and error.

A second difficulty arises when we must decide what constraints should be put on the problem. Care must be taken to include all the important constraints, such as limiting values of the variables as set by availability or safe limits of operation, and by the need to observe the applicable physical or chemical laws, such as the conservation of matter, temperature dependence of reaction rate constant, and the like.

On the other hand, in selecting the constraints we should not be unduly influenced by existing operation practice, as frequently the improvement of this practice is our objective.

Returning to our autoclave, let us assume that both profits and pollution depend only on the temperature and pressure of the reactor, so that

$$I(F, A) = F(T, P) - \lambda A(T, P) \tag{1.1.2}$$

becomes $I(T, P)$, which is plotted in Fig. 1.2 (we note that the contours shown represent lines of constant I). If the problem is only constrained to lie within the bounds of the graph, we see that point A (160 psig, 140°C) is the optimum with $I = 10.3$. Suppose that, because of the danger of explosion, only the area satisfying the *inequality constraint*

$$g_1(P, T) = T + 0.4P - 180 \leq 0 \tag{1.1.3}$$

(shown by area OBC in Fig. 1.2) is safe for operation. Thus point D (120 psig, 130°C) is the new constrained optimum, which has a value of $I = 9$,

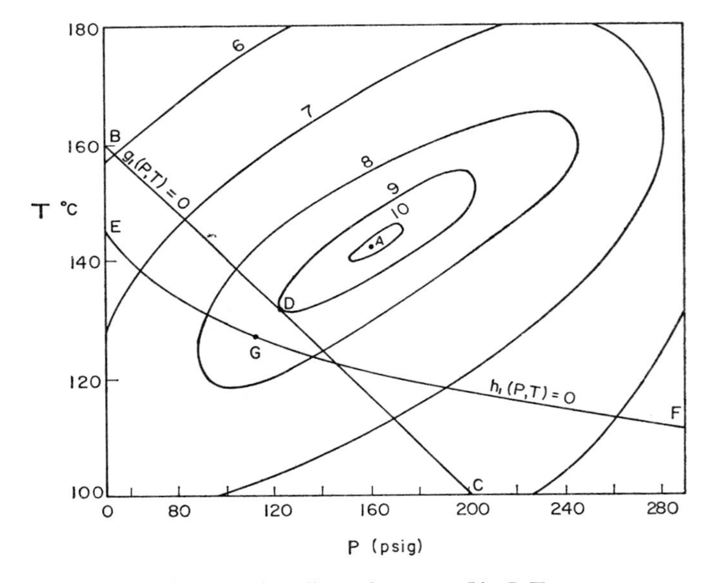

Fig. 1.2 Autoclave Optimization: lines of constant I in P-T space.

so that the explosion hazard reduces profits. Then, let us consider that a plant downstream of the autoclave, which uses the effluent as a feedstock, requires that conversion be kept at 80%. This produces the equality constraint

$$h_1(P, T) = X(P, T) - 0.8 = 0 \tag{1.1.4}$$

shown by curve EF in Fig. 1.2. We see that this constraint gives a minimum at G (115 psig, 125°C) with $I = 8.5$. Thus the conversion constraint also reduces the profits.* So, in general, one would wish to constrain the problem as little as possible.†

Another view of an optimum is that of the separation between what is possible and what is impossible. For example, in our unconstrained autoclave optimization problem, if we consider the pressure P to be a fixed design parameter, and that optimization is done only with respect to temperature, then the optimum is the dividing line between what is possible in profits and what is impossible. This is shown in Fig. 1.3.

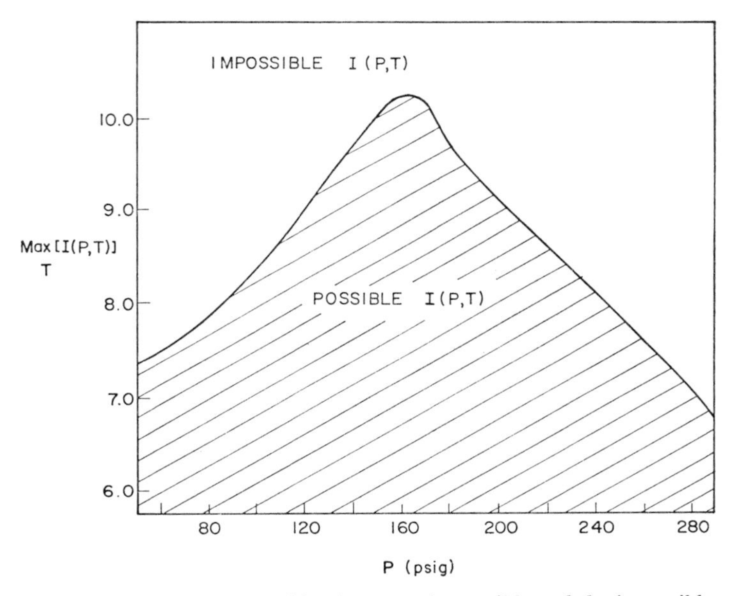

Fig. 1.3 The optimum as a partition between the possible and the impossible.

It is of interest to note that a properly posed optimization problem gives us the often sought condition of getting something for nothing. For example, in our autoclave problem we are able to increase our profits substantially over nonoptimal operation by changing the operating policy. This is possible because our objective function specifies very precisely what we mean by an optimum, and thus we are able to explore the alternative operating policies to find the one which best satisfies our objectives.

* If the last constraint could be removed to allow operation at the higher conversion of point D, the profits would increase in our plant. However the profits in the downstream plant may decrease. This illustrates a difficulty in process optimization (which is discussed in detail later)—the problem of isolating a process from related processes for optimization.

† Note that the conversion constraint makes the explosion constraint unnecessary because point G does not fall on $g_1(P, T) = 0$.

1.2 MAXIMUM OF A FUNCTION

Broadly speaking, optimization is concerned with finding the maximum (or minimum)* of a function. Usually the problem has two main parts.

The Objective Function

We express our objectives in a quantitative form through the formulation of an objective function $I(x_1, x_2, \ldots, x_n)$, which is to be maximized

$$\underset{x_i}{\text{Max}}\, I(x_1, x_2, \ldots, x_n) \qquad i = 1, 2, \ldots, n \tag{1.2.1}$$

Although some recent work has been done to treat problems with more than one objective function, we shall confine our treatment to a single objective function. Here $I(x_1, x_2, \ldots, x_n)$ is a shorthand notation, which will be used extensively in subsequent sections of the text, to denote some functional relationship between the adjustable parameters $x_1, x_2, \ldots, x_n$ and the objective I.

Thus, for example, we may specify our objective function as the cost per unit mass of product in a melting shop, say $\$_{MP}$, which we would of course, wish to minimize rather than to maximize.

For a given system we may say that $\$_{MP}$ is given as

$$\$_{MP} = [C_1V^m + C_2P^n + C_3L^p + \cdots] \tag{1.2.2}$$

where V is the volume of the arc furnace, P is the cost of power, L is the labor cost/hr and C_1, C_2, C_3, m, n, and p are constants. Here the quantity in the square brackets would correspond to $I(x, x_2, \ldots, x_n)$. We stress that Eq. 1.2.2 need not be a simple sum of the individual functions, but may represent a more complex relationship, for example,

$$\$_{MP} = [C_1V^m(1 + C_2P^{n-1})/L + C_3L^p] \tag{1.2.3}$$

Constraints

There are several types of constraints which are normally put on the problem.

(i) Algebraic equality constraints of the type:

$$h_i(x_1, x_2, \ldots, x_n) = 0 \qquad i = 1, 2, \ldots, m_1 \tag{1.2.4}$$

* It is appropriate to point out that a maximization problem

$$\underset{x_i}{\text{Max}}\, I(x_1, x_2, \ldots, x_n)$$

is entirely equivalent to a minimization problem if the sign of the objective function is reversed; that is,

$$\underset{x_i}{\text{Max}}\, I(x_1, x_2, \ldots, x_n) = \underset{x_i}{\text{Min}}\, [-I(x_1, x_2, \ldots, x_n)]$$

The *constraints* represented by Eq. 1.2.4 express the fact that there exist one or more functional relationships connecting the variables $x_1, x_2, \ldots, x_n$. Clearly, each one of these relationships reduces the number of parameters that may be varied independently.

As an example of constraints, in an autoclave of fixed volume and isolated from its surroundings, the pressure and temperature are uniquely related through the equation of state; thus one could vary only one of them independently. Another typical equality constraint one would encounter in slag-metal systems, is that at equilibrium, the composition of the metal phase would be uniquely related to the composition of the slag phase, and so on.

In the example concerned with the operation of an autoclave, Eq. 1.1.4 specifying a fixed level of conversion was another typical equality constraint that one could encounter in optimization problems.

(ii) Algebraic inequality constraints of the type:

$$g_j(x_1, x_2, \ldots, x_n) \leq 0 \qquad j = 1, 2, \ldots, m_2 \tag{1.2.5}$$

The algebraic inequality constraints usually specify the practical operating limits of certain variables within a system. The explosion limit in our example was a typical case of an inequality constraint; other examples may be to assign limits to the ore composition, the temperature, pressure, and perhaps the size of the system.

(iii) Differential or integral equation equality constraints which may have the form:*

$$dx_i/dt = f_i(x_1, x_2, \ldots, x_n) \qquad i = 1, 2, \ldots, m_3 \tag{1.2.6}$$

$$x_i(0) = x_{i0} \qquad t_0 < t \leq t_f \tag{1.2.7}$$

These types of constraints could arise in batch processing problems such as batch distillation, the batch processing of polymers, dyestuffs, the operation of the basic oxygen furnace, and the fuming of zinc where the rate of disappearance or the rate of generation of a component is a function of the concentration, temperature, pressure, and the like. For example,

$$\frac{dc_i}{dt} = f(c_i, T, P, t) \tag{1.2.8}$$

with

$$c_i = c_{i0}, \qquad \text{at} \quad t = 0 \tag{1.2.9}$$

as the initial condition.

Similar equations may arise in optimization problems concerning heat

* Any other form of differential or integral equation constraints is possible; however, Eq. 1.2.6 is a type very commonly found.

conduction problems, for example, the optimal scheduling of a slab reheating furnace, where the constraint would take the form of a partial differential equation relating the temperature and its derivatives to the other parameters in the system.

Since any optimum found must satisfy the constraints, we shall confine our search to the *feasible region*, that is, to the collection of points which satisfies all the constraints. Any point within this *feasible region* is thus termed a *feasible point*.

Traditionally, these maximization problems, detailed under (i) to (iii), have been divided into two main classes.

Mathematical Programming Problems

These are maximization problems which have no constraints or only type (i) and (ii) constraints. The general problem is usually stated as

$$\max_{x_i} I(x_1, x_2, \ldots, x_n) \qquad i = 1, 2, \ldots, n \tag{1.2.1}$$

subject to

$$h_k(x_1, x_2, \ldots, x_n) = 0 \qquad k = 1, 2, \ldots, m_1 \tag{1.2.4}$$

$$g_j(x_1, x_2, \ldots, x_n) \leq 0 \qquad j = 1, 2, \ldots, m_2 \tag{1.2.5}$$

The autoclave optimization problem of the previous section can be formulated in this form if the variables chosen are invariant with time. The variables are sometimes partitioned into *state*, and *control* or *decision* variables. For example, if we solve Eq. 1.2.4 for the first m_1 variables, $x_1, x_2, \ldots, x_{m_1}$ in terms of the last $n - m_1$ variables

$$x_i = F_i(x_{m_1+1}, x_{m_1+2}, \ldots, x_n) \qquad i = 1, 2, \ldots, m_1 \tag{1.2.10}$$

then we could denote the state variables, by the m_1 vector $\mathbf{y}$

$$\mathbf{y} = \begin{bmatrix} y_1 \\ y_2 \\ \cdot \\ \cdot \\ \cdot \\ \cdot \\ y_{m_1} \end{bmatrix} = \begin{bmatrix} x_1 \\ x_2 \\ \cdot \\ \cdot \\ \cdot \\ \cdot \\ x_{m_1} \end{bmatrix}$$

because these are dependent, and the remaining free variables, by the $n - m$ vector $\mathbf{u}$

$$\mathbf{u} = \begin{bmatrix} u_1 \\ u_2 \\ \cdot \\ \cdot \\ \cdot \\ \cdot \\ u_{n-m_1} \end{bmatrix} = \begin{bmatrix} x_{m_1+1} \\ x_{m_1+2} \\ \cdot \\ \cdot \\ \cdot \\ \cdot \\ x_n \end{bmatrix}$$

the control variables because they are independent, and Eq. 1.2.10 is

$$y_i = F_i(u_1, u_2, \ldots, u_{n-m_1}) \qquad i = 1, 2, \ldots, m_1 \tag{1.2.11}$$

In this way our problem can be reformulated as

$$\operatorname*{Max}_{u_j} I(y_1, \ldots, y_{m_1}, u_1, u_2, \ldots, u_{n-m_1}) \qquad j = 1, 2, \ldots, n - m_1 \tag{1.2.12}$$

subject to

$$y_i = F_i(u_1, u_2, \ldots, u_{n-m_1}) \qquad i = 1, 2, \ldots, m_1 \tag{1.2.11}$$

$$g_j(y_1, y_2, \ldots, y_{m_1}, u_1, u_2, \ldots, u_{n-m_1}) \leq 0 \qquad j = 1, 2, \ldots, m_2 \tag{1.2.13}$$

The second formulation has reduced the number of variables to be considered from n to $n - m_1$, and thus is easier to solve. However, the first formulation is usually required in large or complex problems because of the difficulty in solving Eq. 1.2.4 explicitly.

As an illustration of the above procedure, let us consider a hydrometallurgical leaching operation. Suppose that we wish to find the operating conditions which maximize the rate of leaching of the soluble constituent. Let the temperature, pressure, the pH of the solution, and the partitioning of the solute between the solid and the solution at equilibrium be the variables of interest.

Thus we may write

$$\text{Max } I(T, P, \text{pH}, K, C) \tag{1.2.14}$$

as the objective function, which at this stage would contain five variables, that is, temperature, pressure, pH, the equilibrium constant, and the degree of extraction.

However, in the statement of the problem we shall fix the percentage extraction, that is,

$$C = C_1, \text{ i.e., a constant} \tag{1.2.15}$$

Thus thermodynamics would tell us that the equilibrium relationship depends on the temperature, pressure, and the pH, that is,

$$K = f_1(T, P, \text{pH}) \tag{1.2.16}$$

Furthermore, for a system at constant volume, the temperature and pressure are uniquely related, that is,

$$T = f_2(P) \qquad \text{or} \qquad T = f_3(P, C) \tag{1.2.17}$$

if there is a gaseous product.

It follows that in the above problem we have three equality type constraints. Thus we may state that we have *three state variables*, and *two control variables*, and only the control variables may be changed independently. Therefore, in the problem at hand, we may keep altering the temperature and, for example, the pH in order to find the optimum conditions; however, the remaining variables are *not independent*.

Trajectory Optimization Problems

In these optimization problems we must handle type (iii) constraints, that is, equality constraints which take the form of differential or integral equations. These problems may also contain algebraic equality and inequality constraints [i.e., type (i) and (ii) constraints], but these will not be considered at this time. The trajectory optimization problems may then be stated as

$$\underset{x_i(t)}{\text{Max}} \, [I(x_1, x_2, \ldots, x_n)|_{t=t_f}] \tag{1.2.18}$$

subject to

$$\frac{dx_k}{dt} = f_k(x_1, x_2, \ldots, x_n) \qquad k = 1, 2, \ldots, m_3 \quad t_0 < t \le t_f \tag{1.2.6}$$

However, since there is no effort required to partition the variables into dependent and independent ones (no equations to solve) this is nearly always done. When we partition, as before, with the y_i being state variables and u_j the control variables, the problem takes the form

$$\underset{u_j(t)}{\text{Max}} \, I(y_1, y_2, \ldots, y_{m_3}, u_1, u_2, \ldots, u_{n-m_3})|_{t=t_f} \qquad j = 1, 2, \ldots, n - m_3 \tag{1.2.19}$$

subject to

$$\frac{dy_i}{dt} = f_i(y_1, y_2, \ldots, y_{m_3}, u_1, u_2, \ldots, u_{n-m_3}) \qquad i = 1, 2, \ldots, m_3 \quad t_0 \le t \le t_f \tag{1.2.20}$$

where there are suitable boundary conditions on the y_i. We note that when there are differential equation constraints, the objective function must be evaluated at a fixed point in time, t_f. We note further that there is no loss in generality if I is taken only as a function of the y_i, since new y_i can be defined to give the proper type of dependence on u_j.

We have seen in the preceding section an example of a mathematical programming problem; now let us consider an example of a trajectory optimization problem. Let us assume that we wish to determine the optimal temperature and concentration trajectories for the autoclave or chemical reactor.

In the previous section, we assumed that both the state and control variables of the process were constant so that the problem could be formulated as a mathematical program. In this section, we shall consider these variables to be time dependent, and we wish to determine the optimal progression for them. Let us assume that we have a model of the system giving the rate of change of product concentration c and temperature T, as functions of the steam valve setting V. The pressure is assumed to be fixed at the design value. Thus we have the following constraints

$$dc/dt = f_1(c, T) \qquad c(t_0) = c_0 \tag{1.2.21}$$

$$dT/dt = f_2(c, T, V) \qquad T(t_0) = T_0 \tag{1.2.22}$$

Furthermore, we must specify limits on the valve setting V

$$V_* \leq V \leq V^* \tag{1.2.23}$$

where V^* is the highest safe setting of V due to explosion hazard† and V_* is the flow when the valve is closed. Starting from our initial conditions c_0, T_0, we would like to reach our desired final point c_s, T_s, in some optimal fashion. Perhaps we wish to minimize the time it takes to reach this final point, or to minimize the loss of product caused by side reactions. In general, our objective (to be minimized in this instance) is

$$I = \int_{t_0}^{t_f} F(c, T, V, t)\, dt \tag{1.2.24}$$

where for example $F = 1$ if the total time required to reach steady state is to be minimized, and $F =$ (rate of side reactions), if these are to be minimized. Our problem is then formulated as

$$\underset{V(t)}{\text{Min}}\ I = \int_{t_0}^{t_f} F(c, T, V, t)\, dt \tag{1.2.25}$$

subject to

$$dc/dt = f_1(c, T) \qquad c(t_0) = c_0 \tag{1.2.21}$$

$$dT/dt = f_2(c, T, V) \qquad T(t_0) = T_0 \tag{1.2.22}$$

$$V_* \leq V(t) \leq V^* \tag{1.2.23}$$

† If there were no such constraint, then V^* would take the value that corresponds to the valve being full open.

If this problem is then solved for the optimal policy $V(t)$, $t_0 \leq t \leq t_f$, we shall have the optimal trajectory for the system.

In order to see that this problem falls into the class represented by Eqs. 1.2.10 and 1.2.11, let us redefine our variables. Let

$$y_1 = c, y_2 = T, y_3 = -\int_{t_0}^{t} F(c, T, V, t')\, dt', u_1 = V \tag{1.2.26}$$

Then our problem becomes

$$\underset{u_1(t)}{\text{Max}}\ I(y_3)\big|_{t_f} = y_3(t_f) \tag{1.2.27}$$

subject to

$$\frac{dy_i}{dt} = f_i(y_1, y_2, u_1, t) \qquad i = 1, 2, 3, y_i(t_0) = y_{i0} \tag{1.2.28}$$

$$V_* \leq u_1(t) \leq V^* \tag{1.2.29}$$

where we note that Eq. 1.2.29 is the simplest form of inequality constraint of type (ii).

1.3 MORPHOLOGY OF OPTIMIZATION

Now that we have formulated the classes of problems we shall set out to solve, let us examine the techniques available for tackling them. One interpretation of the structure of the field of optimization is shown in Fig. 1.4. A small number of mathematical theories form the basis for all of the optimization techniques. The calculus is used in virtually every technique either in the actual computational procedure, or in the proof of the validity of the method. The theorems of linear vector spaces are used in most of the techniques, while the theory of inequalities has great importance in some of the most recent approaches. The problems which the mathematical programming techniques were designed to solve are listed opposite them.

The solution of trajectory optimization problems is based on the variational trinity of the classical calculus of variations, the maximum principle of Pontryagin, and the dynamic programming of Bellman. It can be shown that, much like the other more familiar trinity, each entity of the triad is equivalent to the others, but each manifests itself in a different way. Also much like the other trinity, there has been great personal prejudice and controversy over the relative merits of each member of the triad. Probably the most commonly used member (from a computational point of view) is the maximum principle of Pontryagin, and most computational techniques available are concerned with satisfying the maximum principle necessary conditions for an optimum.

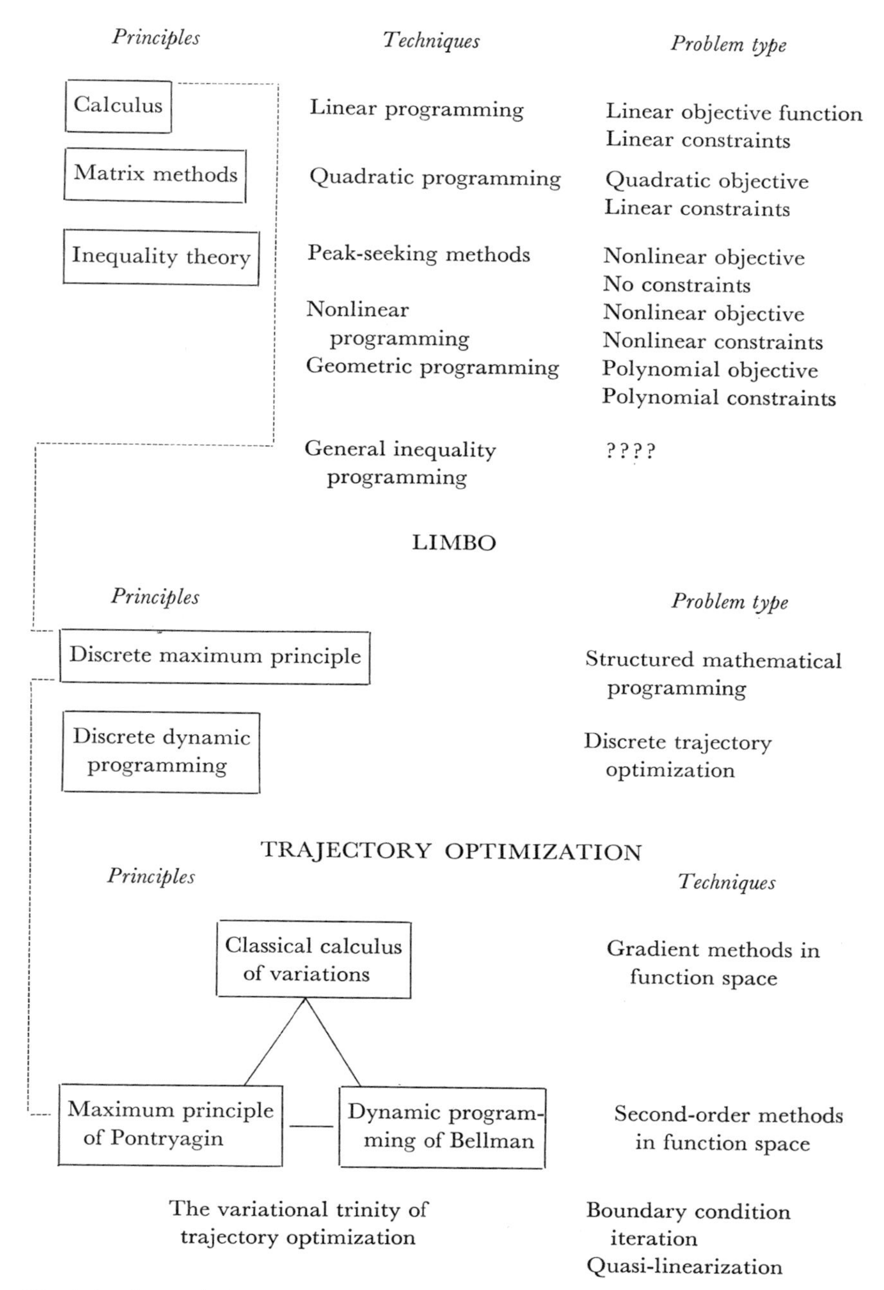

Fig. 1.4 A morphology of optimization.

There are two methods which cannot be fitted easily into either the mathematical programming section or the trajectory optimization section, and thus are in a state of limbo. A discrete form of the maximum principle of Pontryagin has been developed and applied by Katz [1] and Fan and Wang [3]. The original form of the theory was too sweeping, but it has been corrected [2] to a weaker result. Nevertheless the method seems to have computational usefulness [3]. The discrete dynamic programming algorithm is derived from the simplest form of Bellman's "principle of optimality" but has been used mainly for mathematical programming problems [4].

It is hoped that this morphology of optimization has made the material we are to cover more appealing. As the details presented in later chapters are grasped, it may be desirable to return to this outline for more perspective.

1.4 THE PROBLEM DESCRIPTION—PROCESS MODELS

In the preceding three sections we introduced some of the basic concepts of optimization with the ultimate goal of determining the conditions which lead to a maximum or a minimum for the *objective function*, subject to certain equality and inequality constraints.

Two important conclusions must emerge from this discussion:

1. The actual *techniques* for optimization can deal with essentially abstract mathematical functions and are largely independent of the physical nature of the process to be optimized; thus the techniques for devising an optimal distribution system for flat rolled products is analogous, if not identical, to the approaches used in finding the optimal route for a milk truck or a delivery van for a diaper service.
2. In order to perform the optimization, one requires a good description of the process so as to account for all the constraints on the problem.

The success or failure of any optimizing procedure depends, on the sophistication and appropriateness of the mathematical techniques used, but the correctness of the mathematical description employed is also crucial.

For this reason we shall devote some space to the problems involved in the modeling of systems for the purpose of optimization. This discussion will be of greater interest to metallurgical engineers rather than to their chemical counterparts, because there is a paucity of suitable mathematical models of metallurgical systems in contrast to the large number of sophisticated process models that are available for many chemical engineering operations.

From the viewpoint of optimization the mathematical description of a process requires that one should be able to predict the *response* of the system to changes in the *independent* (or control) variables or, in other words, we must be able to predict the effect of the *input variables* (**u** in the previous section) on the *state* or *output* variables (termed **y** previously). Thus, for example, if we wish to describe a basic oxygen furnace in mathematical terms, *we would have as input variables*:

> scrap composition, quantity and composition of the flux additions, hot metal composition, hot metal temperature.

As output variables we would have:

> the instantaneous bath composition, the slag composition, bath temperature, offgas composition, and heat loss through the wall.

By the same token, in the operation of a stirred batch reactor, the *input variables* could include:

> initial batch composition, initial temperature, the rate of agitation and the extent of external heat transfer.

As output variables we could have the following:

> instantaneous composition, instantaneous temperature, and possibly the instantaneous pressure

There exist several forms of process descriptions and in many respects these forms are complementary. Thus the dividing line between them is drawn somewhat arbitrarily in order to aid the classification.

Theoretical or Semi-Theoretical Mathematical Representation

In certain simple physical situations it is possible to describe the system by means of well-established basic physical laws. Thus, for example, the temperature distribution in a cylindrical billet receiving purely radiant heat in a billet reheating furnace can be described by combining Fourier's law of heat conduction and the Stefan-Boltzmann radiation law. Thus for radial symmetry and in the absence of axial conduction we have:

$$\frac{\partial T}{\partial t} = \kappa\left(\frac{\partial^2 T}{\partial r^2} + \frac{1}{r}\cdot\frac{\partial T}{\partial r}\right); \qquad 0 \leq r \leq R \tag{1.4.1}$$

with

$$T = T_i, \qquad t = 0 \tag{1.4.2}$$

$$\frac{\partial T}{\partial r} = 0, \qquad r = 0 \tag{1.4.3}$$

and

$$-k\frac{\partial T}{\partial r} = F_{EC}\sigma[T_E^4 - T^4], \qquad r = R \tag{1.4.4}$$

where κ is the thermal diffusivity, T is the (absolute) temperature, r is the radial coordinate, F_{EC} is the view factor, σ is Boltzmann's constant, and T_E is the temperature of the environment within the furnace.

We may wish to optimize this system by finding the optimal time dependence of T_E, such that we minimize the overall cost of preheating the billet.

Equations 1.4.1 to 1.4.4 provide an accurate theoretical representation of the system, since no gross simplifying assumptions were made in their statement. Such situations, however, are quite rare in the representation of metallurgical situations. In the majority of cases one has to make rather gross simplifying assumptions either to describe poorly understood physical phenomena or to keep the mathematics tractable.

The choice of the approximations made is dictated by the nature of the system, the availability of data, the sophistication warranted, and by the time-honored engineering compromise between the faithful representation of physical reality and the need for mathematical simplicity. In one approach, based on transport laws, one constructs a certain simplified physical picture, which is then faithfully represented by the appropriate expressions based on chemical kinetics and transport theory.

As an example we may represent heat transfer in a continuous slab [5] casting machine by assuming that:

1. The liquid pool is completely mixed.
2. Heat conduction may be neglected in the direction of the casting velocity.
3. Heat transfer between the mold wall and the outer surface of the solidified slab may be described in terms of a simple heat transfer coefficient.

This system is sketched in Fig. 1.5 and the "governing equations" are given by

$$\kappa_s\frac{\partial^2 T_s}{\partial y^2} = u_z\frac{\partial T_s}{\partial z}; \qquad 0 \leq y \leq Y(z) \tag{1.4.5}$$

where T_s is the temperature within the solidified crust, u_z is the casting velocity, and $Y(z)$ is the position of the solidified crust, which is, of course, a function of z.

The boundary conditions are given as:

$$T_s = T_{mp}, \qquad \text{at} \quad y = Y(z) \tag{1.4.6}$$

$$-k_s\frac{\partial T_s}{\partial y} = -\rho\,\Delta H_T\frac{dY(z)}{dz}, \qquad \text{at} \quad y = Y(z) \tag{1.4.7}$$

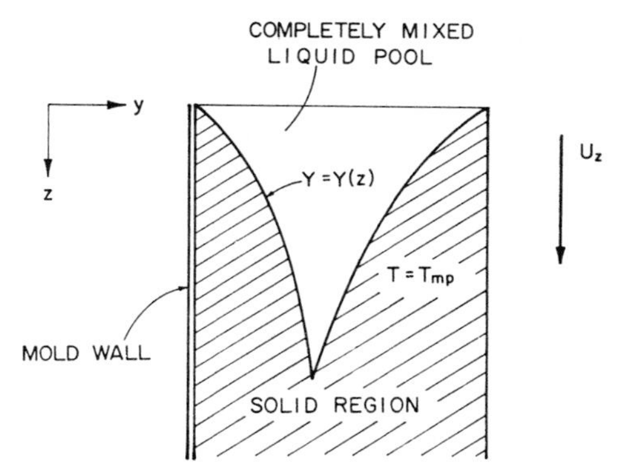

Fig. 1.5 Sketch of a continuous slab casting operation.

and

$$k_s \frac{\partial T_s}{\partial y} = h_m(T_s - T_m) \tag{1.4.8}$$

where T_{mp} is the melting point, ΔH_T is the sum of the latent heat and sensible heat content of the incoming hot metal, h_m is the mold heat transfer coefficient, and T_m is the temperature of the mold surface.

The system of Eqs. 1.4.5 to 1.4.8 is *a fundamental transport based mathematical model* of the continuous casting operation, the solution of which allows us to predict the position of the solidification front $Y(z)$, as well as the temperature distribution in the solid, $T_s(y, z)$, as a function of the parameters, u_z, h_m, and T_m.

For a given metal T_m would be fixed, but in principle one could examine varying u_z or h_m with a view of maximizing the output from the system.

It is noted that from a practical viewpoint, we would wish to have u_z, the casting speed, as high as possible, subject to the constraint that we must be able to abstract sufficient heat through the mold wall, so that the solidified metal crust of the metal exiting the mold is sufficiently thick to contain the molten core.

One can envision numerous transport based models for the representation of metallurgical systems, for example, the description of the stack region of the blast furnace as a packed bed [6], the modeling of a reverbatory furnace in copper smelting in terms of "plug flow," completely mixed and dead water regions [7], and the modeling of the basic oxygen furnace (BOF) by a set of simultaneous linear differential equations [8].

While the mathematical modeling of metallurgical systems is a relatively new discipline and one must quote recent journal articles for representative examples, much work has been done on the construction of mathematical models relating to the chemical and petroleum industry. Several excellent text-books are available on the modeling of chemical reactors [9, 10], and for the broader area of process simulation in the chemical industry the reader is referred to the texts by Himmelblau and Bischoff [11] and Crowe et al. [12].

Whenever approximations are used to minimize mathematical complexity, these often change the description of the problem quite drastically. In a *distributed parameter* representation one deals with partial differential equations and thus has more than one independent variable. For example, the transient temperature distribution in the stack region of the blast furnace, such as a response to a step change in the blast temperature, would require a distributed parameter representation, as both length and time would appear as the independent variables.

Similarly problems involving the transient behavior of any packed bed reactor would constitute a distributed parameter problem. Hansen et al. [13] give an excellent example of the great sensitivity of the results to the packed bed reactor model selected.

A typical approximation in both chemical and metallurgical process models is to assume that a given property, such as temperature, or concentration is spatially uniform throughout the system, thus allowing the system to be represented by a set of ordinary differential equations—a *lumped parameter model*. For example, in modeling a BOF or a copper converter one may assume that there are no spatial variations in temperature of the metal phase, or that the concentration of some impurity would depend only on time, thus reducing the description to a lumped parameter problem.

Similarly, a catalyst pellet is often assumed to have uniform concentration and temperature profiles even though spatial variations are known to exist.

The billet reheating problem mentioned earlier could be regarded as a lumped parameter problem if we assume that there are no radial gradients within the cylinder—this may, or may not, be a good assumption depending on the geometry and the heat flux incident on the surface.

As a lumped parameter problem, Eqs. 1.4.1 to 1.4.4 reduce to

$$R^2 \pi \rho C_p \frac{dT}{dt} = 2\pi R \sigma F_{EC}(T_E{}^4 - T^4) \tag{1.4.9}$$

with

$$T = T_i \qquad \text{at} \quad t = 0 \tag{1.4.10}$$

Needless to say, the use of this ordinary differential equation allows considerable simplification, but at the expense of our ignoring some aspects of physical reality.

Empirical (Black Box) Input-Output Models

These models use only data on the input and output of the system and fit the most convenient empirical relationships between these. Because of their isolation from theory, these models provide no physical insight into the basic phenomena observed, and should not be extrapolated beyond the range of the data examined.

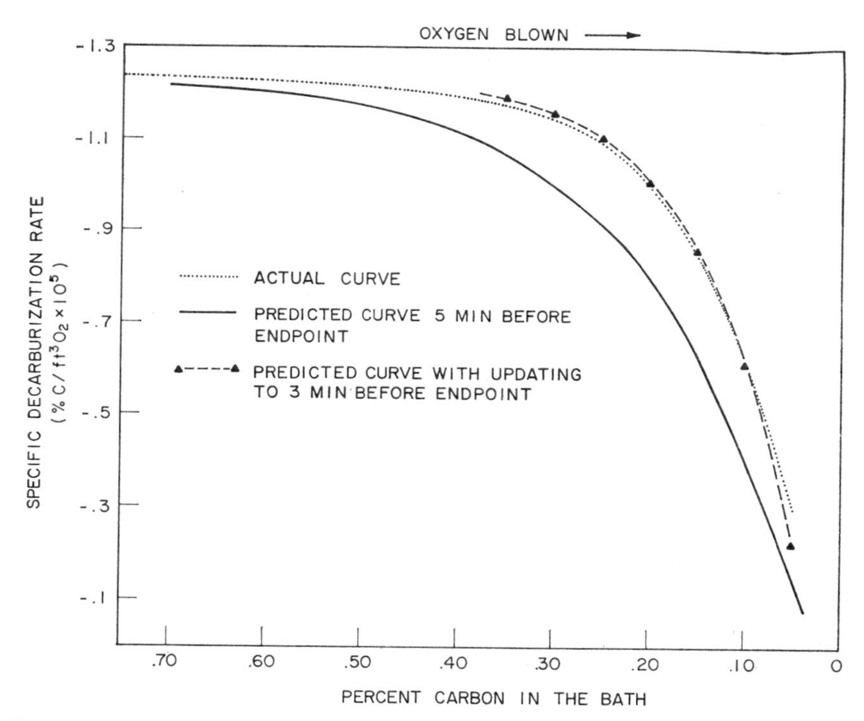

Fig. 1.6 A comparison of measured and predicted specific decarburization rates for a BOF, after Meyer et al. [14].

As a good, successful example of curve fitting, we may quote the work of Meyer and associates [14] who represented the rate of carbon oxidation in the BOF by the following expression:

$$\phi = \alpha - \beta e^{\gamma X_C} \tag{1.4.11}$$

Here ϕ is the rate of carbon oxidation per unit volume of oxygen blown, X_C is the weight percent of carbon in the metal bath, α, β, and γ are empirical constants, which depend on the mode of operation.

The use of this technique is illustrated in Fig. 1.6 where it is seen, that by the updating of the coefficients, it is possible to predict the "carbon end point" quite accurately.

It is noted that this curve fitting representation of the BOF is an alternative or complementary technique to a transport based model for the system, such as described by Muchi [8].

Implicit Process Models

In some cases, even the information required for an empirical input-output "black box model" is not available. For example, the process description may only be deduced from the noise-corrupted measurements from the process itself, so that optimization must be done in an "evolutionary" manner by a sequence of process experiments. An excellent monograph [15] by Box and Draper treats this problem in some detail.

An even less quantitative description may have to be sufficient if one is dealing with a number of interacting complex operations each run by a manager and for which no mathematical description is available. In this case, the manager has a "gut" feeling for the process which determines his operating decisions. The optimizer in this case must provide the appropriate incentives for each manager to optimize his operation so as to maximize the profits from the entire system. We note that this may result in some managers actually reducing their profits so that others may have theirs appreciably increased. These problems will be discussed in Chapter 5.

In this section we have emphasized that a necessary prerequisite of any successful optimization is the ability to represent the system in question in some descriptive form. A very brief review was presented of the various types of representations available for the modeling of processes. We note that "ready-made" process models are far more likely to be available in optimization problems dealing with chemical processes, whereas in many metallurgical operations the development of the model must necessarily precede any optimization studies. Because of the many nonlinearities inherent in the equations that represent materials processing at high temperatures, the modeling of metallurgical systems is a complex subject, well worthy of study by those interested in the optimization of metallurgical processes.

Just as the most powerful digital computer is powerless to improve poorly written programs or algorithms, the most sophisticated optimization techniques will be wasted if the basic model for the process is incorrect or inappropriate.

Ideally, process optimization should be a team effort, with significant inputs from those familiar with the process, from those with experience in mathematical modeling, and from those having expertise in optimization. This is particularly important in situations where the process models are not available and where their development must also be included in the overall optimization effort.

REFERENCES

1. S. Katz, *I and EC Fundam*, **1,** 276 (1962).
2. R. Jackson and F. Horn, *I and EC Fundam*, **4,** 487 (1965).
3. L. T. Fan and C. S. Wang, *Discrete Maximum Principle*, Wiley, 1964.
4. R. Aris, *Discrete Dynamic Programming*, Blaisdell, 1964.
5. A. W. D. Hills, *J Iron Steel Inst* **203,** 18 (1965).
6. J. Yagi and I. Muchi, *Trans ISIJ* **10,** 5, 392 (1970).
7. N. J. Themelis and P. Spira, *Trans Met Soc. AIME* **236,** 821 (1966).
8. S. Asai and I. Muchi, *Trans ISIJ* **10,** 4, 250 (1970).
9. R. Aris, *Elementary Chemical Reactor Analysis*, Prentice-Hall, New York, 1969.
10. K. G. Denbigh and J. C. R. Turner, *Chemical Reactor Theory*, Cambridge, 1971.
11. D. M. Himmelblau and K. B. Bischoff, *Process Analysis and Simulation*, Wiley, 1968.
12. C. M. Crowe, A. E. Hamielec, T. W. Hoffman, A. I. Johnson, P. T. Shannon, and D. R. Woods, *Chemical Plant Simulation*, Prentice-Hall, 1971.
13. K. W. Hansen, H. Livbjerg, and J. Villadsen, *Proc. DISCOP Symp* Gyor, Hungary, 1971.
14. H. W. Meyer, E. Aukrust, and W. F. Porter, "Heat and Mass Transfer in Process Metallurgy," A. W. D. Hills, Ed., *The Institution of Mining and Metallurgy*, London, 1967 p. 173.
15. G. E. P. Box and N. R. Draper, *Evolutionary Operation*, Wiley, 1969.

2 The Maxima of Functions Through Use of the Calculus

2.1 RELATIVE MAXIMA AND ABSOLUTE MAXIMA

In the preceding chapter we introduced the concepts of the objective function, state variables, and control variables, together with the concepts of equality and inequality constraints.

It was shown that our purpose in an optimization problem is to find the maximum of the objective function, subject to the constraints which are usually posed by the physical nature of the problem. This maximum is then found by the appropriate manipulation of the control variables.

In this chapter we shall discuss the mathematical techniques available to us for finding the maxima or minima of functions, through the use of the calculus.

The use of calculus for finding the maxima or minima of simple analytical functions will be recalled by the reader. As an example, let us consider the

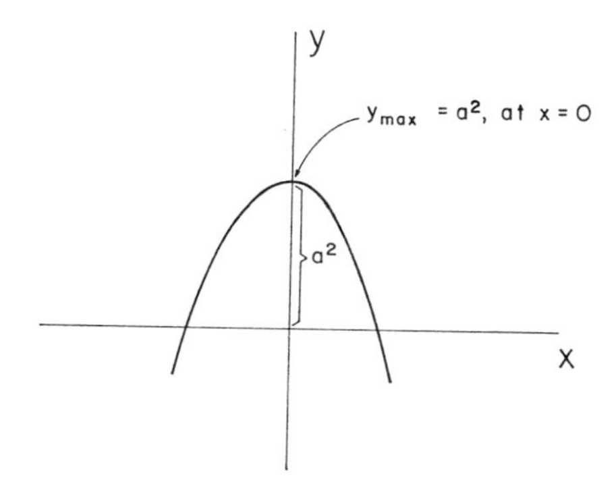

Fig. 2.1 Plot of $y = a^2 - x^2$.

function

$$y = a^2 - x^2 \tag{2.1.1}$$

sketched in Fig. 2.1. It may be shown that the necessary condition for an extremum (i.e., maximum or minimum) is that the first derivative is zero. Furthermore, this extreme value is a maximum, if the second derivative is negative. By applying these considerations to Eq. 2.1.1, we obtain

$$\frac{dy}{dx} = -2x \tag{2.1.2}$$

and

$$\frac{d^2y}{dx^2} = -2 \tag{2.1.3}$$

It follows that the extreme value of the function occurs at $x = 0$, and the value $y = a^2$ is indeed a maximum, because the second derivative is negative.

The above example is a very simple illustration of the use of calculus for finding the maximum of a function; the majority of real life problems involve rather more complex objective functions and, therefore, we need to express our previously used method in a more general form.

As an example, let us consider the objective function $I(x)$ shown in Fig. 2.2. It is seen that $I(x)$ shows several peaks within the interval considered.

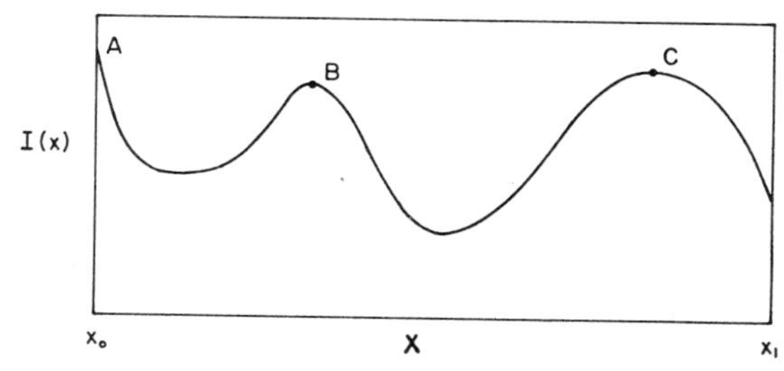

Fig. 2.2 Multiple extreme points.

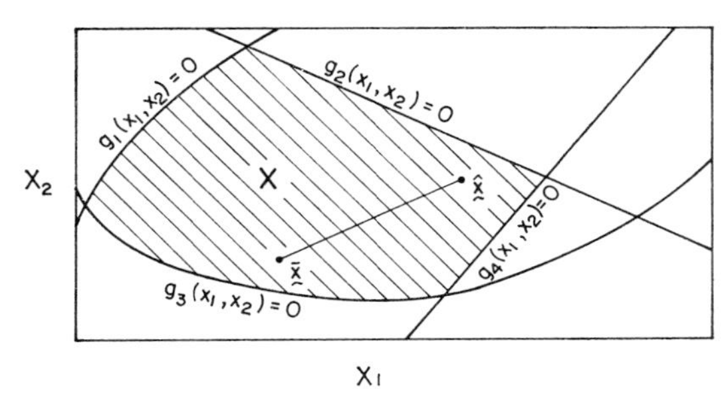

Fig. 2.3 A convex set.

The *absolute maximum* is located at point A, whereas points B and C correspond to what are termed *local* or *relative maxima*.

Clearly, care has to be exercised in any search procedure, so that one is not satisfied with finding local or relative maxima, when the absolute maximum is required.

In principle the simplest and safest way of finding the absolute maximum of a function is to locate all the maxima and then to select the largest of these.

However, this procedure may not be practical under some circumstances, and therefore it is desirable to develop alternative techniques and theorems concerning the nature of functions, their maxima and minima.

The concepts described in this chapter play an important role in the optimization procedures to which the bulk of this text is devoted. The derivation of the theorems is given for the sake of completeness and may be skipped by those who are interested only in acquiring the various optimization techniques.

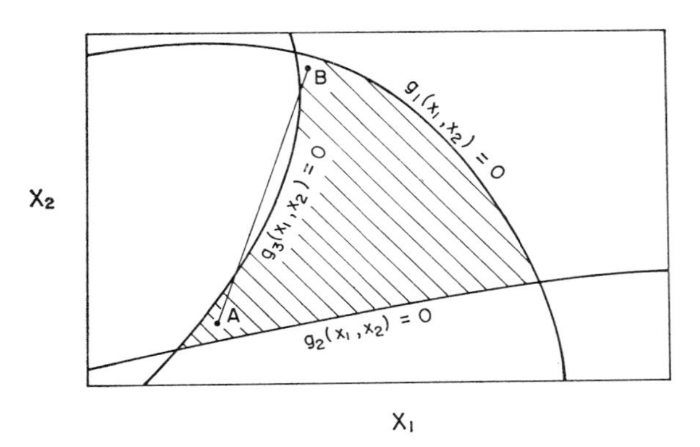

Fig. 2.4 A nonconvex set.

Let us define the following notions:

1. *Convex set* X*: A set of points* X *is convex if for any two points* $\bar{\mathbf{x}}$ *and* $\hat{\mathbf{x}}$ *in* X *and for all* λ, $0 \leq \lambda \leq 1$ *the point* $\mathbf{x} = \lambda\hat{\mathbf{x}} + (1 - \lambda)\bar{\mathbf{x}}$ *lies in the set* X (cf. Fig. 2.3). Basically one can see that convex sets "bulge out." Figure 2.4 shows a nonconvex set in the two variables x_1, x_2, where it is clear that some points on the line AB lie outside the set.

2. *Concave function* $I(x_1, x_2, \ldots, x_n)$*: A function* $I(x_1, x_2, \ldots, x_n)$ *is concave over a convex set* X *if for any two points* $\hat{\mathbf{x}}$ *and* $\bar{\mathbf{x}}$ *in the set* X *and for all* λ, $0 \leq \lambda \leq 1$,

$$I(\lambda\hat{\mathbf{x}} + (1 - \lambda)\bar{\mathbf{x}}) \geq \lambda I(\hat{\mathbf{x}}) + (1 - \lambda)I(\bar{\mathbf{x}}) \tag{2.1.4}$$

Figure 2.5 shows a concave function over the convex set of numbers $x_0 \leq x \leq x_1$ on the real line.

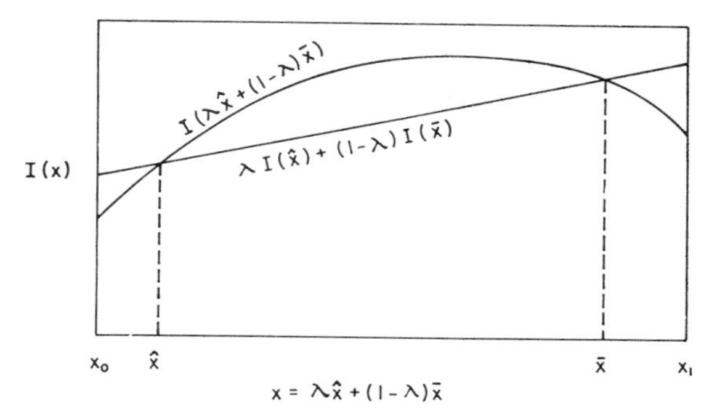

Fig. 2.5 A concave function.

3. *Convex function* $I(x_1, x_2, \ldots, x_n)$*:* A function* $I(x_1, x_2, \ldots, x_n)$ *is convex over a convex set* X *if for any two points* $\hat{\mathbf{x}}$ *and* $\bar{\mathbf{x}}$ *in the set* X *and for all* λ, $0 \leq \lambda \leq 1$,

$$I(\lambda\hat{\mathbf{x}} + (1 - \lambda)\bar{\mathbf{x}}) \leq \lambda I(\hat{\mathbf{x}}) + (1 - \lambda)I(\bar{\mathbf{x}}) \tag{2.1.5}$$

Figure 2.6 shows a convex function over the convex set $x_0 \leq x \leq x_1$.

Figures 2.5 and 2.6 provide a clear graphical illustration of the definitions given for concave and convex functions in Eqs. 2.1.4 and 2.1.5, respectively.

In simple graphical terms a function is concave over an interval, if it passes above a straight line drawn between any two points in the interval.

The opposite is true for a convex function, which has to pass below the straight line drawn between any two points in the interval. These illustrations were given in one dimension, but can be readily generalized to higher

* Note that when the equality holds in Eqs. 2.1.4 and 2.1.5, the functions are both convex and concave. The definitions with the inequality sign would read "strictly" convex or concave.

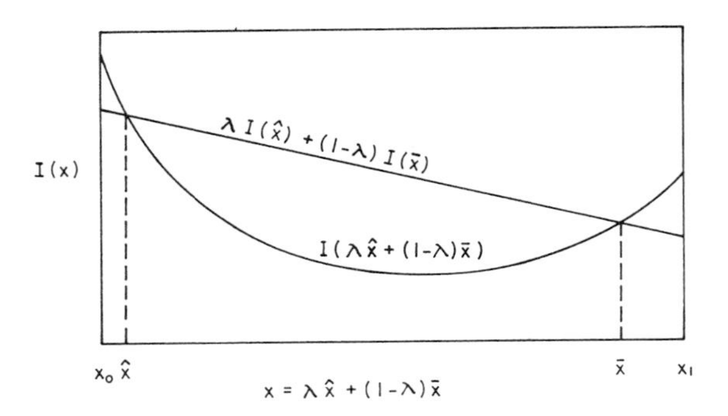

Fig. 2.6 A convex function.

dimensional systems. We note that the definitions of concave and convex functions given above specify, in effect, that *a function which is concave or convex within a given interval can have only one peak* (maximum for a concave function and minimum for a convex function) within this interval. We note that the function shown in Fig. 2.2 is neither convex nor concave within the interval $x_0 \leq x \leq x_1$, so that our finding is not inconsistent with the behavior seen there.

Now that we have some fundamental definitions we can state a theorem which is derived in Ref. 1, page 93.

THEOREM 1. *If* $\mathrm{I}(\mathrm{x}_1, \mathrm{x}_2, \ldots, \mathrm{x}_n)$ *is a concave function over a convex set of allowable values of* $(\mathrm{x}_1, \mathrm{x}_2, \ldots, \mathrm{x}_n)$, *then a relative maximum of* $\mathrm{I}(\mathrm{x}_1, \mathrm{x}_2, \ldots, \mathrm{x}_n)$ *over this convex set,* X, *will be an absolute maximum.*

A similar theorem applies also to the minimization of convex functions over convex sets.

This formal theorem is, of course, the precise statement of the well-known fact that the local maximum of a function having only one peak is a global maximum.

The importance of performing the optimization over a convex set can be illustrated by the function shown in Fig. 2.7. The function $I(x)$ is certainly concave, but it is defined over the nonconvex set of numbers

$$I(x) = \begin{cases} I(x) & \text{over } x_0 \leq x \leq x_a \\ & \text{and } x_b \leq x \leq x_c \\ & \text{and } x_d \leq x \leq x_1 \\ 0 & \text{elsewhere} \end{cases} \tag{2.1.6}$$

Thus this concave function over a nonconvex set has three local maxima (A, B, C), only one of which (B) is a global maximum.

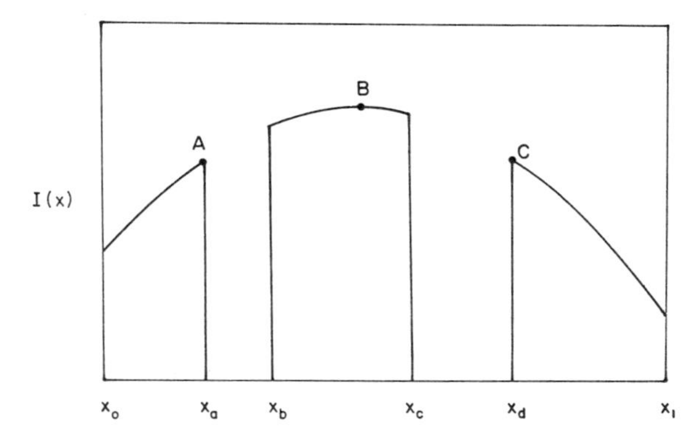

Fig. 2.7 A concave function defined over a nonconvex set.

In general, Theorem 1 gives us a test we can make for an absolute maximum. However, the requirement of concavity is such a stringent test that it may not be helpful for offering guidance on many problems encountered in practice; thus, in many practical situations we may have to check all the relative maxima.

The discussion concerning concave and convex sets may well have appeared to be rather abstract; thus, it may be worthwhile to illustrate the application of these concepts with a practical example.

Let us consider that we wish to find the optimum size for a tube and shell heat exchanger. While in the process calculations the heat transfer area could be regarded as a continuous function, practical considerations, such as the need to use standard tube and shell dimensions may limit us to certain ranges of this variable.

The previously discussed Fig. 2.7 may be used as an illustration of this, where

$$x_0 \leq x \leq x_a; \quad x_b \leq x \leq x_c; \qquad \text{and} \quad x_d \leq x \leq x_1$$

represent the "permissible values of x" (the heat transfer area); thus our search has to be confined within this region. As shown earlier, this problem provides a nonconvex set of x values, so that in our search care must be taken to find the abscissa which gives us $I(x)$ corresponding to point B, the true maximum, rather than to points A and C, which are only local maxima.

2.2 NECESSARY AND SUFFICIENT CONDITIONS FOR UNCONSTRAINED RELATIVE MAXIMA

We have seen that our search is for relative maxima which can be examined subsequently to determine whether one or more of these constitute the global maximum. It is desirable therefore, to develop the necessary and sufficient conditions for the existence of local maxima.

Let us assume that the maximum is found at an interior point of the set X and that there are no constraints to consider. (The problem with constraints requires different strategies which will be treated later.) Thus we wish to discover the maximum of $I(x_1, x_2, \ldots, x_n)$. Now let us suppose that the point

$$\mathbf{x}^* = \begin{bmatrix} x_1^* \\ x_2^* \\ \cdot \\ \cdot \\ \cdot \\ x_n^* \end{bmatrix}$$

is the maximum sought and there is another point, $\mathbf{x} = \mathbf{x}^* + \delta\mathbf{x}$ slightly displaced from it. Since $\mathbf{x}^*$ is the relative maximum,

$$\delta I = I(\mathbf{x}^* + \delta\mathbf{x}) - I(\mathbf{x}^*) \leq 0 \qquad (2.2.1)$$

for *all* displacements

$$\delta\mathbf{x} = \begin{bmatrix} \delta x_1 \\ \delta x_2 \\ \cdot \\ \cdot \\ \cdot \\ \delta x_n \end{bmatrix}$$

Let us expand Eq. 2.2.1 in a Taylor series about the maximum to obtain

$$\delta I = \sum_{i=1}^{n} \left(\frac{\partial I}{\partial x}\right)_{\mathbf{x}^*} \delta x_i + \frac{1}{2} \sum_{j=1}^{n} \sum_{i=1}^{n} \delta x_j \left(\frac{\partial^2 I}{\partial x_i \, \partial x_j}\right)_{\mathbf{x}^*} \delta x_i + 0(\delta\mathbf{x}^3) \qquad (2.2.2)$$

If we consider *very* small displacements (e.g., $|\delta x_i| < \epsilon$), then the second-order terms are negligible and

$$\delta I = \sum_{i=1}^{n} \left(\frac{\partial I}{\partial x_i}\right)_{\mathbf{x}^*} \delta x_i + 0(\delta\mathbf{x}^2) \qquad (2.2.3)$$

The only way that $\delta I \leq 0$ can hold for *all* possible small values of $\delta\mathbf{x}$, is for

$$\left(\frac{\partial I}{\partial x_i}\right)_{\mathbf{x}^*} = 0 \qquad i = 1, 2, \ldots, n \qquad (2.2.4)$$

[Points satisfying Eq. 2.2.4 are called *extreme* points of the function $I(\mathbf{x})$.] Thus we have stated the first necessary condition for a maximum to occur at the point $\mathbf{x}^*$. On substituting Eq. 2.2.4 into Eq. 2.2.2 we obtain

$$\delta I = \tfrac{1}{2}\delta\mathbf{x}^T(\mathbf{H})_{\mathbf{x}^*}\,\delta\mathbf{x} + 0(\delta\mathbf{x}^3) \tag{2.2.5}$$

where $\mathbf{H}$ is the symmetric matrix of second derivatives

$$\mathbf{H} = \begin{bmatrix} \dfrac{\partial^2 I}{\partial x_1{}^2} & \dfrac{\partial^2 I}{\partial x_1\,\partial x_2} & \cdots\cdots & \dfrac{\partial^2 I}{\partial x_1\,\partial x_n} \\ \dfrac{\partial^2 I}{\partial x_2\,\partial x_1} & \dfrac{\partial^2 I}{\partial x_2{}^2} & \cdots\cdots & \dfrac{\partial^2 I}{\partial x_2\,\partial x_n} \\ & & \ddots & \\ \dfrac{\partial^2 I}{\partial x_n\,\partial x_1} & \cdots\cdots & \cdots\cdots & \dfrac{\partial^2 I}{\partial x_n{}^2} \end{bmatrix} \tag{2.2.6}$$

known as the *Hessian matrix*. From matrix algebra we know that in Eq. 2.2.5 δI will be negative for all $\delta\mathbf{x}$ if the matrix $\mathbf{H}$ is negative definite. For a symmetric matrix this means that all eigenvalues must be negative.† If $\mathbf{H}$ is only negative semidefinite (no positive eigenvalues), then $\delta I \leq 0$ for all $\delta\mathbf{x}$. This last condition is necessary for a maximum, but is not sufficient because at a nonzero displacement $\delta\mathbf{x}$ where $\delta I = 0$ for a second-order expansion, the character of the extreme point must be determined from the higher order derivatives. However, negative definiteness allows no nonzero $\delta\mathbf{x}$ where $\delta I = 0$ and thus is sufficient (but not necessary) for a relative maxima. In summary, then, the conditions

$$\left(\frac{\partial I}{\partial x_i}\right)_{\mathbf{x}^*} = 0 \qquad i = 1, 2, \ldots, n \tag{2.2.4}$$

$$\delta\mathbf{x}^T\mathbf{H}\,\delta\mathbf{x} \leq 0 \qquad (H \text{ negative-definite or semidefinite}) \tag{2.2.7}$$

may be used to specify a relative maximum at $\mathbf{x} = \mathbf{x}^*$.

The reader will note that the criteria set out by Eqs. 2.2.4 and 2.2.7 represent a generalization of the one-dimensional case discussed at the beginning of the chapter, that is, Eqs. 2.1.1 to 2.1.2.

† See Appendix A for a discussion of quadratic forms.

At this stage it may be worthwhile to illustrate the use of Eqs. 2.2.4 and 2.2.7 for finding the maxima of some more complex functions.

EXAMPLE 2.2.1. Find the maximum of

$$I = 3x_1 + 2x_2 - 3x_1x_2 - 4x_1^2 - 5x_2^2$$

Solution. From Eq. 2.2.4 the necessary condition for an extreme value is that

$$\frac{\partial I}{\partial x_1} = 0, \qquad \frac{\partial I}{\partial x_2} = 0 \tag{2.2.8}$$

Thus we have

$$\frac{\partial I}{\partial x_1} = 3 - 3x_2 - 8x_1 = 0 \tag{2.2.9}$$

and

$$\frac{\partial I}{\partial x_2} = 2 - 3x_1 - 10x_2 = 0 \tag{2.2.10}$$

On solving Eqs. 2.2.9 and 2.2.10 for x_1^* and x_2^*, that is, the location of the extreme value, we have

$$x_1^* = \tfrac{24}{71}, \qquad x_2^* = \tfrac{7}{71}$$

In order to decide whether this location corresponds to a maximum or a minimum, we must form the *Hessian*, defined in Eq. 2.2.6.

Since

$$\frac{\partial^2 I}{\partial x_1^2} = -8, \qquad \frac{\partial^2 I}{\partial x_2^2} = -10$$

and

$$\frac{\partial^2 I}{\partial x_1\,\partial x_2} = -3, \qquad \frac{\partial^2 I}{\partial x_2 \partial x_1} = -3$$

$$\mathbf{H} = \begin{bmatrix} -8 & -3 \\ -3 & -10 \end{bmatrix}$$

On noting the criteria for a maximum, expressed by Eq. 2.2.7 and the definition of the eigenvalues for a symmetric matrix, it is readily seen that the eigenvalues of Eq. 2.2.6 are indeed negative, so that the location (24/71, 7/71) is a maximum. (Some basic definitions in matrix algebra are given in Appendix A for those needing a review.)

The reader is urged to plot contours of $I(x_1, x_2)$ as given in this example in the vicinity of the maximum, in order to acquire a "feel" for the nature of this problem.

Now that we understand the basic conditions for a maximum, let us investigate other extreme points. On examining Fig. 2.8 showing lines of

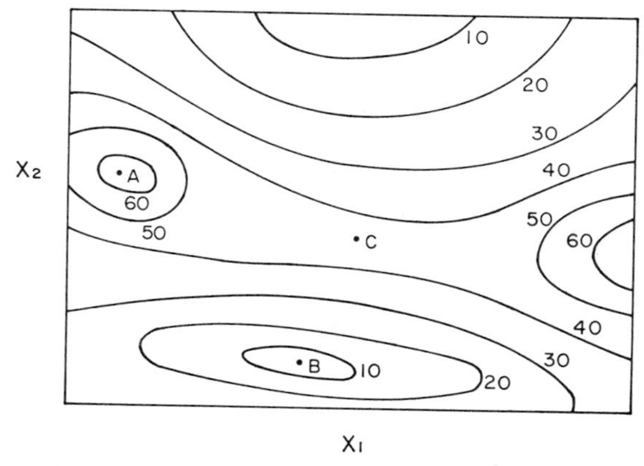

Fig. 2.8 A function with multiple extreme points.

constant $I(x_1, x_2)$, we see that points A, B, C are all extreme points of the function. Point A is a *relative maximum*, point B is a *relative minimum*, and point C is what is known as a *saddle point*. If one goes through the development of Eqs. 2.2.1 to 2.2.7 again while seeking a minimum, the character of each of these extreme points should become clear; thus

1. For a relative maximum, Eqs. 2.2.4 and 2.2.7 must be satisfied.
2. For a relative minimum, Eqs. 2.2.4 and

$$\delta \mathbf{x}^T \mathbf{H}\, \delta \mathbf{x} \geq 0 \qquad (H \text{ positive definite or semidefinite}) \tag{2.2.11}$$

must be satisfied.

3. For a saddle point, Eq. 2.2.4 and **H** indefinite must be satisfied.

The computational technique one normally uses from these theorems is to find all the solutions to Eq. 2.2.4 and then check the eigenvalues of H at each of the extreme points or simply compare values of $I(x)$ at each of the extreme points.

2.3 NECESSARY AND SUFFICIENT CONDITIONS FOR MAXIMA HAVING EQUALITY CONSTRAINTS

Let us now consider the case where we have equality constraints so that following the discussion in Chapter 1, the problem may be written as:

$$\underset{x_i}{\text{Max}}\ I(x_1, x_2, \ldots, x_n) \qquad i = 1, 2, \ldots, n \tag{1.2.1}$$

where

$$h_j(x_1, x_2, \ldots, x_n) = 0 \qquad j = 1, 2, \ldots, m_1 \tag{1.2.2}$$

In principle one could tackle optimization problems with equality constraints, by combining Eqs. 1.2.1 and 1.2.2 and thus eliminating all but the *control variables*. Having done this, one could proceed in the manner described in the preceding section.

Although the above procedure is attractive in concept, it is impractical in the majority of cases, because the form taken by the objective function and the constraints renders direct substitution impossible.

A convenient way of tackling these problems is through the use of the *Lagrange multiplier*, which will be illustrated in a very simple example, and then given subsequently in a more general form.

EXAMPLE 2.3.1. Find the minimum of

$$I = 4x_1^2 + 5x_2^2$$

subject to the constraint

$$h = 2x_1 + 3x_2 - 6 = 0$$

Solution. Let us form a new objective function, called the *Lagrangian*:

$$L(\mathbf{x}, \lambda) = I - \lambda h$$

or

$$L(\mathbf{x}, \lambda) = 4x_1^2 + 5x_2^2 - \lambda(2x_1 + 3x_2 - 6)$$

Here λ is an as yet unknown constant, termed the *Lagrange multiplier*.

We may proceed by forming the derivatives of $L(\mathbf{x}, \lambda)$ with respect to x_1 and x_2 and set the result equal to zero to find the stationary point, that is,

$$\frac{\partial L}{\partial x_1} = 8x_1 - 2\lambda = 0$$

and

$$\frac{\partial L}{\partial x_2} = 10x_2 - 3\lambda = 0$$

from which we find that the location of the stationary value in L is

$$x_1 = \tfrac{1}{4}\lambda, \qquad x_2 = \tfrac{3}{10}\lambda$$

All that remains is to find the value of λ which satisfies the constraint, $h = 0$.

On substituting these values for x_1 and x_2 into the original equation for h, we obtain

$$2(\tfrac{1}{4}\lambda) + 3(\tfrac{3}{10}\lambda) - 6 = 0$$

that is

$$\lambda = \tfrac{30}{7} \qquad \text{and} \quad x_1 = \tfrac{15}{14}, \quad x_2 = \tfrac{9}{7}$$

The reader is urged to verify this result by solving $h = 0$ for x_1 and substituting into the original equation, and solving for optimal x_2.

Example 2.3.1 provided a purely "mechanical" illustration of the use of the Lagrange multiplier, without justification, and for a particularly simple case. In the following, we shall develop these concepts in a more general manner.

Let us form the Lagrangian, L, by *adjoining* Eq. 1.2.2 to Eq. 1.2.1 through the m_1 Lagrange multipliers λ_j. The procedure followed here is the same as used in the previous example, in that we form the Lagrangian by taking the objective function and subtracting from it the product of the equality constraint and the as yet unknown Lagrange multiplier.

Thus we have

$$L(\mathbf{x}, \boldsymbol{\lambda}) \equiv I(\mathbf{x}) - \sum_{j=1}^{m_1} \lambda_j h_j(\mathbf{x}) \tag{2.3.1}$$

For any small displacements from the optimum $\mathbf{x}^*$, $\boldsymbol{\lambda}^*$, on expanding Eq. 2.3.1 in a Taylor series, we obtain

$$\begin{aligned} \delta L(\mathbf{x}^*, \boldsymbol{\lambda}^*) &= \left[\sum_{i=1}^{n} \left(\frac{\partial I}{\partial x_i}\right)_{\mathbf{x}^*} - \sum_{j=1}^{m_1} \lambda_j^* \sum_{i=1}^{n} \left(\frac{\partial h_j}{\partial x_i}\right)_{\mathbf{x}^*}\right] \delta x_i \\ &\quad - \sum_{j=1}^{m_1} h_j(\mathbf{x}^*)\, \delta\lambda_j + 0(\delta\mathbf{x}^2, \delta\boldsymbol{\lambda}^2) \\ &= \sum_{i=1}^{n} \frac{\partial L(\mathbf{x}^*, \boldsymbol{\lambda}^*)}{\partial x_i}\, \delta x_i + \sum_{j=1}^{m_1} \frac{\partial L(\mathbf{x}^*, \boldsymbol{\lambda}^*)}{\partial \lambda_j}\, \delta\lambda_j + 0(\delta\mathbf{x}^2, \delta\boldsymbol{\lambda}^2) \end{aligned} \tag{2.3.2}$$

We note that for any small variation from the maximum $I(\mathbf{x}^*)$, we must have

$$\delta I(\mathbf{x}^*) \leq 0$$

and since the constraints must be satisfied, then $\delta h_j(\mathbf{x}^*) = 0$ must also hold. Thus at the optimum we must have

$$\delta L(\mathbf{x}^*, \boldsymbol{\lambda}^*) \leq 0 \tag{2.3.3}$$

But the only way this can be true for *all* small values of δx_i, and $\delta \lambda_j$ is for

$$\frac{\partial L}{\partial x_i} = 0 \qquad i = 1, 2, \ldots, n \tag{2.3.4}$$

$$\frac{\partial L}{\partial \lambda_j} = 0 \qquad j = 1, 2, \ldots, m_1 \tag{2.3.5}$$

to hold at the optimum $\mathbf{x}^*$, $\boldsymbol{\lambda}^*$. This can be formalized as follows:

THEOREM. In order for the problem given in Eqs. 1.2.1 and 1.2.2 to have a relative maximum (minimum) at $\mathbf{x}^*$, $\boldsymbol{\lambda}^*$, it is necessary that Eqs. 2.3.4 and 2.3.5 hold there.

Now let us work an example to see how these rules may be applied to a simple constrained problem.

EXAMPLE 2.3.2. We wish to maximize the profit from the operation of a continuous agitated leaching unit, where stirring is provided by the momentum of the incoming stream; thus the cost of agitation may be taken as proportional to the throughput. The conversion, that is, the fraction of soluble *solute* extracted, is given as

$$X = \frac{K\theta}{1 + K\theta} = \frac{x_1 x_2}{1 + x_1 x_2} \tag{2.3.6}$$

here $\theta = x_1$ is the holding time and $K = x_2$ is, say, a mass transfer coefficient which is a function of the rate of agitation.

Equation 2.3.6 above is also the familiar expression for conversion in a continuous stirred reactor, in which a first-order reaction (or mass transfer controlled leaching operation) is being performed. Clearly, the longer the holding time, the higher the conversion, and it is reasonable to assume that the higher the rate of agitation, the larger is the value of K.

Thus a simple, physically realistic objective function could take the following form:

$$I(\theta, K) = I(x_1, x_2) = \frac{X}{\theta} - \frac{\beta(K)}{\theta} = \frac{x_2}{1 + x_1 x_2} - \frac{\beta(x_2)}{x_1} \tag{2.3.7}$$

where $I(\theta, K)$ is the profit from the operation, X/θ is the net value of the product produced and $\beta(x_2)/x_1$ is cost of power consumption for agitation.

If for the sake of simplicity we assume that $\beta(x_2)$ is a linear function of x_2, say

$$\beta(x_2) = 2x_2 \tag{2.3.8}$$

then Eq. 2.3.7 may be written as

$$\operatorname*{Max}_{x_1 x_2} \left[I(x_1, x_2) = \frac{x_2}{1 + x_1 x_2} - \frac{2x_2}{x_1} \right] \tag{2.3.9}$$

which expresses the *conflict* inherent in most optimization problems; in this particular case the conflict is between the shorter contact time achievable with a higher degree of agitation and the increased cost involved in meeting the power requirements for greater agitation.

Let us assume that the fractional extraction must be maintained at 0.9, thus

$$h_1(x_1, x_2) = \frac{x_1 x_2}{1 + x_1 x_2} - 0.9 = 0 . \tag{2.3.10}$$

Solution. The Lagrangian for this problem is given as

$$L(x_1, x_2, \lambda_1) = \frac{x_2}{1 + x_1 x_2} - \frac{2x_2}{x_1} - \lambda_1\left(\frac{x_1 x_2}{1 + x_1 x_2} - 0.9\right) \tag{2.3.11}$$

and our necessary conditions for a constrained maximum become

$$\frac{\partial L}{\partial x_1} = \frac{-(x_2)^2}{(1 + x_1 x_2)^2} + \frac{2x_2}{(x_1)^2} - \lambda_1\left[\frac{x_2}{1 + x_1 x_2} - \frac{x_1 (x_2)^2}{(1 + x_1 x_2)^2}\right] = 0 \tag{2.3.12}$$

and

$$\frac{\partial L}{\partial x_2} = \frac{1}{1 + x_1 x_2} - \frac{x_2 x_1}{(1 + x_1 x_2)^2} - \frac{2}{x_1} - \lambda_1\left[\frac{x_1}{1 + x_1 x_2} - \frac{(x_1)^2 x_2}{(1 + x_1 x_2)^2}\right] = 0 \tag{2.3.13}$$

Equations 2.3.10, 2.3.12, and 2.3.13 contain three unknowns, x_1, x_2, and λ_1.

The solution is readily obtained by multiplying Eq. 2.3.12 by x_1 and Eq. 2.3.13 by x_2 and subtracting to yield

$$\frac{x_1 x_2}{1 + x_1 x_2} - 4x_2 = 0 \tag{2.3.14}$$

By using Eq. 2.3.10 one immediately obtains the optimum at $x_1 = 40$, $x_2 = 0.225$. The optimality of this stationary point can be tested by substitution into Eq. 2.3.9, and the value of I compared with the value at the boundary points that satisfy Eq. 2.3.10.

Clearly one may envision objective functions with a larger number of variables and constraints, which take rather more involved forms. The actual technique for tackling these problems would, of course, be identical, but the use of computers would be highly desirable for solving the set of simultaneous nonlinear algebraic equations that would result. As we shall see in the next chapter, there may be more practical ways of proceeding in the case of very complex problems, because of the difficulty in solving these nonlinear equations.

2.4 INEQUALITY CONSTRAINTS—KUHN-TUCKER CONDITIONS

We shall now consider inequality constraints of the form

$$g_j(\mathbf{x}) \leq 0 \qquad j = 1, 2, \ldots, m_2 \tag{1.2.5}$$

Upon considering the inequality constraints (Eq. 1.2.5) we note that at any feasible point $\mathbf{x}$ (and, in particular, at the maximum $\mathbf{x}^*$) some of the constraints will be equalities $g_j^1(\mathbf{x}^*) = 0$ and the remainder strict inequality constraints $g_j^2(\mathbf{x}^*) < 0$. Thus at the maximum $\mathbf{x}^*$ we need only be concerned about the equality constraints $g_j^1(\mathbf{x}^*)$, since the strict inequalities $g_j^2(\mathbf{x}^*) < 0$ do not affect the problem.

We can form a Lagrangian

$$L(\mathbf{x}, \boldsymbol{\lambda}) = I(\mathbf{x}) - \boldsymbol{\lambda}^T\mathbf{g} \tag{2.4.1}$$

where $\boldsymbol{\lambda}$ is an m vector Lagrange multiplier. At the maximum if we partition $\boldsymbol{\lambda}$ and $\mathbf{g}(\mathbf{x})$

$$\boldsymbol{\lambda}^* = \begin{bmatrix} \boldsymbol{\lambda}_1^* \\ \hline \mathbf{0} \end{bmatrix}, \qquad \mathbf{g}(\mathbf{x}) = \begin{bmatrix} \mathbf{g}^1 \\ \hline \mathbf{g}^2 \end{bmatrix} \tag{2.4.2}$$

so that the multipliers for the inequality constraints at $\mathbf{x}^*$ are zero and only equality constraints are included in the Lagrangian,† then we may use the

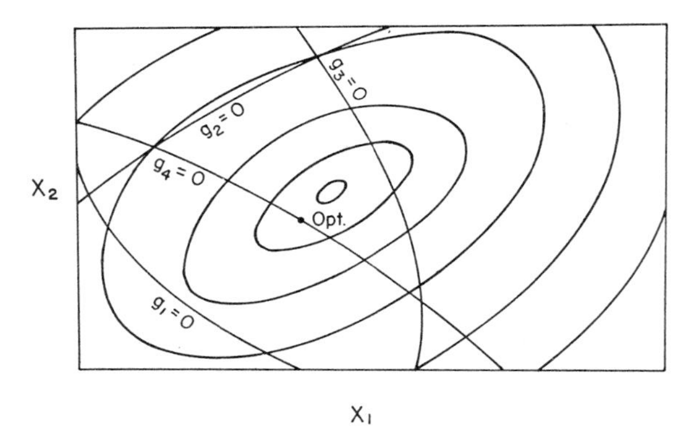

Fig. 2.9 An optimum constrained by $g_4(x_1, x_2) \leq 0$.

† These considerations are illustrated graphically in Fig. 2.9 where it is seen that g_1, g_2, g_3, are strict inequalities at the optimum (or need not be considered), while g_4 becomes an equality constraint.

results of the previous section, that is,

$$\left(\frac{\partial L}{\partial x_i}\right)_{\mathbf{x}^*} = \left(\frac{\partial I}{\partial x_i}\right)_{\mathbf{x}^*} - \boldsymbol{\lambda}^{*T}\left(\frac{\partial \mathbf{g}}{\partial x_i}\right)_{\mathbf{x}^*} = 0 \tag{2.4.3}$$

If we add the condition that $\mathbf{x}^* \geq 0$ must hold, (usually required for physical reasons), then our Lagrangian could possibly take its maximum at the boundary $x_i^* = 0$ where

$$\left(\frac{\partial L}{\partial x_i}\right)_{\mathbf{x}^*} < 0 \qquad i = 1, 2, \ldots, n \tag{2.4.4}$$

must hold. In this case, we could write our conditions for a maximum in the form:

$$\boxed{x_i^* \geq 0, \qquad \frac{\partial L}{\partial x_i} \leq 0, \qquad x_i \frac{\partial L}{\partial x_i} = 0 \qquad i = 1, 2, \ldots, n} \tag{2.4.5}$$

We can also establish similar conditions on $\boldsymbol{\lambda}$. Since we know that all small deviations about $\mathbf{x}^*$ which satisfy the constraints must satisfy

$$\delta I = \sum_{i=1}^{n} \delta x_i \left(\frac{\partial I}{\partial x_i}\right)_{\mathbf{x}^*} \leq 0 \tag{2.4.6}$$

and from Eq. 2.4.3 we have

$$\left(\frac{\partial I}{\partial x_i}\right)_{\mathbf{x}^*} = \sum_{j=1}^{m_2} \lambda_j^* \left(\frac{\partial g_j}{\partial x_i}\right)_{\mathbf{x}^*} \qquad i = 1, 2, \ldots, n \tag{2.4.7}$$

so that

$$\delta I = \sum_{i=1}^{n} \delta x_i \sum_{j=1}^{m_2} \lambda_j^* \left(\frac{\partial g_j}{\partial x_i}\right)_{\mathbf{x}^*} \leq 0 \tag{2.4.8}$$

is required. However, in order for x_i to satisfy the constraints for all possible small variations δx_i, we must have the following

$$\delta g_j = \sum_{i=1}^{n} \left(\frac{\partial g_j}{\partial x_i}\right)_{\mathbf{x}^*} \delta x_i \leq 0 \qquad j = 1, 2, \ldots, m_2 \tag{2.4.9}$$

But this implies that

$$\delta I = \sum_{j=1}^{m_2} \lambda_j^* \left(\sum_{i=1}^{n} \frac{\partial g_j}{\partial x_i} \delta x_i\right) \leq 0 \tag{2.4.10}$$

can be true for all possible small values of δx_i only if at $\mathbf{x}^*$, the vector λ_j^* satisfies

$$\lambda_j^* \geq 0 \qquad j = 1, 2, \ldots, m_2 \tag{2.4.11}$$

Thus we can summarize conditions on $\boldsymbol{\lambda}$ as

$$\lambda_j^* \geq 0, \qquad \left(\frac{\partial L}{\partial \lambda_j}\right)_{\mathbf{x}^*} = -g_j \geq 0, \qquad \lambda_j^* \frac{\partial L}{\partial \lambda_j} = 0, \qquad j = 1, 2, \ldots, m_2 \tag{2.4.12}$$

These conditions have been derived much more elegantly by Kuhn and Tucker in their first theorem [2]:

FIRST KUHN-TUCKER THEOREM (stationarity condition). In order that $\mathbf{x}^* \geq 0$ be the maximum of $I(\mathbf{x})$ subject to $g_j(\mathbf{x}) \leq 0 \quad j = 1, 2, \ldots, m_2$ it is necessary that Eqs. 2.4.5 and 2.4.12 hold at $\mathbf{x}^*$.

This is the first necessary condition for a maximum, subject to inequality constraints. As we have pointed out it is an extension of the treatment of equality constraints in the previous section.

An interesting point to note is that the first Kuhn-Tucker theorem requires† that the Lagrangian be a maximum with respect to $\mathbf{x}$ and a minimum with respect to $\boldsymbol{\lambda}$ at $\mathbf{x}^*$, $\boldsymbol{\lambda}^*$; that is,

$$L(\mathbf{x}, \boldsymbol{\lambda}^*) \leq L(\mathbf{x}^*, \boldsymbol{\lambda}^*) \leq L(\mathbf{x}^*, \boldsymbol{\lambda})$$

For this reason the theorem is sometimes called the "saddle point" theorem.

Let us pause now, and reflect on what we have done here. We have seen that all that is required for meeting the equality constraints is that the Lagrangian be stationary with respect to $\mathbf{x}$ and $\boldsymbol{\lambda}$ at $\mathbf{x}^*$ and $\boldsymbol{\lambda}^*$, and these two conditions together can be used to determine the maximum. The analysis can be extended to cover inequality constraints. To do this, we consider that the inequality constraints that *are binding* must become equality constraints at the maximum, that is, at $\mathbf{x}^*$. It follows that the previously developed criteria for equality constraints may be applied to the problem and that the nonbinding inequality constraints can be safely ignored. The form of the theorem for finding a minimum of a function follows trivially after a sign change in $I(x)$. The usefulness of the Kuhn-Tucker theory will be shown in Chapter 4 where we discuss the various nonlinear programming algorithms.

EXAMPLE 2.4.1. Let us conclude this section by a simple example, demonstrating an application of the Kuhn-Tucker theorem.

Let us maximize the profit from the operation of a casting shop, by selecting the optimum ingot size. The following relationships are thought to hold:

$$M = C_1 l^3, \quad A = C_2 l^2, \quad t_c = C_3 l^2$$

† Under the proper concavity assumptions.

where M is the mass of an ingot, l is a characteristic dimension of the ingot, A is the floor area required for handling an ingot, t_c is the time required for achieving a given, desired degree of solidification, and C_1, C_2, and C_3 are constants.

Let us consider that the overall profit from the operation, over a given time period may be described by the following relationship:

$$P = \frac{C_4 M}{t_c + C_5} - AC_6 \qquad (2.4.13)$$

where C_1, C_2, C_3, C_4, C_5, and C_6 are constants, and more specifically, C_5 refers to the "dead time" required for the loading and unloading of the molds. Equation 2.4.13 above may be written in terms of l, as follows:

$$P = \frac{C_1 C_4 l^3}{C_3 l^2 + C_5} - C_2 C_6 l^2 \qquad (2.4.14)$$

Equation 2.4.14 is now our objective function.

Regarding constraints, let us consider that for practical reasons, the permissible ingot size must fall between two prescribed limits, that is,

$$C_7 \leq l \leq C_8 \qquad (2.4.15)$$

so that our problem is now defined.

For purposes of illustration let us set, $C_1 = 5$; $C_2 = C_3 = C_4 = C_5 = 1$; and $C_6 = \frac{1}{3}$.

Then on writing $l = x$ we have that

$$P = \frac{5x^3}{1 + x^2} - \tfrac{1}{3}x^2 \qquad (2.4.16)$$

A plot depicting Eq. 2.4.16 is shown in Fig. 2.10 where it is seen that the maximum of P would occur at $x = 7.5$.

If $C_7 < 7.5$, $C_8 > 7.5$, then these constraints would not become operable and the problem could be solved as if it were unconstrained. However, let us consider the following cases:

(i) $C_8 = 6$. In this case the upper bound becomes binding and the Kuhn-Tucker condition (Eq. 2.4.12) tells us that the Lagrangian

$$L = P - \lambda_1(C_7 - l) - \lambda_2(l - C_8) \qquad (2.4.17)$$

has

$$\lambda_1 = 0, \text{ lower bound inoperable}$$

$$\lambda_2 > 0, \text{ upper bound binding}$$

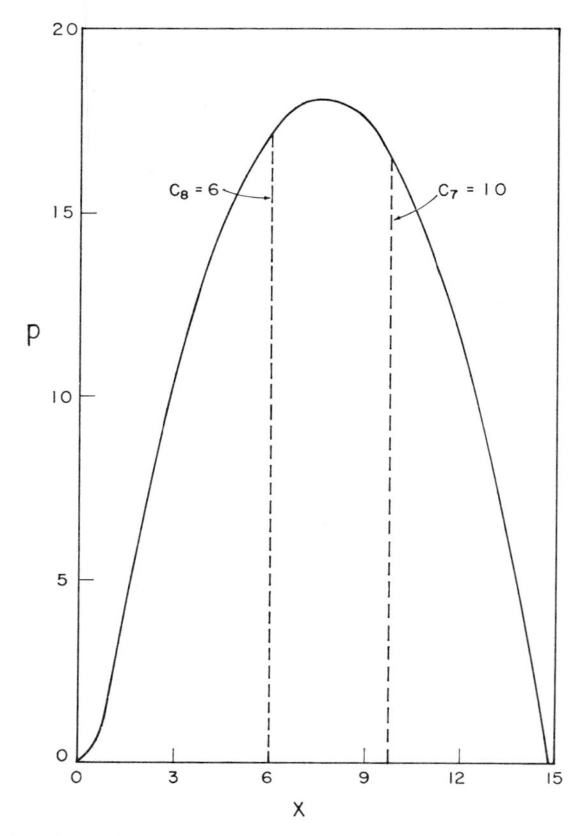

Fig. 2.10 Plot of P against x for Example 2.4.1.

and

$$\frac{\partial L}{\partial l} = \frac{\partial P}{\partial l} - \lambda_2 = 0 \tag{2.4.18}$$

which implies $\partial P/\partial l > 0$ is necessary at the constrained optimum $l = 6$. Inspection of Fig. 2.10 confirms this.

(ii) $C_7 = 10$. In this case the lower bound becomes binding and

$$\lambda_1 > 0, \qquad \lambda_2 = 0$$

and

$$\frac{\partial L}{\partial l} = \frac{\partial P}{\partial l} + \lambda_1 = 0 \tag{2.4.19}$$

which implies $\partial P/\partial l < 0$ is necessary at the constrained optimum $l = 10$. Again, inspection of Fig. 2.10 confirms this condition.

2.5 DUALITY

At this stage it is appropriate to discuss the concept of duality because of its importance in later chapters. We shall first illustrate the concept of duality for convex programming problems and then indicate how these ideas may be generalized.

Consider the "primal" problem

$$\underset{\mathbf{x}}{\text{Max}}\ [I(\mathbf{x})] \tag{2.5.1}$$

subject to

$$g_j(\mathbf{x}) \leq 0 \qquad j = 1, 2, \ldots, m \tag{2.5.2}$$

where $I(\mathbf{x})$ is a concave function and the constraints (Eq. 2.5.2) form a convex set. Through the use of Lagrange multipliers we can form the Lagrangian

$$L(\mathbf{x}, \boldsymbol{\lambda}) = I(\mathbf{x}) - \boldsymbol{\lambda}^T \mathbf{g}(\mathbf{x}) \tag{2.5.3}$$

which (according to the Kuhn-Tucker theorem of the last section) has the saddle point at the optimum $(\mathbf{x}^*, \boldsymbol{\lambda}^*)$

$$L(\mathbf{x}, \boldsymbol{\lambda}^*) \leq L(\mathbf{x}^*, \boldsymbol{\lambda}^*) \leq L(\mathbf{x}^*, \boldsymbol{\lambda}) \tag{2.5.4}$$

Note that we can consider the following problem

$$\underset{\boldsymbol{\lambda}}{\text{Min}} \left\{ \phi(\boldsymbol{\lambda}) = \underset{\mathbf{x}}{\text{Max}}\ [L(\mathbf{x}, \boldsymbol{\lambda}) = I(\mathbf{x}) - \boldsymbol{\lambda}^T \mathbf{g}(\mathbf{x})] \right\} \tag{2.5.5}$$

which also leads to the optimal pair $(\mathbf{x}^*, \boldsymbol{\lambda}^*)$. The problem given by Eq. 2.5.5 is called the *dual* of the primal problem (Eqs. 2.5.1 and 2.5.2). The dual can be rewritten

$$\left[\underset{\boldsymbol{\lambda}}{\text{Min}}\ [L(\mathbf{x}, \boldsymbol{\lambda}) = I(\mathbf{x}) - \boldsymbol{\lambda}^T \mathbf{g}(\mathbf{x}) \right] \tag{2.5.6}$$

subject to the Kuhn-Tucker conditions (Eq. 2.4.5)

$$x_i \frac{\partial L}{\partial x_i} = 0, \qquad \frac{\partial L}{\partial x_i} \leq 0, \quad x_i \geq 0 \tag{2.5.7}$$

We note that because $L(\mathbf{x}^*, \boldsymbol{\lambda}^*) = I(\mathbf{x}^*)$, and $L(\mathbf{x}^*, \boldsymbol{\lambda}) = \phi(\boldsymbol{\lambda})$, the *dual function*, $\phi(\boldsymbol{\lambda})$ provides an upper bound to the objective, that is,

$$I(\mathbf{x}) \leq I(\mathbf{x}^*) \leq \phi(\boldsymbol{\lambda}) \tag{2.5.8}$$

This, in fact, is one of the most important properties of the dual problem—*it provides a readily available upper bound in the objective at each stage of the calculations.*

The concept of duality can be generalized to a wide range of optimization problems (e.g., cf. [3]). In many situations the dual problem may be easier to solve than the primal problem and thus affords a computational shortcut to the optimum (cf. Sections 4.4, 4.5, and 4.7 for illustrations).

PROBLEMS

1. For the function

$$I = \sin^2 (x_1^2 + x_2) + x_2^2 \qquad x_1, x_2 \in [-\infty, \infty]$$

find
(a) All stationary points.
(b) All relative maxima.
(c) All relative minima.
(d) The global minimum; is it unique?

2. For the function

$$I = (x_2 - x_1)^2 + 2x_1^2 \qquad x_1, x_2 \in [-\infty, \infty]$$

and for x_1, x_2 satisfying

$$x_1^2 + x_2^2 = 1$$

find
(a) All stationary points of the Lagrangian.
(b) All relative maxima of I.
(c) All relative minima of I.
(d) The global minimum; is it unique?

3. The rate of production from a continuously operating hydrometallurgical reactor is given by the following expression:

$$\text{Production rate} = vC_I\left[1 - \frac{v^2}{Vk_0e^{a(T-T_0)} + v}\right]$$

where v = volumetric flow rate
V = volume of reactor
k_0 = frequency factor in reaction rate expression
T = absolute temperature
C_I = inlet concentration of soluble constituent

We wish to find the optimum values of v and T, for a fixed reactor volume. A realistic expression for the objective function could take the following form:

$$I = vC_I\alpha_1\left[1 - \frac{v^2}{Vk_0e^{a(T-T_0)} + v}\right] - \alpha_2(T - T_0)^2 - vC_I\alpha_3$$

Here α_1, α_2, and α_3 are the various cost factors. For the set of values indicated below

(i) Find the values of v and T which maximize I in an unconstrained situation.

(ii) Maximize I for the case when $250 \leq T \leq 400°K$,

$$0 \leq v \leq 1000 \text{ ft}^3/\text{hr}$$

$V = 1000 \text{ ft}^3$
$C_I = 2 \text{ lb/ft}^3$
$\alpha_1 = 0.2$
$k_0 = 0.2 \text{ hr}^{-1}$
$a = 0.1°K^{-1}$
$T_0 = 300°K$
$\alpha_2 = 0.005 \text{ \$/(°K)}^2 \text{ hr}$
$\alpha_3 = 0.05 \text{ \$/lb}$

REFERENCES

1. G. Hadley, *Non-linear and Dynamic Programming*, Addison-Wesley, 1964.
2. H. W. Kuhn and A. W. Tucker, *Second Berkeley Symposium on Mathematical Statistics and Probability*, California, 1951.
3. D. G. Luenberger, *Optimization by Vector Space Methods*, Wiley, 1969, p. 223.

3 Unconstrained Peak Seeking Methods

In the preceding chapter we developed the analytical criteria for the existence of constrained and unconstrained maxima or minima. We noted that in special cases, when relatively simple functions were involved, the analytical techniques described there would provide a straightforward way of finding the maximum of the objective function.

In the majority of practical cases, however, such procedures cannot be applied and, therefore, the main practical usefulness of the techniques discussed in Chapter 2 is to *test whether a given point is indeed the maximum*. The actual methods required for searching for such optima form the subject matter of Chapters 3 and 4. In this chapter we shall restrict ourselves to unconstrained problems, that is, we shall seek the optimum of an objective function, in the absence of any equality or inequality constraints.

Regarding the contents of this chapter, Section 1 deals with functions of a single variable, whereas Sections 2 and 3 deal with multivariable search methods. *Gradient methods*, that is, problems where information is available on the derivatives of the function, are discussed in Section 2, and *Nongradient methods*, that is, systems where there is no information on the derivatives are discussed in Section 3. The chapter is then concluded with a comparison of the various unconstrained peak seeking methods.

3.1 SINGLE VARIABLE ELIMINATION METHODS

In this section we discuss problems that involve just one independent variable, under conditions where there is no information available on the derivatives. Basically, all the techniques that are presented for dealing with these problems involve trying various values of the independent variable, with a view of ultimately selecting the optimum.

From a practical viewpoint, problems of this type may arise as follows:

Plant Scale Experiments

Here if one can regard one operating parameter as the independent variable, such as blast temperature, degree of preheat, feed composition, and so on, then we can perform a number of experiments, with a view of finding the optimum value of this *input variable*. The appropriate search techniques will allow us to reach this optimum in a relatively small number of steps.

Mathematical Model Experiments

Single variable elimination methods may also be used, when we have a fairly *complex mathematical model* for a given process, so that "internal" manipulation of the various functions forming the model would be too cumbersome. Under these conditions, the mathematical model is treated like a "black box" and, here again, we are just concerned with finding the particular value of the *input variable* which gives us the optimum value for the output.

To summarize the preceding discussion, all the methods dealing with a single variable will be concerned with performing "experiments" such that an input variable is selected, which then results in the generation of an output variable $I(x)$, as sketched in Fig. 3.1.

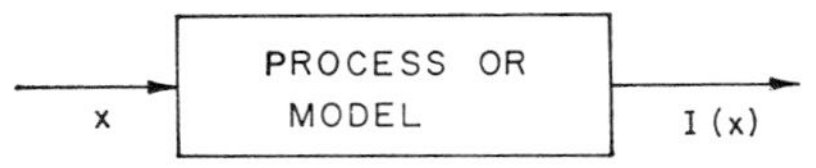

Fig. 3.1 Relationship between the input, x and the output $I(x)$ in an experimental apparatus, or in a process model.

The treatment that follows relies on the assumption that $I(x)$ is a unimodal function of x, that is, it has only one peak or valley. Figures 3.2*a* and *b* show functions with unimodal maxima and minima, respectively, while the curve in Fig. 3.2*c* is multimodal. The techniques to be discussed in this

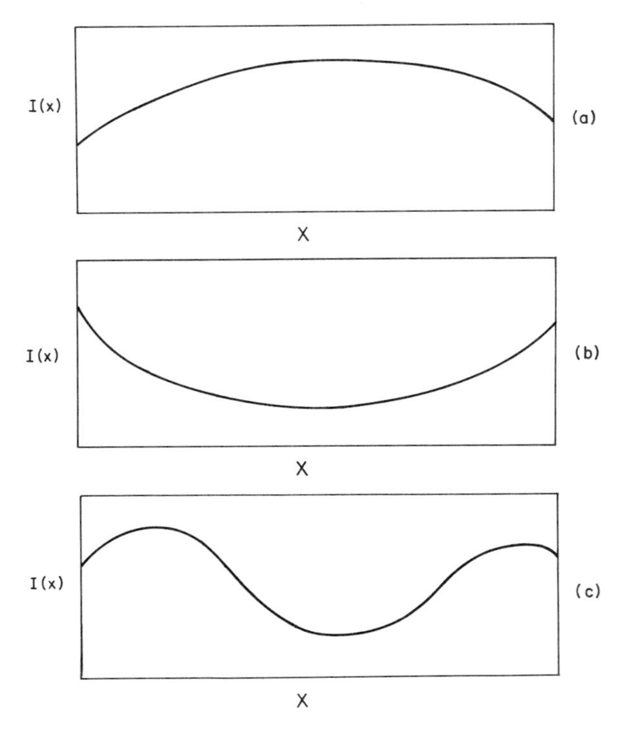

Fig. 3.2 Illustration of unimodal and multimodal functions: (*a*) unimodal maximum, (*b*) unimodal minimum, and (*c*) multimodal function.

section all rely on the unimodality of the objective function, which enables us to confine our search to a progressively smaller interval by eliminating certain portions of the domain from further consideration.

In the following, the search techniques will be developed on the assumption that $I(x)$ is a unimodal maximum; analogous procedures could be presented for curves having a unimodal minimum.

Dichotomous Search

Let us consider that the interval to be searched lies within $0 \leq x \leq 1$ and our objective is to find the maximum located at x^*. In concept the simplest way of narrowing the interval within which x^* is located is to use a procedure which halves the remaining interval at each step. This *dichotomous search* technique can be described as follows:

1. We select two points separated by a small distance, ϵ, in the center of the interval. Since we have assumed unimodality, then if $I(x_1) > I(x_2)$ (cf. Fig. 3.3), the area $x \geq x_2$ can be eliminated from further consideration.
2. We select two points, ϵ apart in the center of the remaining interval and if $I(x_4) > I(x_3)$, then area $x \leq x_3$ can be eliminated.
3. We continue halving the interval until the remaining interval is within the desired tolerance of the maximum. The procedure is illustrated in Fig. 3.3.

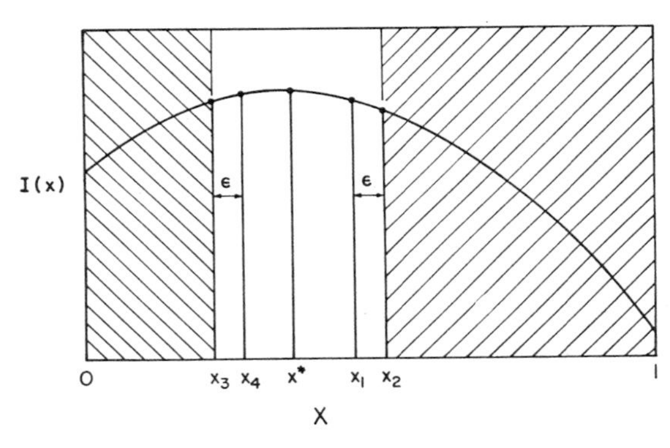

Fig. 3.3 The dichotomous search.

Since (neglecting ϵ) the interval is halved for every two experiments, the interval remaining after n experiments* is

$$L_n = 2^{-n/2} \tag{3.1.1}$$

This technique is very simple, but not very efficient as can be seen by considering the methods to be discussed subsequently.

Fibonacci Search

This search technique has the property that it furnishes the smallest possible interval L_n after n experiments, that is, it is the *optimum* elimination procedure. To see why this is true, let us consider the search problem in reverse. Let us assume that we have performed all but two of the n experiments. With only two experiments left, it is easy to show that their best placement is in the center of the interval ϵ apart so that if we neglect ϵ (cf. Fig. 3.4)

$$L_n = \tfrac{1}{2}L_{n-1} \tag{3.1.2}$$

where L_{n-1} is the length remaining when only one of the experiments is in the center. Now to insure that we obtain the desired L_n in n experiments,

* This expression contains a small error because ϵ was neglected; the true expression is $L_n = (1 + [2^{n/2} - 1]\epsilon)/2^{n/2}$.

we need to have L_{n-1} take the value $2L_n$ no matter which end is discarded after experiment $n - 1$. Thus we must have (as shown in Fig. 3.4)

$$L_{n-2} = L_n + L_{n-1} = 3L_n \tag{3.1.3}$$

This type of reasoning can be carried back to

$$L_{j-1} = L_j + L_{j+1} \qquad j = 2, 3, \ldots, n-1 \tag{3.1.4}$$

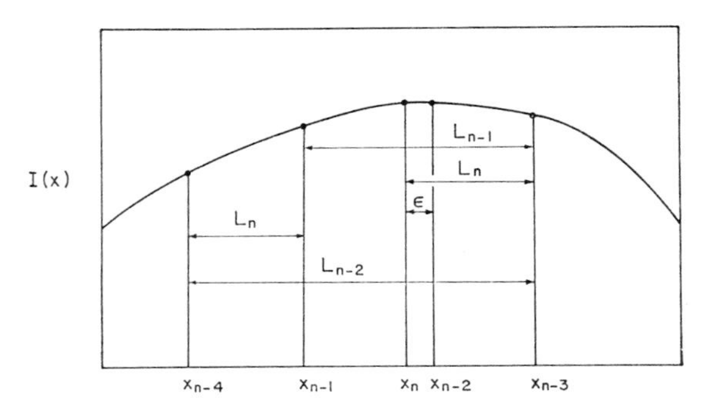

Fig. 3.4 The Fibonacci search.

which leads to

$$\begin{aligned} L_{n-3} &= 5L_n \\ L_{n-4} &= 8L_n \\ L_{n-5} &= 13L_n \\ &\vdots \end{aligned} \tag{3.1.5}$$

At this stage it is convenient to define the Fibonacci† numbers F_j as:

$$F_j = F_{j-1} + F_{j-2}, \quad F_0 = F_1 = 1, \qquad j = 2, 3 \cdots \tag{3.1.6}$$

Thus Eq. 3.1.3 may be generalized to

$$L_{n-j} = F_{j+1}L_n \qquad j = 2, 3, \ldots, n-1 \tag{3.1.7}$$

so that for $j = n - 1$

$$L_n = \frac{1}{F_n} \tag{3.1.8}$$

† Named for the Italian mathematician Fibonacci of Pisa who used them to estimate the reproduction rate of rabbits.

Thus after n steps the reduction in the interval is the reciprocal of F_n. In order to locate the position of the first experiment let us use Eqs. 3.1.7 and 3.1.8.

$$L_2 = F_{n-1}L_n = \frac{F_{n-1}}{F_n} \tag{3.1.9}$$

In order to illustrate the application of the method let us solve the following problem:

EXAMPLE 3.1.1. Find the maximum of $I(x)$, an arbitrary curve, sketched in Fig. 3.5, by reducing the initial interval $0 \leq x \leq 1$ to less than 0.05.

Solution. Since L_n must be 0.05 or less, then from Eq. 3.1.8, $F_n \geq 20$ and since $F_7 = 21$ is the next largest Fibonacci number, then we need seven experiments to do this. Let us divide $0 \leq x \leq 1$ into 21 equal spaces and number them 1 to 21. Then from Eq. 3.1.9, $L_2 = 13/21$, so x_1 must be at $13/21$, x_2 at $8/21$. Since $I(x_2) > I(x_1)$, then $x > x_1$ is eliminated.

Then from Eq. 3.1.4 we must place x_3 symmetrically with x_2 so that $x_3 = 5/21$. We continue by placing each experiment symmetrically in

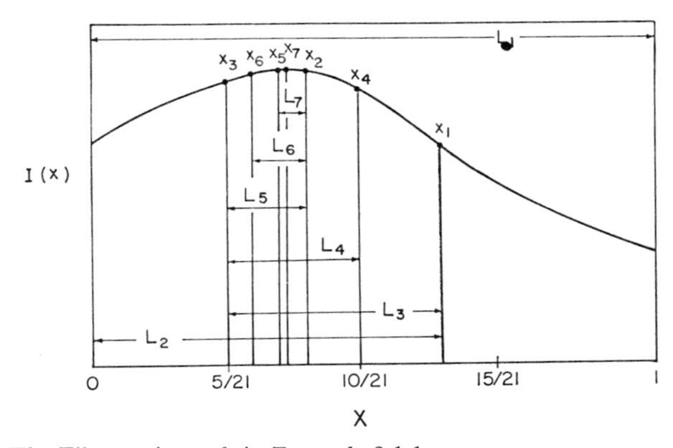

Fig. 3.5 The Fibonacci search in Example 3.1.1.

the remaining interval and obtain $x_4 = 10/21$, $x_5 = 7/21$, $x_6 = 6/21$, $x_7 = 7/21 + \epsilon$ to give the optimum between 7/21 and 8/21. The sequence of experiments is shown in Fig. 3.5.

We note that the starting point depends on the total number of experiments, so that we must decide on the number of experiments before beginning the search. This is, in fact, the principal disadvantage of the method. However, a technique not having this disadvantage is described in the following.

The Golden Section

An important property of the Fibonacci search is embodied in Eq. 3.1.4 which requires that succeeding experiments be selected symmetrically in the remaining interval. If we require further that each step should reduce the interval by the same factor, that is,

$$\tau = \frac{L_{j-1}}{L_j} = \frac{L_j}{L_{j+1}} \tag{3.1.10}$$

Then Eq. 3.1.4 becomes

$$\tau = -1 + \tau^2 \tag{3.1.11}$$

which may be solved for τ to obtain:

$$\tau = \frac{1 + \sqrt{5}}{2} = 1.618033989 \cdots \dagger$$

Then

$$L_n = \tau^{1-n} \tag{3.1.12}$$

and the first two experiments are selected so that $L_2 = 1/\tau$.

Thus we have a method which embodies many of the attractive properties of the Fibonacci search, without requiring that the number of experiments be known in advance.

The relative efficiency of these methods are shown in Table 3.1.1. As can be seen, the method of the golden section is nearly as efficient as the Fibonacci search while it does not require that the number of experiments be specified in advance. The dichotomous search is much less efficient than either of the others. Some further illustrations of the relative efficiencies of these three search techniques may be found in the homework problems given at the end of the chapter.

As a very real illustration of the value of these techniques in choosing optimization experiments for actual processes or process models, we apply the golden section to the following problem.

† This number τ was first derived by Euclid and a rectangle whose sides were in the ratio τ was so pleasing to the eye of the Greeks that doors, windows, and other parts of Greek structures were made of these "golden sections." A more immediate verification to the beauty of the number lies in the fact that the numbers 36-22-36 are approximately in the ratio τ.

TABLE 3.1.1 REDUCTION L_1/L_n

Experiments	Dichotomous	Fibonacci	Golden section
1	1	1	1
2	2	2	1.62
3	2	3	2.62
4	4	5	4.24
5	4	8	6.85
6	8	13	11.09
7	8	21	17.94
8	16	34	29.0
9	16	55	47.0
10	32	89	76.0
11	32	144	123
12	64	233	199
13	64	377	322
14	128	610	521
15	128	987	843
16	256	1,597	1,364
17	256	2,584	2,207
18	512	4,181	3,570
19	512	6,765	5,778
20	1,024	10,946	9,346

EXAMPLE 3.1.2

The Optimal Bath Depth in a Ladle Degassing Process [1–3]

Let us evaluate the optimal bath depth in a ladle degassing process by using the golden section technique applied to a mathematical model of the process to be described subsequently.

Basically, the operation consists of placing molten steel, containing the undesirable impurities of dissolved hydrogen, nitrogen, and oxygen, into a ladle, the freeboard of which is then evacuated. As a result, the steel becomes supersaturated with respect to these constituents, a fraction of which is then transferred into the gas phase and is thus removed from the steel. A schematic sketch of the system is given in Fig. 3.6.

If no external "purge gas" is used, then the process takes place in the following three sequential steps:

(i) Gas bubbles nucleate at the bottom of the container; these bubbles then grow, until they reach their detachment diameter.

(ii) On detachment the bubbles rise through the molten metal in the

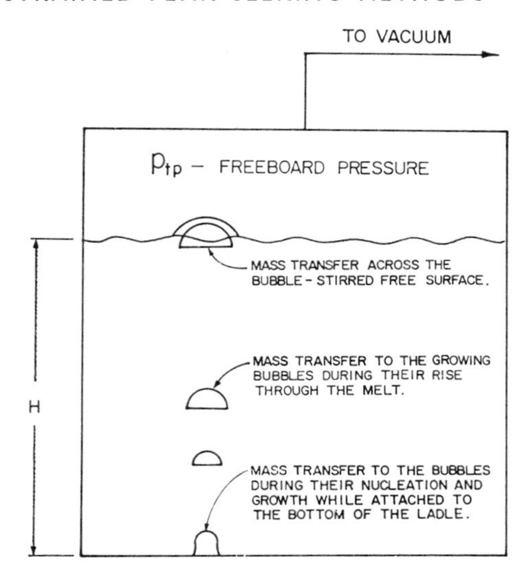

Fig. 3.6 Sketch of the vacuum degassing operation discussed in Example 3.1.2.

course of which growth will occur as a result of mass transfer into the bubble and the gradual reduction in the absolute pressure as the bubble ascends.

(iii) Finally, when the bubble reaches the free surface of the melt, the agitation provided by the passage of the bubble allows the direct transfer of the gaseous impurities from the metal into the gas space.

Our objective in this example is to find the optimal bath depth, which provides the maximum rate of mass transfer for a given nucleation site at the bottom of the bath; this quantity should be proportional to the mass transfer rate per unit cross sectional area of the bath.

Before doing any calculations, it may be worthwhile to outline the physical reasoning which suggests the existence of such an optimum. The rate at which bubbles are nucleated, or the frequency of their production at a given site, shows some form of proportionality to the effective driving force; thus, for a given solute gas concentration, the lower the absolute pressure at the bottom of the ladle, the higher the frequency of nucleation. It follows that this frequency will tend to decrease with the depth of the bath. For progressively increasing bath depths, eventually a stage will be reached when bubble nucleation is completely suppressed.

Regarding the rise of the gas bubbles through the molten steel, the greater the depth of the metal, the larger the contact time and, therefore, the greater the amount of solute that can be transferred into the bubble.

The *conflicting roles* played by the metal depth must be apparent from the foregoing discussion: a high bubble nucleation frequency would require a shallow metal depth, whereas effective mass transfer between the rising bubble and the melt would need a large metal depth.

Finally, the rate at which solute is transferred across the free metal surface will depend on both the frequency and the size of the bubbles [4].

In order to find the optimal bath depth for a given top pressure, P_{tp} and solute concentration, the problem can be formulated in the following way.

Formulation

Growth While Attached to the Bottom Solid Surface

Here we need a statement of the diffusion equation, and of the equations of continuity and motion, describing the movement of the surface of the bubble. These two sets of equations are coupled through the terms containing the pressure within the bubble.

The equations of continuity and motion may be combined and written as [2]:

$$R\ddot{R} + \tfrac{3}{2}\dot{R}^2 = \frac{1}{\rho_L}\left[P_G(R) - \frac{2\sigma}{R} - P_L(\infty)\right] \qquad R_0 \le R \le R_f \quad (3.1.13)$$

where R is the radius of the bubble, R_0 and R_f are the initial and final (detachment) radii of the bubble, respectively,

$$\dot{R} = \frac{dR}{dt}, \qquad \ddot{R} = \frac{d^2R}{dt^2}$$

ρ_L is the density of the melt, $P_G(R)$ is the pressure inside the bubble at a given time, and σ is the interfacial tension.

Equation 3.1.13 expresses the fact that the pressure within the bubble must be larger than $P_L(\infty)$, which is the pressure within the melt at the same depth as the bubble, in order that growth may be sustained.

The initial conditions specify the initial bubble size (R_0) together with the fact that the initial growth rate is zero:

$$R = R_0, \qquad \frac{dR}{dt} = 0, \qquad \text{at} \quad t = 0 \qquad (3.1.14\text{–}15)$$

The conservation of the diffusing species (i.e., the solute) may be written as

$$\frac{\partial C}{\partial t} = D\left(\frac{\partial^2 C}{\partial r^2} + \frac{2}{r}\frac{\partial C}{\partial r}\right) - \left(\frac{R}{r}\right)^2 \dot{R}\frac{\partial C}{\partial r}, \qquad r > R, \qquad R_0 \le R \le R_f \quad (3.1.16)$$

The boundary conditions for Eq. 3.1.16 are written as

$$C = C_0 \qquad \text{at} \quad r = \infty \tag{3.1.17}$$

$$D\frac{\partial C}{\partial r} = \rho_G \dot{R}, \qquad r = R \tag{3.1.18}$$

and

$$P_G(R) = f(C) \qquad \text{at} \quad r = R \tag{3.1.19}$$

Here Eq. 3.1.17 expresses the fact that at some distance from the bubble surface the solute is maintained at its original concentration level; Eq. 3.1.18 states that the growth rate of the bubble multiplied by the molar density of the gas (ρ_G) equals the molar flux of the solute to the bubble surface.

Finally, Eq. 3.1.19 expresses the existence of thermodynamic equilibrium at the bubble surface. For the particular system considered in this example, namely the transfer of hydrogen from the melt to the gas phase, Eq. 3.1.19 takes the following form:

$$C = K_G^{-1/2}(R) \qquad \text{at} \quad r = R \tag{3.1.20}$$

which reflects the fact that the reaction at the bubble surface involves

$$2H_{\text{metal}} = H_{2(\text{gas})}$$

The above Eqs. 3.1.13 to 3.1.20 describe the growth of the bubble up to its detachment radius, the value of which may be estimated by the expression

$$R_f = 7.4 \times 10^{-3}\theta \left(\frac{2\sigma}{g(\rho_L - \rho_G)}\right)^{1/2} \tag{3.1.21}$$

where θ is the contact angle in degrees.

The system of Eqs. 3.1.13 to 3.1.20 may be solved numerically, and as a result we can calculate the time required for the bubble to grow to its detachment radius, for a given set of conditions.

Let us designate this time as t_f, the reciprocal of which is the bubble frequency, designated f.

Bubble Growth During Its Rise Through the Melt

Let us assume, as a first approximation, that the translational motion will not affect the pressure within the bubble, except for the fact that during its rise the gas bubble will be exposed to a progressively lower environmental pressure.

Under these conditions the behavior of the bubble may be described by a set of equations, almost identical to that given earlier, except that

(*a*) the initial radius is R_f, and
(*b*) the equation of motion must take the following form:

$$R\ddot{R} + \tfrac{3}{2}\dot{R}^2 = \frac{1}{\rho_L}\left[P_G(R) - P_{L\infty}(h)\right] \tag{3.1.22}$$

where the term $P_{L\infty}(h)$ provides allowance for the progressively lower environmental pressure exerted on the bubble during its rise. The term containing the surface tension, which was important during initial bubble growth (due to the small radius of curvature), may be neglected at this stage.

The equations describing bubble growth during its rise may also be solved numerically (using the known relationship between bubble rising velocity and bubble volume), and as a result we obtain R_t and $\rho_G{}^t$, that is, the radius and the gas density within the bubble exiting at the free melt surface.

Mass Transfer from the Free Melt Surface Augmented by Bubble Agitation

It has been shown that mass transfer from a free liquid surface, agitated by rising gas bubbles, may be described by the following expression [4]:

$$\bar{N} = \sqrt{\frac{Df}{\pi}}\,(C_0 - C_{eq}) \tag{3.1.23}$$

where $\bar{N}$ is the molar flux of the diffusing species, and C_{eq} corresponds to the concentration of the solute in equilibrium with that in the gas space above the melt. The flux $\bar{N}$ is computed over the area projected by the rising bubble as it meets the free surface.

On combining the effects detailed above, we may calculate the total mass transfer rate for a given nucleation site, which is given as

$$N_T = N_B + \bar{N}$$

where N_B is the mass transfer to the bubble during its initial growth and subsequent rise through the metal, and $\bar{N}$ is the secondary mass transfer induced by agitation of the free surface. The total mass transfer can then be written

$$N_T(H) = f\frac{4\pi}{3}R_t{}^3\rho_G{}^t + \sqrt{\frac{Df}{\pi}}\,(C_0 - C_{eq}) \tag{3.1.24}$$

Solution. Let us apply the golden section technique to the problem by using the following algorithm:

(i) Let us restrict our search of the optimal depth H to $0 < H \leq 5$ cm. Our first guess of H then is $H_1 = 5/1.618 = 3.09$ cm.

(ii) Solve the modeling Eqs. 3.1.13 to 3.1.24 numerically to produce the mass transfer rate, $N_T(H)$.

(iii) Place the next experiment symmetrically in the interval (i.e., $H_2 = 5-3.09 = 1.91$ cm) and go back to step (ii) and iterate until the optimal H is found to the desired accuracy.

This procedure was applied to the problem of removing hydrogen from molten steel [3], with the resulting experiments shown in Fig. 3.7. By using

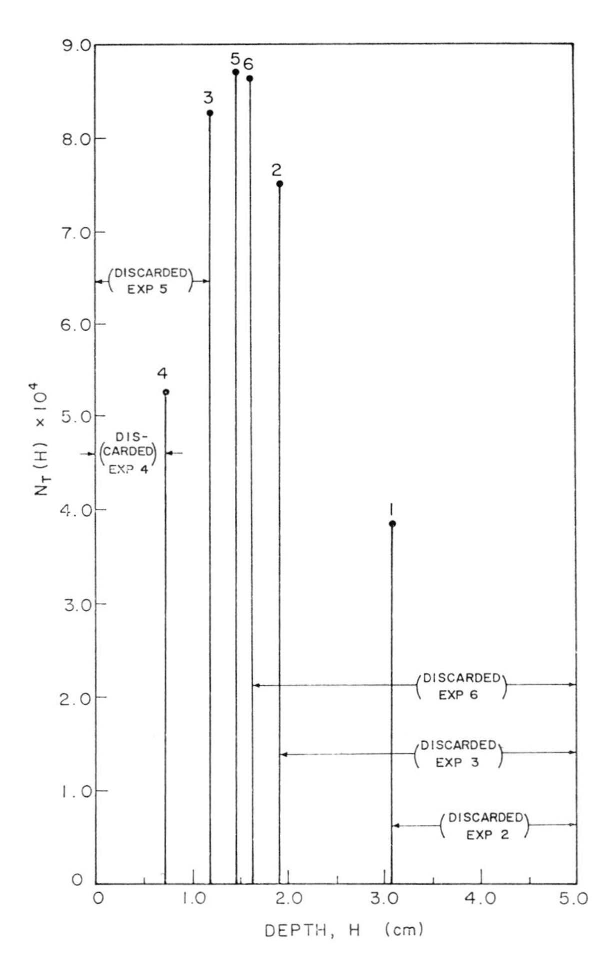

Fig. 3.7 Determination of the optimal depth in Example 3.1.2.

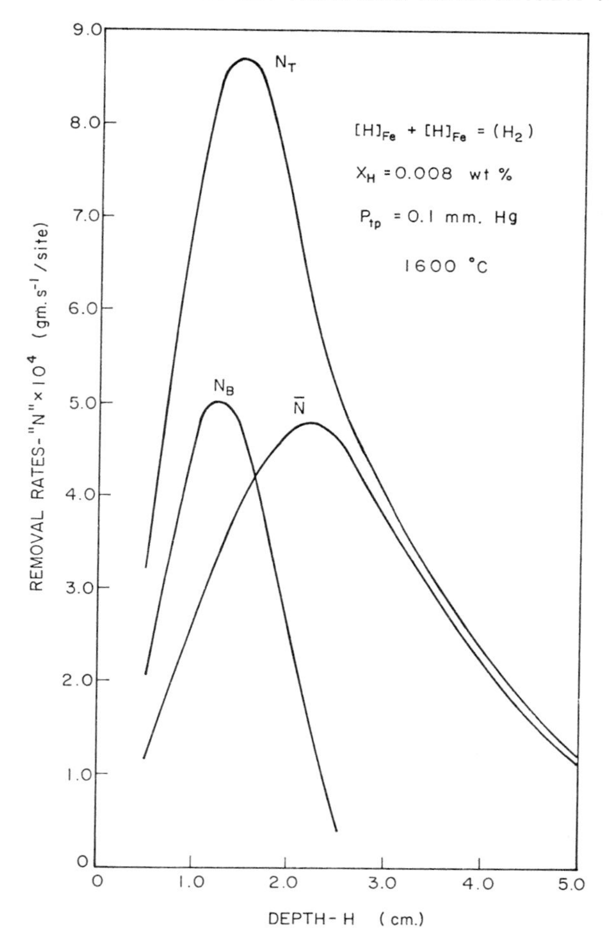

Fig. 3.8 The optimal depths for N_B, $\bar{N}$, and N_T in Example 3.1.2.

only six experiments we have found the optimal value of H to lie within $1.18 \text{ cm} \leq H \leq 1.63$ cm.

An added benefit of choosing experiments by this approach is that only these six experiments are needed to draw a very accurate, smooth curve of $N_T(H)$. Thus these elimination methods are also very helpful in choosing the minimum number of points needed to plot functions having a maximum or a minimum.

For the interested reader, plots of $\bar{N}$, N_B, and N_T are given in Fig. 3.8 and show optimal depths for the individual terms N_B and $\bar{N}$.

It is stressed that this example uses a simplified model of a fairly complex process; in a practical situation it would be desirable to perform a number of physical experiments with a view of defining the optimum. The number and actual placement of these experiments would be best undertaken with the aid of an elimination technique—perhaps, with consideration of the experimental errors involved [5].

The preceding example was taken from the metallurgical field; a further problem dealing with a single variable and involving the optimum temperature in a continuous stirred tank reactor is posed as a homework problem at the end of the chapter.

3.2 MULTIVARIABLE SEARCH PROCEDURES—GRADIENT METHODS

Let us consider a multivariable maximization problem, such that

$$\underset{x_i}{\text{Max}}\ I(\mathbf{x}) \tag{3.2.1}$$

Here $\mathbf{x}$ is a vector, with the components $x_1, x_2, \ldots, x_n$ and $I(\mathbf{x})$ is the output function, as illustrated in Fig. 3.9.

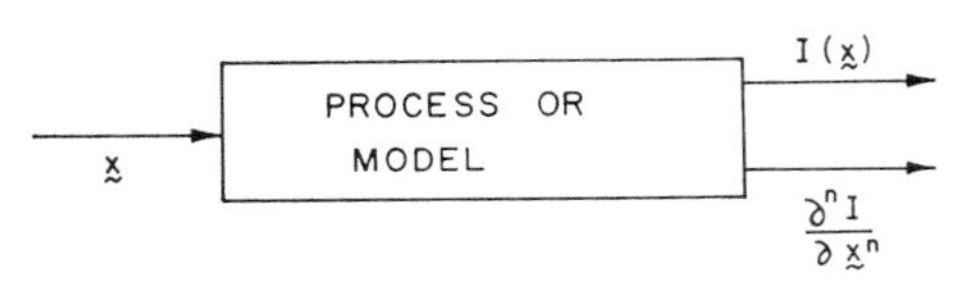

Fig. 3.9 Relationship between input, output, and the derivatives of the output for a process or a process model.

When the partial derivatives of $I(\mathbf{x})$ are available, we may proceed by performing an experiment on the process at the point $\mathbf{x}$ and calculating $I(\mathbf{x})$ and perhaps its derivatives, $\left(\frac{\partial^k I}{\partial x_i^k}\right)_{\mathbf{x}}$ (cf. Fig. 3.9). From this information we determine the correction

$$\delta\mathbf{x} = \begin{bmatrix} \delta x_1 \\ \delta x_2 \\ \cdot \\ \cdot \\ \cdot \\ \delta x_n \end{bmatrix}$$

which will produce the maximum increase in I to obtain the new point $\mathbf{x} = \bar{\mathbf{x}} + \delta\mathbf{x}$. We shall now discuss a number of techniques available for accomplishing this objective.

Gradient and Steepest Ascent Methods

These techniques use the local gradient $\partial I/\partial \mathbf{x}$ defined as

$$\frac{\partial I}{\partial \mathbf{x}} = \begin{bmatrix} \dfrac{\partial I}{\partial x_1} \\ \dfrac{\partial I}{\partial x_2} \\ \cdot \\ \cdot \\ \cdot \\ \dfrac{\partial I}{\partial x_n} \end{bmatrix}$$

in order to make the best choice of the correction $\delta\mathbf{x}$. For a small perturbation, $\delta\mathbf{x}$, about our first estimate, $\mathbf{x} = \bar{\mathbf{x}}$, we obtain the first-order approximation

$$\delta I = I(\bar{\mathbf{x}} + \delta\mathbf{x}) - I(\bar{\mathbf{x}}) = \left(\frac{\partial I}{\partial \mathbf{x}}\right)_{\bar{\mathbf{x}}} \delta\mathbf{x} + 0(\delta\mathbf{x}^2) \tag{3.2.2}$$

Thus if we correct our estimate $\bar{\mathbf{x}}$ in the direction of *steepest ascent*

$$\delta\mathbf{x} = \epsilon\left(\frac{\partial I}{\partial \mathbf{x}}\right)_{\bar{\mathbf{x}}}, \qquad \epsilon > 0 \tag{3.2.3}$$

we obtain the greatest local improvement in I and $\delta I > 0$ so that convergence always is assured, although this may require an infinite number of steps.

The procedure outlined here in quite general terms is perhaps best elucidated if we examine the problem in two dimensions.

Let us consider a function $I(\mathbf{x})$, where the vector $\mathbf{x}$ has two components, that is,

$$\mathbf{x} = \begin{bmatrix} x_1 \\ x_2 \end{bmatrix}$$

Figure 3.10 shows $I(x) = I(x_1, x_2)$ as a plot of the contours of $I =$ constant, in the $x_1 - x_2$ plane.

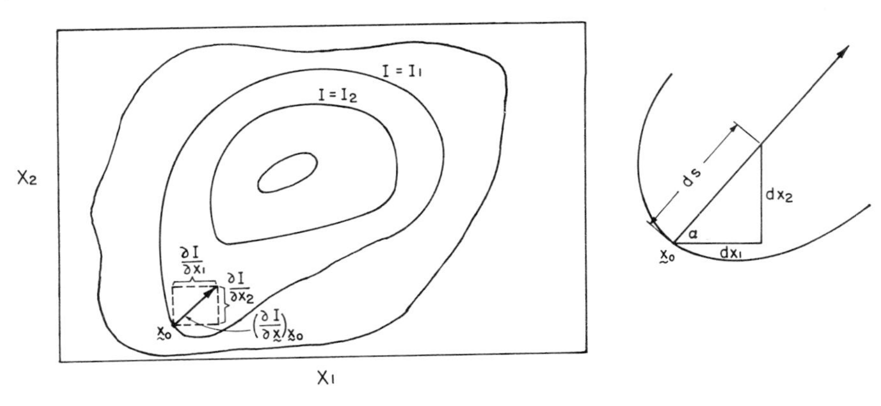

Fig. 3.10 An illustration of the direction of the steepest ascent.

Let us now consider a fixed point, on the $I = I_1$ contour, corresponding to $\mathbf{x}_0$, that is,

$$\mathbf{x}_0 = \begin{bmatrix} x_{10} \\ x_{20} \end{bmatrix}$$

and let us form the gradient

$$\frac{\partial I}{\partial \mathbf{x}} \equiv \begin{bmatrix} \dfrac{\partial I}{\partial x_1} \\ \dfrac{\partial I}{\partial x_2} \end{bmatrix}$$

at this location as illustrated in Fig. 3.10.

For a small change about $\mathbf{x}_0$ in any direction, we may write that the corresponding change in $I(\mathbf{x})$ is given by

$$dI = \left(\frac{\partial I}{\partial x_1}\right)_{\mathbf{x}_0} dx_1 + \left(\frac{\partial I}{\partial x_2}\right)_{\mathbf{x}_0} dx_2 \tag{3.2.4}$$

Let us now assume, that we select increments in x_1 and x_2, that is, dx_1 and dx_2 and that we are forced to move along the line dS, which forms an angle of α with the x_1 axis, as indicated on the inset of Fig. 3.10. If dS is the distance along the line of the search, then by the chain rule of differentiation we have that

$$\frac{dI}{dS} = \left(\frac{\partial I}{\partial x_1}\right)_{\mathbf{x}_0} \frac{dx_1}{dS} + \left(\frac{\partial I}{\partial x_2}\right)_{\mathbf{x}_0} \frac{dx_2}{dS} \tag{3.2.5}$$

The quantity dI/dS is called the *directional derivative of* I and represents (within a linear approximation) the increase in I per unit movement S in the direction described by the angle α. Let us recall that our objective in this procedure is to find the direction, that is, the value of α or the values of dx_1 and dx_2, which will give the maximum increment in dI. In other words, we wish *to maximize the directional derivative.*

In seeking to maximize the directional derivative, we may proceed as follows: from standard trigonometric arguments we have that

$$\frac{dx_2}{dx_1} = \tan \alpha, \quad \frac{dx_1}{dS} = \cos \alpha \qquad \text{and}$$

$$\frac{dx_2}{dS} = \sin \alpha$$

By using these definitions, Eq. 3.2.5 may be written as

$$\frac{dI}{dS} = \left(\frac{\partial I}{\partial x_1}\right)_{\mathbf{x}_0} \cos \alpha + \left(\frac{\partial I}{\partial x_2}\right)_{\mathbf{x}_0} \sin \alpha \tag{3.2.6}$$

Equation 3.2.6 may be interpreted as a relationship between the *directional derivative* and α.

It may be shown, by a simple trigonometric argument, that there exist two values of α for which $dI/dS = 0$.

On setting $dI/dS = 0$, after some rearrangement Eq. 3.2.6 may be written as

$$\tan \alpha = -\frac{(\partial I/\partial x_1)_{\mathbf{x}_0}}{(\partial I/\partial x_2)_{\mathbf{x}_0}} \tag{3.2.7}$$

The value of α satisfying Eq. 3.2.7 defines the position of the *contour tangent.*

Between the pairs of values of α where dI/dS is zero, the directional derivative will have a stationary point. The location of this stationary point is defined by

$$\frac{d}{d\alpha}\left(\frac{dI}{dS}\right) = 0$$

which after some manipulation may be written as

$$\tan \alpha = \frac{dx_2}{dx_1} = \frac{(\partial I/\partial x_2)_{\mathbf{x}_0}}{(\partial I/\partial x_1)_{\mathbf{x}_0}} \tag{3.2.8}$$

A comparison of Eqs. 3.2.5 and 3.2.8 shows that the direction corresponding to the maximum of the directional derivative (and thus the greatest local increase in I) is precisely the gradient direction

$$\frac{\partial I}{\partial \mathbf{x}} = \begin{bmatrix} \dfrac{\partial I}{\partial x_1} \\ \dfrac{\partial I}{\partial x_2} \end{bmatrix}$$

Thus a move in the direction of the gradient (Eq. 3.2.2)

$$\delta \mathbf{x} = \begin{bmatrix} \delta x_1 \\ \delta x_2 \end{bmatrix} = \epsilon \begin{bmatrix} \dfrac{\partial I}{\partial x_1} \\ \dfrac{\partial I}{\partial x_2} \end{bmatrix}$$

gives the largest improvement in I in the region about $\mathbf{x}_0$ where a linear approximation is valid. Because our corrections often extend beyond this region where the linear representation is valid, the gradient direction is not always the best for large corrections $\delta \mathbf{x}$ as illustrated in the subsequent Example 3.2.1.

After this preliminary discussion, the main purpose of which was to clarify the concepts of the directional derivative, the gradient, and the contour tangent, let us return to a more general treatment of the methods available for finding the optimum of multivariable functions, where information is available on the derivatives.

There are basically two types of algorithms that can be used here.

1. *"Gradient" Methods*
   (a) Calculate $[\partial I/\partial \mathbf{x}]_{\bar{\mathbf{x}}}$ and $\delta \mathbf{x}$.
   (b) Move to the highest value of I in the direction $\delta \mathbf{x}$.
   (c) Go back to (*a*) and continue until the maximum is found.

or

2. *"Steepest Ascent" Methods*
   (a) Calculate $[\partial I/\partial \mathbf{x}]_{\bar{\mathbf{x}}}$ and $\delta \mathbf{x}$.
   (b) Move a step length ϵ.
   (c) Go back to (*a*) and continue until the maximum is found.

Scheme 1 (shown in Fig. 3.11a) does not require derivative information at each step and tends to follow ridges. Scheme 2 shown in Fig. 3.11b always corrects the direction at each move, but it may require more steps because

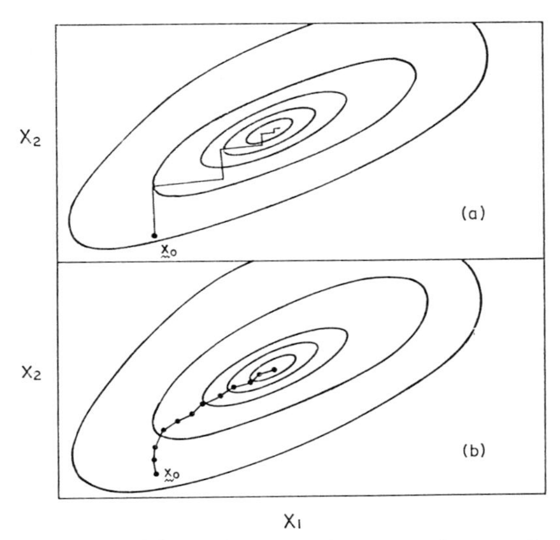

Fig. 3.11 An illustration of first-order methods: (*a*) gradient method and (*b*) steepest ascent method.

of the small step size. Usually scheme 1 is to be preferred, because it will give more improvement per function evaluation.

There is a basic difficulty with such gradient methods which arises because of scaling problems. The efficiency of the method is very sensitive to the scale factors (i.e., dimensions) chosen for the x_i. An engineering example provides perhaps the best illustration of this problem.

EXAMPLE 3.2.1. Let us consider a processing unit (which could be a CSTR, a leaching tank, or a hydrometallurgical reactor) whose performance we wish to optimize by maximizing the yield, through the appropriate adjustment of the operating temperature T and the flow rate Q of the unit. We use the present operating conditions, corresponding to T_s, Q_s as the base point, define $x_1 = T - T_s$, $x_2 = Q - Q_s$ and then seek the maximum of the objective $I(x_1, x_2)$.

For argument's sake let us assume that (unknown to us) the objective function has the form

$$I(x_1, x_2) = 125 - \{(x_1 - 10)^2 + (x_2 - 5)^2\} \tag{3.2.9}$$

when T has units °F and Q units gal/min. The contours of this function are shown in Fig. 3.12 with the optimum at $x_1 = 10$°F, $x_2 = 5$ gal/min. As the contours are circular, the gradient method finds the optimum in one step.

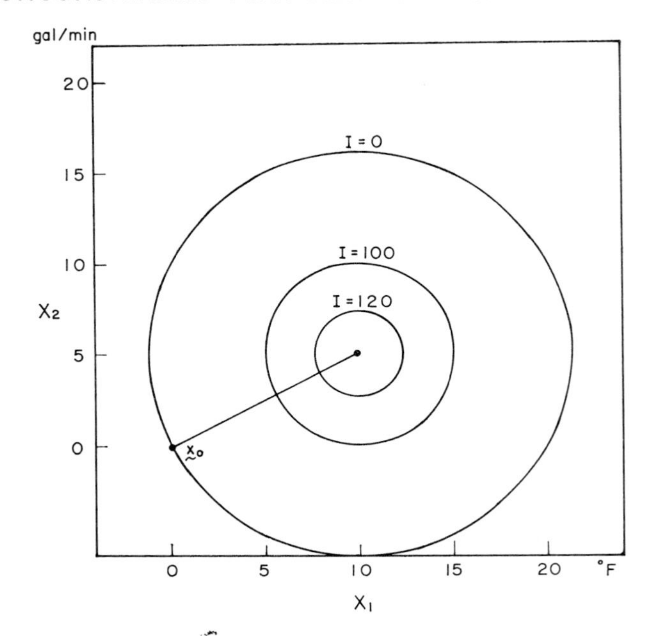

Fig. 3.12 Circular contours in Example 3.2.1.

Now let us change our units to °C and ft³/min; then $I(x_1, x_2)$ becomes $I(\hat{x}_1, \hat{x}_2)$ which may be written as

$$I(\hat{x}_1, \hat{x}_2) = -[(1.8\hat{x}_1)^2 + (7.5\hat{x}_2)^2] + 20(1.8)\hat{x}_1 + 10(7.5)\hat{x}_2$$

$$= -[3.24\hat{x}_1{}^2 + 56.2\hat{x}_2{}^2] + 36\hat{x}_1 + 75\hat{x}_2$$

$$I(\hat{x}_1, \hat{x}_2) = -[3.24(\hat{x}_1 - 5.55)^2 + 56.2(\hat{x}_2 - 0.67)^2] + 125 \tag{3.2.10}$$

due to the conversion factors: 1.8°F/°C, and 7.5 gal/ft³.

The optimum is at $\hat{x}_1 = 5.55$°C, $\hat{x}_2 = 0.67$ ft³/min, but the contours are now elliptical as shown in Fig. 3.13. It is seen that under these conditions many steps are required to find the maximum by the gradient method. Thus an arbitrary change in the units has increased the difficulty of the problem considerably.

To circumvent this difficulty, what is really required is that the correction be of the form

$$\delta \mathbf{x} = \mathbf{W} \frac{\partial I}{\partial \mathbf{x}}$$

where $\mathbf{W}$ is a matrix of scale factors known as the "*weighting matrix*." In the following we shall describe a technique for the evaluation of the best weighting matrix.

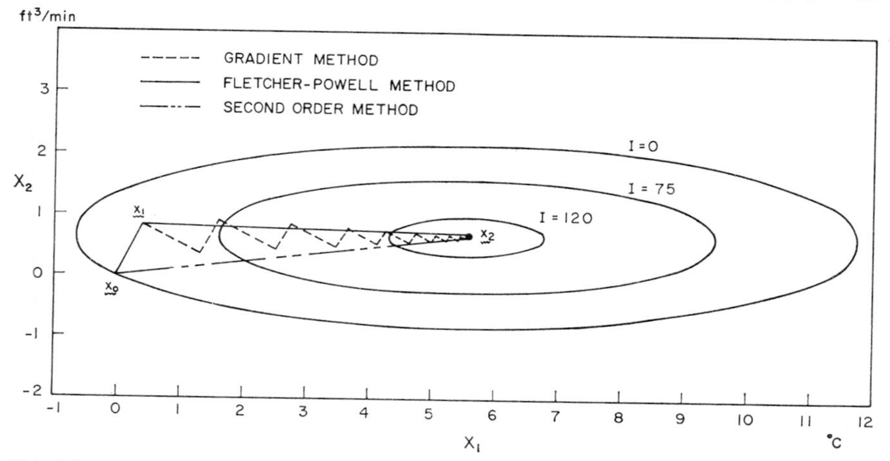

Fig. 3.13 A comparison of gradient type search methods, showing the elliptical contours in Example 3.2.1—on changing the units.

Second-Order Methods

Let us assume that we can calculate the second derivatives of our objective function so that a Taylor series about our present point $\bar{\mathbf{x}}$ gives

$$I(\mathbf{x}) = I(\mathbf{x}) + \sum_{i=1}^{n} \left(\frac{\partial I}{\partial x_i}\right)_{\bar{\mathbf{x}}} (x_i - \bar{x}_i)$$

$$+ \frac{1}{2} \sum_{i=1}^{n} \sum_{j=1}^{n} \left(\frac{\partial^2 I}{\partial x_i \partial x_j}\right)_{\bar{\mathbf{x}}} (x_i - \bar{x}_i)(x_j - \bar{x}_j) + 0(\mathbf{x} - \bar{\mathbf{x}})^3$$

or

$$\delta I = I(\mathbf{x}) - I(\bar{\mathbf{x}}) = \left(\frac{\partial I}{\partial \mathbf{x}}\right)_{\bar{\mathbf{x}}}^T \delta\mathbf{x} + \frac{1}{2}\delta\mathbf{x}^T \left(\frac{\partial^2 I}{\partial \mathbf{x}^2}\right)_{\bar{\mathbf{x}}} \delta\mathbf{x} + 0(\mathbf{x} - \bar{\mathbf{x}})^3 \quad (3.2.11)$$

where

$$\frac{\partial^2 I}{\partial \mathbf{x}^2} = \begin{bmatrix} \dfrac{\partial^2 I}{\partial {x_1}^2} & \dfrac{\partial^2 I}{\partial x_1\, \partial x_2} & \cdots & \dfrac{\partial^2 I}{\partial x_1\, \partial x_n} \\ \dfrac{\partial^2 I}{\partial x_2\, \partial x_1} & \dfrac{\partial^2 I}{\partial {x_2}^2} & \cdots & \vdots \\ \dfrac{\partial^2 I}{\partial x_n\, \partial x_1} & & \cdots & \dfrac{\partial^2 I}{\partial {x_n}^2} \end{bmatrix}$$

is the Hessian matrix of second derivatives.

If we wish to find the correction $\delta\mathbf{x}$ that makes δI a maximum (i.e., greatest increase in I), then we set

$$\frac{\partial(\delta I)}{\partial(\delta\mathbf{x})} = \left(\frac{\partial I}{\partial\mathbf{x}}\right)_{\bar{\mathbf{x}}} + \left(\frac{\partial^2 I}{\partial\mathbf{x}^2}\right)\delta\mathbf{x} = 0 \tag{3.2.12}$$

which gives the optimal correction

$$\delta\mathbf{x} = -\left(\frac{\partial^2 I}{\partial\mathbf{x}^2}\right)_{\bar{\mathbf{x}}}^{-1}\left(\frac{\partial I}{\partial\mathbf{x}}\right)_{\bar{\mathbf{x}}} \tag{3.2.13}$$

Now we see that the optimal weighting matrix is the inverse of the Hessian matrix with a minus sign prefixed. Thus if we know the second derivatives, then we can select the optimal direction and step length at each point. A suitable computational procedure would then involve the following steps:

(*a*) Compute $\partial I/\partial\mathbf{x}$ and $\partial^2 I/\partial\mathbf{x}^2$ at $\bar{\mathbf{x}}$ and the correction $\delta\mathbf{x}$ from Eq. 3.2.13, and

(*b*) Move to $\bar{\mathbf{x}} + \delta\mathbf{x}$ and go back to (*a*) and iterate until the optimum is found.

It should be noted that if the function $I(\mathbf{x})$ is a quadratic, then the Hessian matrix is a constant and the optimum will be found in one step. If the function is not quadratic, then for trial points $\bar{\mathbf{x}}$ far from the optimum, the procedure may not converge. This is because the local Hessian may not be a good approximation to the best quadratic fit to the function. For this reason, it is usually better to use a pure gradient method to get close to the optimum and then to switch to the second-order method in order to complete the search procedure.

This two-part procedure works well, because the pure gradient methods tend to be efficient in the region far from the optimum, whereas the second-order methods are more appropriate in the vicinity of the optimum (where a quadratic approximation tends to be very good).

The use of the weighting matrix is illustrated by the following example.

EXAMPLE 3.2.2. Evaluate the weighting matrix and use a second-order procedure to determine the maximum for the system discussed in Example 3.2.1.

Solution. When we use the units of °F and gal/min, then $I(x_1, x_2)$ is given by Eq. 3.2.9, and it follows that

$$\frac{\partial^2 I}{\partial x_1{}^2} = -2, \quad \frac{\partial^2 I}{\partial x_2{}^2} = -2, \quad \frac{\partial^2 I}{\partial x_1\,\partial x_2} = 0, \quad \frac{\partial^2 I}{\partial x_1 x_2} = 0$$

therefore, the Hessian is

$$\frac{\partial^2 I}{\partial \mathbf{x}^2} = \begin{bmatrix} -2 & 0 \\ 0 & -2 \end{bmatrix}$$

The weighting matrix becomes

$$\mathbf{W} = -\left[\frac{\partial^2 I}{\partial \mathbf{x}^2}\right]^{-1} = \begin{bmatrix} \frac{1}{2} & 0 \\ 0 & \frac{1}{2} \end{bmatrix}$$

and the correction, that is, the increment in $\mathbf{x}$ (in the same direction as the gradient)

$$\delta\mathbf{x} = \mathbf{W}\left(\frac{\partial I}{\partial \mathbf{x}}\right)_{\mathbf{x}=0} = \begin{bmatrix} \frac{1}{2} & 0 \\ 0 & \frac{1}{2} \end{bmatrix}\begin{bmatrix} 20 \\ 10 \end{bmatrix} = \begin{bmatrix} 10 \\ 5 \end{bmatrix}$$

allows us to reach the optimum in one step.

Now for the change in units to °C, ft^3/min

$$\frac{\partial^2 I}{\partial \hat{\mathbf{x}}^2} = \begin{bmatrix} -6.48 & 0 \\ 0 & -112.4 \end{bmatrix}$$

so that

$$\mathbf{W} = -\left[\frac{\partial^2 I}{\partial \hat{\mathbf{x}}^2}\right]^{-1} = \begin{bmatrix} 0.154 & 0 \\ 0 & 0.0089 \end{bmatrix}$$

and the correction

$$\delta\hat{\mathbf{x}} = \mathbf{W}\left(\frac{\partial I}{\partial \hat{\mathbf{x}}}\right)_{\hat{\mathbf{x}}=0} = \begin{bmatrix} 0.154 & 0 \\ 0 & 0.0089 \end{bmatrix}\begin{bmatrix} 36 \\ 75 \end{bmatrix} = \begin{bmatrix} 5.55 \\ 0.67 \end{bmatrix}$$

allows us to obtain the optimum in one step.† We notice that the use of $\mathbf{W}$ has changed the direction of our search in Fig. 3.13 by about 60° from the direction of the gradient.

It is stressed that Example 3.2.2 involved an objective function with a very simple quadratic form; for more complex objective functions, the Hessian will not be a constant but will be a function of $\mathbf{x}$. Thus, in general, several steps may be required before the maximum is reached. The requirement of this method, that second derivative information be available, is very stringent; therefore, an approximation due to Fletcher and Powell‡ has been found very useful.

† This is because $I(\mathbf{x})$ is a quadratic function.

‡ Actually, Davidon first developed this result, but it was improved and popularized by Fletcher and Powell.

Method of Fletcher and Powell [6]

This method constructs an approximation to the Hessian matrix at each step based just on gradient information. Let us assume that our weighting matrix, $\mathbf{W}$, which is the negative of the inverse of the Hessian

$$\mathbf{W} = \left(\frac{\partial^2 I}{\partial \mathbf{x}^2}\right)_{\bar{\mathbf{x}}}^{-1} \tag{3.2.14}$$

is approximated at each step until at some point $\mathbf{W}$ is exact. Now what we wish to do is to move in a sequence of directions $\delta\mathbf{x}_1, \delta\mathbf{x}_2, \ldots, \delta\mathbf{x}_n$ so that the gradient $[\partial I/\partial \mathbf{x}]_{\mathbf{x}_k}$ at each point $\mathbf{x}_k = \mathbf{x}_{k-1} + \delta\mathbf{x}_k$ is orthogonal (i.e., perpendicular) to each of the previous corrections, $\delta\mathbf{x}_j$, that is,

$$\left(\frac{\partial I}{\partial \mathbf{x}}\right)_{\mathbf{x}_k}^T \delta\mathbf{x}_j = 0 \qquad j = 1, 2, \ldots, k \tag{3.2.15}$$

Then at the point $\mathbf{x}_n = \mathbf{x}_{n-1} + \delta\mathbf{x}_n$, the relation

$$\left(\frac{\partial I}{\partial \mathbf{x}}\right)_{\mathbf{x}_n}^T \delta\mathbf{x}_j = 0 \qquad j = 1, 2, \ldots, n \tag{3.2.16}$$

must hold, but since the dimension of the space is n, and each of the $\delta\mathbf{x}_j$ are independent, then the only way the gradient $(\partial I/\partial \mathbf{x})_{\mathbf{x}_n}$ can be orthogonal to all $\delta\mathbf{x}_j$ is for it to vanish, and therefore we have

$$\left(\frac{\partial I}{\partial \mathbf{x}}\right)_{\mathbf{x}_n} = 0 \tag{3.2.17}$$

which is a condition for a stationary point of I. Thus for a quadratic function (where $\mathbf{W}$ is always constant), we can find the optimum in n trials $\delta\mathbf{x}_j$ if we can construct the proper sequence of directions $\delta\mathbf{x}_j$. To develop the method, let us assume that in the region of search, a quadratic approximation to $I(\mathbf{x})$ is valid, so that the Hessian matrix is a constant, that is,

$$I = a + \mathbf{b}^T\mathbf{x} + \tfrac{1}{2}\mathbf{x}^T\mathbf{Q}\mathbf{x} \tag{3.2.18}$$

Here a, $\mathbf{b}$, and $\mathbf{Q}$ are unknown at this stage and will be determined as the result of the subsequent calculations. Let us start with a first estimate of the weighting matrix $\mathbf{W}_0$. We note that it should be positive definite, since the inverse of the Hessian, $\mathbf{Q}^{-1}$, should be negative definite for a maximum. Then

$$\delta\mathbf{x}_1 = \mathbf{W}_0\left(\frac{\partial I}{\partial \mathbf{x}}\right)_{\mathbf{x}_0} \tag{3.2.19}$$

is our first step, and we need to decide how to choose $\mathbf{W}_1, \ldots, \mathbf{W}_n$ for the succeeding steps. From Eq. 3.2.18 we have that

$$\left(\frac{\partial I}{\partial \mathbf{x}}\right)_{\mathbf{x}_k} = \mathbf{b} + \mathbf{Q}\mathbf{x}_k \qquad k = 1, 2, \ldots, n \tag{3.2.20}$$

and the difference in gradients $\mathbf{d}_k$ is given as

$$\mathbf{d}_k = \left(\frac{\partial I}{\partial \mathbf{x}}\right)_{\mathbf{x}_k} - \left(\frac{\partial I}{\partial \mathbf{x}}\right)_{\mathbf{x}_{k-1}} = \mathbf{Q}\,\delta\mathbf{x}_k \qquad k = 1, 2, \ldots, n \tag{3.2.21}$$

We see that after n independent directional moves $\delta\mathbf{x}_k$, we would have $\mathbf{d}_k$ and $\delta\mathbf{x}_k$ in the n^2 Eqs. 3.2.21 and could determine the n^2 elements of $\mathbf{Q}$. However, we would like to use what information we have at each step; thus, let us choose the following updating scheme

$$\mathbf{W}_k = \mathbf{W}_{k-1} + \mathbf{A}_k + \mathbf{B}_k \qquad k = 1, 2, \ldots, n \tag{3.2.22}$$

where we wish $\mathbf{W}_n \doteq -\mathbf{Q}^{-1}$ and our initial guess is $\mathbf{W}_0$. Then

$$\mathbf{W}_n = \mathbf{W}_0 + \sum_{k=1}^{n} \mathbf{A}_k + \sum_{k=1}^{n} \mathbf{B}_k \tag{3.2.23}$$

and it would be convenient to select $\mathbf{A}_k$ and $\mathbf{B}_k$ so that

$$\sum_{k=1}^{n} \mathbf{A}_k = -\mathbf{Q}^{-1} \tag{3.2.24}$$

and

$$\sum_{k=1}^{n} \mathbf{B}_k = -\mathbf{W}_0 \tag{3.2.25}$$

Thus our computational procedure would be to find corrections

$$\begin{aligned} \delta\mathbf{x}_1 &= \mathbf{W}_0\left(\frac{\partial I}{\partial \mathbf{x}}\right)_{\mathbf{x}_0} \\ \delta\mathbf{x}_2 &= \mathbf{W}_1\left(\frac{\partial I}{\partial \mathbf{x}}\right)_{\mathbf{x}_1} \\ &\;\;\vdots \\ \delta\mathbf{x}_k &= \mathbf{W}_{k-1}\left(\frac{\partial I}{\partial \mathbf{x}}\right)_{\mathbf{x}_{k-1}} \end{aligned} \tag{3.2.26}$$

where $\mathbf{W}_k$ is updated by Eq. 3.2.23. Equations 3.2.14 to 3.2.22 describe the essential features of the method proposed by Fletcher and Powell. Readers who wish to use this technique, without any further proof, may skip the material that follows, and may proceed to the bottom of the next page for a "recipe" for the procedure and a worked example.

Now let us determine what $\mathbf{A}_k$ and $\mathbf{B}_k$ should be. Fletcher and Powell have shown by induction that if the directions $\delta\mathbf{x}_k$ are chosen optimally, then

$$\delta\mathbf{x}_i^T\mathbf{Q}\,\delta\mathbf{x}_j = \delta\mathbf{x}_i^T\,\mathbf{d}_j = 0 \qquad i \neq j \tag{3.2.27}$$

that is, the directions are all orthogonal with respect to Q. Also

$$-\mathbf{W}_k\mathbf{Q}\,\delta\mathbf{x}_i = \delta\mathbf{x}_i \qquad i \leq k \tag{3.2.28}$$

Equation 3.2.27 implies that

$$\mathbf{Q} = (\mathbf{S}\mathbf{\Lambda}^{-1}\mathbf{S}^T)^{-1} \tag{3.2.29}$$

where $\mathbf{S}$ is the matrix of directions $\delta\mathbf{x}_i$

$$\mathbf{S} = (\delta\mathbf{x}_1\ \delta\mathbf{x}_2 \cdots \delta\mathbf{x}_n) \tag{3.2.30}$$

and $\mathbf{\Lambda}$ is the diagonal matrix

$$\mathbf{\Lambda} = \begin{bmatrix} \delta\mathbf{x}_1^T\mathbf{Q}\,\delta\mathbf{x}_1 & 0 & \cdots & 0 \\ 0 & \delta\mathbf{x}_2^T\mathbf{Q}\,\delta\mathbf{x}_2 & \cdots & 0 \\ \cdot & & & \\ \cdot & & & \\ \cdot & & & \\ 0 & & & \delta\mathbf{x}_n^T\mathbf{Q}\,\delta\mathbf{x}_n \end{bmatrix} \tag{3.2.31}$$

Thus

$$\begin{aligned} \mathbf{Q}^{-1} &= \mathbf{S}\mathbf{\Lambda}^{-1}\mathbf{S}^T \\ &= \sum_{k=1}^{n}\left(\frac{\delta\mathbf{x}_k\,\delta\mathbf{x}_k^T}{\delta\mathbf{x}_k^T\mathbf{Q}\,\delta\mathbf{x}_k}\right) \end{aligned} \tag{3.2.32}$$

which from Eq. 3.2.21 becomes

$$\mathbf{Q}^{-1} = \sum_{k=1}^{n}\left(\frac{\delta\mathbf{x}_k\,\delta\mathbf{x}_k^T}{\delta\mathbf{x}_k^T\,\mathbf{d}_k}\right) \tag{3.2.33}$$

Thus $\mathbf{A}_k$ should take the value

$$\mathbf{A}_k = \frac{-\delta\mathbf{x}_k\,\delta\mathbf{x}_k^T}{\delta\mathbf{x}_k^T\,\mathbf{d}_k} \tag{3.2.34}$$

Now from Eq. 3.2.28

$$\begin{aligned} -\mathbf{W}_k\mathbf{Q}\,\delta\mathbf{x}_k &= \delta\mathbf{x}_k \\ &= -(\mathbf{W}_{k-1}\mathbf{Q}\,\delta\mathbf{x}_k + \mathbf{A}_k\mathbf{Q}\,\delta\mathbf{x}_k + \mathbf{B}_k\mathbf{Q}\,\delta\mathbf{x}_k) \end{aligned} \tag{3.2.35}$$

Since from Eqs. 3.2.21 and 3.2.34 we have

$$-\mathbf{A}_k\mathbf{Q}\,\delta\mathbf{x}_k = -\mathbf{A}_k\,\mathbf{d}_k = \frac{\delta\mathbf{x}_k\,\delta\mathbf{x}_k^{\,T}\,\mathbf{d}_k}{\delta\mathbf{x}_k^{\,T}\,\mathbf{d}_k} = \delta\mathbf{x}_k \tag{3.2.36}$$

then from Eqs. 3.2.35 and 3.2.21 we obtain

$$\mathbf{B}_k\,\mathbf{d}_k = -\mathbf{W}_{k-1}\,\mathbf{d}_k \tag{3.2.37}$$

We would like to use this expression to select $\mathbf{B}_k$; however, $\mathbf{B}_k = -\mathbf{W}_{k-1}$ is not useful, since it cancels the correction in Eq. 3.2.35. Thus we will proceed in the following manner. Let $\mathbf{z}$ be any nonzero vector, so that

$$\mathbf{B}_k\,\mathbf{d}_k = -\mathbf{W}_{k-1}\,\mathbf{d}_k\left(\frac{\mathbf{z}^T\,\mathbf{d}_k}{\mathbf{z}^T\,\mathbf{d}_k}\right) \tag{3.2.38}$$

or

$$\mathbf{B}_k = -\frac{\mathbf{W}_{k-1}\,\mathbf{d}_k\mathbf{z}^T}{\mathbf{z}^T\,\mathbf{d}_k} \tag{3.2.39}$$

Since $\mathbf{B}_k$ must be symmetric, we can choose

$$\mathbf{z} = \mathbf{W}_{k-1}\,\mathbf{d}_k \tag{3.2.40}$$

to give the required expression for $\mathbf{B}_k$

$$\mathbf{B}_k = -\frac{\mathbf{W}_{k-1}\,\mathbf{d}_k\,\mathbf{d}_k^{\,T}\mathbf{W}_{k-1}^T}{\mathbf{d}_k^{\,T}\mathbf{W}_{k-1}\,\mathbf{d}_k} \tag{3.2.41}$$

Thus on combining our results we are led to the following computational procedure:

(i) Guess $\mathbf{W}_0$ and move to the high point in the direction $\mathbf{W}_0(\partial I/\partial \mathbf{x}_0)_{\mathbf{x}_0}$; the first step* then is $\delta\mathbf{x}_1 = \epsilon_1\mathbf{W}_0(\partial I/\partial\mathbf{x})_{\mathbf{x}_0}$

(ii) Update the weighting matrix $\mathbf{W}_k$ from Eq. 3.2.22. Move in the direction $\mathbf{W}_k(\partial I/\partial\mathbf{x})_{\mathbf{x}_k}$ to find the optimal step $\delta\mathbf{x}_{k+1} = \epsilon_{k+1}\mathbf{W}_k(\partial I/\partial\mathbf{x}_0)_{\mathbf{x}_k}$

(iii) Keep iterating until the maximum is found. If I is a quadratic, both $\mathbf{Q}$ and the maximum will be found in n steps. If I is not a quadratic, then proceed past n iterations until $\delta\mathbf{x}_n$ has reached a certain small tolerance.

Although the method has been developed for quadratic functions only, it was found to work well for some very difficult nonquadratic functions.

* ϵ_k is defined as the optimal length to move in the direction

$$\mathbf{W}_{k-1}\left(\frac{\partial I}{\partial \mathbf{x}}\right)_{\mathbf{x}_{k-1}}$$

This behavior may be explained by the fact that almost all functions can be approximated by a quadratic expression close to the optimum. This method is one of the most efficient ones available at the present.

Recently, a number of other conjugate gradient procedures have been developed which do not require an accurate minimum in each search direction and can show increased rates of convergence over the Fletcher-Powell algorithm. Murtagh and Sargent [7] have discussed these and have pointed out their value in constrained optimization problems.

EXAMPLE 3.2.3. Let us now apply this conjugate gradient method to the problem defined by Eq. 3.2.10 and which was solved previously by gradient and second order methods. If we omit the "^" on the variables for convenience, our objective is

$$I(x_1, x_2) = 125 - 3.24(x_1 - 5.55)^2 - 56.2(x_2 - 0.67)^2 \quad (3.2.42)$$

which had the contours plotted in Fig. 3.13.

Let us choose

$$\mathbf{W}_0 = \mathbf{I} = \begin{bmatrix} 1 & 0 \\ 0 & 1 \end{bmatrix} \quad (3.2.43)$$

as our initial guess of $\mathbf{W}$. Following the procedure developed above, the steps are:

(i) From the initial point $\mathbf{x}_0 = \begin{bmatrix} 0 \\ 0 \end{bmatrix}$, $I(\mathbf{x}_0) = 0$ we move in the direction

$$\delta\mathbf{x}_1 = \epsilon_1 \mathbf{W}_0 \left(\frac{\partial I}{\partial \mathbf{x}}\right)_{\mathbf{x}_0} = \epsilon_1 \begin{bmatrix} 1 & 0 \\ 0 & 1 \end{bmatrix} \begin{bmatrix} 36 \\ 75 \end{bmatrix} = \epsilon_1 \begin{bmatrix} 36 \\ 75 \end{bmatrix} \quad (3.2.44)$$

until I is maximized. It is found by several experiments (cf. Fig. 3.13 for progress) that

ϵ_1	I
0.001	6.898
0.004	22.910
0.008	35.290
0.012	37.421
0.016	29.301

and we fit a *quadratic* $I(\epsilon_1)$ to these data to produce the optimum $\epsilon_1 = 0.011$. Thus our new point $\mathbf{x}_1$ is

$$\mathbf{x}_1 = \mathbf{x}_0 + \delta\mathbf{x}_1 = \begin{bmatrix} 0 \\ 0 \end{bmatrix} + 0.011 \begin{bmatrix} 36 \\ 75 \end{bmatrix} = \begin{bmatrix} 0.396 \\ 0.825 \end{bmatrix} \quad (3.2.45)$$

with

$$I(\mathbf{x}_1) = 37.849$$

(ii) Now we see that

$$\left(\frac{\partial I}{\partial \mathbf{x}}\right)_{\mathbf{x}_1} = \begin{bmatrix} 33.29 \\ -17.42 \end{bmatrix}; \qquad \mathbf{d}_1 = \left(\frac{\partial I}{\partial \mathbf{x}}\right)_{\mathbf{x}_1} - \left(\frac{\partial I}{\partial \mathbf{x}}\right)_{\mathbf{x}_0} = \begin{bmatrix} -2.71 \\ -92.42 \end{bmatrix} \tag{3.2.46}$$

so that from Eqs. 3.2.22, 3.2.34, and 3.2.42 our new approximation $\mathbf{W}_1$ becomes

$$\mathbf{W}_1 = \mathbf{W}_0 - \frac{\delta\mathbf{x}_1\, \delta\mathbf{x}_1^T}{\delta\mathbf{x}_1^T\, \mathbf{d}_1} - \frac{\mathbf{W}_0\, \mathbf{d}_1\, \mathbf{d}_1^T \mathbf{W}_0^T}{\mathbf{d}_1^T \mathbf{W}_0\, \mathbf{d}_1} \tag{3.2.47}$$

or

$$\mathbf{W}_1 = \begin{bmatrix} 1 & 0 \\ 0 & 1 \end{bmatrix} - \epsilon_1 \frac{\begin{bmatrix} 36 \\ 75 \end{bmatrix} [36 \quad 75]}{[36 \quad 75] \begin{bmatrix} -2.71 \\ -92.42 \end{bmatrix}} - \frac{\begin{bmatrix} -2.71 \\ -92.42 \end{bmatrix} [-2.71 \quad -92.42]}{[-2.71 \quad -92.42] \begin{bmatrix} -2.71 \\ -92.42 \end{bmatrix}}$$

$$= \begin{bmatrix} 1.00143 & -0.0251 \\ -0.0251 & 0.0088 \end{bmatrix} \tag{3.2.48}$$

Thus

$$\delta\mathbf{x}_2 = \epsilon_2 \mathbf{W}_1 \left(\frac{\partial I}{\partial \mathbf{x}}\right)_{\mathbf{x}_1} = \epsilon_2 \begin{bmatrix} 1.00143 & -0.0251 \\ -0.0251 & 0.0088 \end{bmatrix} \begin{bmatrix} 33.29 \\ -17.42 \end{bmatrix} = \epsilon_2 \begin{bmatrix} 33.735 \\ -0.983 \end{bmatrix} \tag{3.2.49}$$

where again we must perform experiments in the direction $\mathbf{W}_1(\partial I/\partial \mathbf{x})_{\mathbf{x}_1}$ to obtain the data

ϵ_2	$I(\epsilon_2)$
0.001	38.986
0.04	77.494
0.12	120.973
0.16	124.808
0.18	122.249

Fitting a quadratic $I(\epsilon_2)$ to these data yields the optimal $\epsilon_2 = 0.153$. Thus our new point, $\mathbf{x}_2$

$$\mathbf{x}_2 = \mathbf{x}_1 + \delta\mathbf{x}_2 = \begin{bmatrix} 0.396 \\ 0.825 \end{bmatrix} + 0.153 \begin{bmatrix} 33.735 \\ -0.983 \end{bmatrix} = \begin{bmatrix} 5.5575 \\ 0.6746 \end{bmatrix} \tag{3.2.50}$$

with $I(\mathbf{x}_2) = 124.9986$

(iii) Since our function $I(x_1, x_2)$ is a quadratic with only two independent variables, in principle only two iterations are needed to produce the optimum. However, because of small inaccuracies in determining ϵ_1 and ϵ_2 exactly, we see that $\mathbf{x}_2$ deviates only slightly from the true optimum shown in Fig. 3.13. Finally let us carry out one more iteration to improve matters. Thus

$$\left(\frac{\partial I}{\partial \mathbf{x}}\right)_{\mathbf{x}_2} = \begin{bmatrix} -0.0482 \\ -0.517 \end{bmatrix}; \qquad \mathbf{d}_2 = \left(\frac{\partial I}{\partial \mathbf{x}}\right)_{\mathbf{x}_2} - \left(\frac{\partial I}{\partial \mathbf{x}}\right)_{\mathbf{x}_1} = \begin{bmatrix} -33.338 \\ 16.903 \end{bmatrix} \tag{3.2.51}$$

and

$$\mathbf{W}_2 = \mathbf{W}_1 + \mathbf{A}_2 + \mathbf{B}_2$$

$$= \begin{bmatrix} 1.00143 & -0.0251 \\ -0.0251 & 0.088 \end{bmatrix} - \epsilon_2 \frac{\begin{bmatrix} 33.735 \\ -0.983 \end{bmatrix} [33.735 \quad -0.983]}{[33.735 \quad -0.983] \begin{bmatrix} -33.338 \\ 16.903 \end{bmatrix}}$$

$$- \frac{\begin{bmatrix} 1.00143 & -0.0251 \\ -0.0251 & 0.0088 \end{bmatrix} \begin{bmatrix} -33.338 \\ 16.903 \end{bmatrix} [-33.338 \quad 16.903] \begin{bmatrix} 1.00143 & -0.0251 \\ -0.0251 & 0.0088 \end{bmatrix}}{[-33.338 \quad 16.903] \begin{bmatrix} 1.00143 & -0.0251 \\ -0.0251 & 0.0088 \end{bmatrix} \begin{bmatrix} -33.338 \\ 16.903 \end{bmatrix}}$$

$$= \begin{bmatrix} 0.15462 & -0.00041 \\ -0.00041 & 0.0081 \end{bmatrix} \tag{3.2.52}$$

and

$$\delta \mathbf{x}_3 = \mathbf{W}_2 \left(\frac{\partial I}{\partial \mathbf{x}}\right)_{\mathbf{x}_2} = \begin{bmatrix} 0.15462 & -0.00041 \\ -0.00041 & 0.0081 \end{bmatrix} \begin{bmatrix} -0.0482 \\ -0.517 \end{bmatrix} = \begin{bmatrix} -0.00906 \\ -0.00418 \end{bmatrix} \tag{3.2.53}$$

Thus a closer approximation to the exact optimum is at

$$\mathbf{x}_3 = \mathbf{x}_2 + \delta \mathbf{x}_3 = \begin{bmatrix} 5.5485 \\ 0.6704 \end{bmatrix} \tag{3.2.54}$$

To conclude this section, let us consider another application of the Fletcher and Powell conjugate gradient method, now perhaps on a more realistic metallurgical example.

EXAMPLE 3.2.4. Let us use the conjugate gradient method to maximize the profit/time for a batch packed reactor of fixed size, in which a solid state reaction is being carried out at high temperatures.

Let us consider that the reaction is endothermic, and that the rate depends primarily on radiative heat transfer within the packed bed assembly. Then the rate of production could be described, in an approximate fashion, by an equation of the following type:

$$P = \frac{d_p[(1 - \epsilon)\epsilon]^2 T^3 e^{c_4 T}}{1 + c_5\, d_p^{\,2}} \tag{3.2.55}$$

where P = production (lb converted/hr)
d_p = particle size
ϵ = void fraction
T = temperature (°R)
c_3, c_4, c_5 = constants

Furthermore, let us consider that the rate at which the reactor walls are eroded may be expressed by

$$\text{Erosion (corrosion) rate} = c_1 e^{c_2 T} \tag{3.2.56}$$

Let the costs of preparing the granular material in a suitable form for reaction (crushing, grinding, sizing, and subsequent pellet formation) be given by

$$\text{Cost of burden preparation: } c_6[(1 - \epsilon)\epsilon]^2 \tag{3.2.57}$$

Finally, let heat losses be given by

$$\text{Heat losses: } c_7 T^4 \tag{3.2.58}$$

Then the profit from the operation of the reactor may be given as

$$\text{Profit (\$/hr)} = \frac{\alpha_1\, d_p\, c_3[(1 - \epsilon)\epsilon]^2 T^3 e^{c_4 T}}{1 + c_5\, d_p^{\,2}} - \alpha_2 c_1 e^{c_2 T} - \alpha_3 c_6[(1 - \epsilon)\epsilon]^2 - \alpha_4 c_7 T^4 \tag{3.2.59}$$

Here α_1, α_2, α_3, and α_4 represent the appropriate multipliers for profit per unit production and the cost of erosion, pellet preparation, and heat losses, respectively.† Let us consider ϵ_1, d_p, and T as the variables and use the Fletcher and Powell technique.

Before proceeding further, let us define our new variables as

$$T = x_1, \quad d_p = x_2 \qquad \text{and} \qquad \epsilon = x_3$$

† We note that Eq. 3.2.55 is not a rigorous transport based model of the process, but should be regarded either as an empirical expression, or as a relationship which may be consistent with intuition. We note specifically that the term $(1 - \epsilon)\epsilon$ has been squared for mathematical convenience to avoid search in the region of $\epsilon < 0$ and $\epsilon > 1.0$.

Systematic techniques for dealing with inequality constraints are surveyed in Chapter 4 and a more rigorous formulation of this specific packed bed problem will be presented in Chapter 8.

and set the constants at the following levels,

$$\alpha_1 c_3 = 10^{-15}, \quad c_2 = 0.02, \quad \alpha_2 c_1 = 10^{-15}, \quad c_4 = 0.01,$$
$$\alpha_3 c_6 = 10^{-17}, \quad c_5 = 0.01, \quad \alpha_4 c_7 = 10^{-16}$$

Then Eq. 3.2.59 may be written as

$$P \times 10^{15} = \frac{x_2(1 - x_3)x_3 x_1^3 e^{0.01x_1}}{1 + 0.01x_2^2} - e^{+0.02x_1} - 0.01[(1 - x_3)x_3]^2 - 0.1x_1^4 \tag{3.2.60}$$

Solution. Expressions for P, $\partial P/\partial x_1$, $\partial P/\partial x_2$, $\partial P/\partial x_3$ were furnished to a standard general purpose subroutine "FMFP" available from the IBM Scientific Subroutine Library. This program then determined the optimum of $P = \$2210.68/\text{hr}$ at $x_1 = T = 2126.2°\text{R}$, $x_2 = d_p = 10$ mm, $x_3 = \epsilon = 0.5$ in about 20 iterations of the Fletcher-Powell method from the initial guess $x_1 = 1800$, $x_2 = 5$, $x_3 = 0.25$. This required approximately $\frac{1}{2}$ sec on the CDC 6400 computer.

This problem is interesting in that there are multiple optima present, which can be found from other starting points. For example, if initial values of $x_1 \sim 500°\text{R}$ are chosen, a local optimum close to $x_1 \sim -0.1$ is found. Similarly if initial values $x_3 > 1$ or $x_3 < 0$ are chosen, the optimum is found at $x_3 \to \infty$ or $x_3 \to -\infty$. Fortunately, the true feasible optimum can be found from a number of starting points without invoking constraints to suppress these "unwanted" optima.

At the conclusion of this example it may be worthwhile to reemphasize that a wide range of problems in chemical, metallurgical, or ceramics engineering would be amenable to solution by this technique, provided that the objective function is a continuous differentiable function, information is available on the derivatives, and there are no constraints on the values that can be taken by the control variables. From a practical viewpoint the absence of constraints implies that the actual form of the objective function would rule out optima that would correspond to physically impossible values of the control variables (e.g., negative particle diameters, temperatures below absolute zero, etc).

Techniques suitable for handling problems where no information is available on the derivatives are discussed in the next section, and a general treatment of constrained optimization problems is presented in Chapter 4.

3.3 MULTIVARIABLE SEARCH TECHNIQUES—NO DERIVATIVE INFORMATION

If we have a profit function $I(\mathbf{x})$ for which derivatives $\partial I/\partial x_i$ are not readily available (a very common case for models of processing units), then we need search techniques not requiring derivatives. A very large number of these methods have been developed and tested and for this reason we will mention only a few of the most efficient ones. The first method to be discussed makes use of some of the ideas developed by Fletcher and Powell which were described in the previous section.

Method of Conjugate Directions

Powell [8] has developed a method using the properties of conjugate directions. As in the previous case, this technique is based on the assumption that the objective function can be approximated by a quadratic in the interval of interest, so that

$$I = a + \mathbf{b}^T\mathbf{x} + \tfrac{1}{2}\mathbf{x}^T\mathbf{Q}\mathbf{x} \tag{3.2.18}$$

From our initial point $\mathbf{x}_0$, we would like to determine search directions, $\boldsymbol{\xi}_k$, $k = 1, 2, \ldots, n$ so that we need to search in each direction only once to reach the optimum. These directions $\boldsymbol{\xi}_k$ are called *conjugate directions*, and they are defined by the condition

$$\boldsymbol{\xi}_k^T\mathbf{Q}\boldsymbol{\xi}_j = 0 \qquad k \neq j \tag{3.3.1}$$

which must be satisfied for each direction $\boldsymbol{\xi}_k$.†

It is interesting to note that the directions of search found by using the Fletcher-Powell method were all conjugate directions, because Eqs. 3.3.1 and 3.2.27 are identical. It is easy to see that searching in conjugate directions means that we need to search in each direction only once because we are seeking the maximum of Eq. 3.2.18, which can be rewritten as

$$I(\mathbf{x}) = I\left(\mathbf{x}_0 + \sum_{i=1}^{n} \alpha_i\boldsymbol{\xi}_i\right) = \sum_{i=1}^{n} [(\alpha_i)^2\boldsymbol{\xi}_i^T\mathbf{Q}\boldsymbol{\xi}_i + \alpha_i\boldsymbol{\xi}_i^T(\mathbf{Q}\mathbf{x}_0 + \mathbf{b})] + I(\mathbf{x}_0) \tag{3.3.2}$$

Because there are no cross terms $\alpha_i\alpha_j$, $I(\mathbf{x})$ is separable, and we can search for each optimal step length α_i independently, and thus search in each direction only once.

Now that we have established that we wish to search in conjugate directions, let us find a way to obtain a set of conjugate directions without the a priori knowledge of $\mathbf{Q}$ and without derivatives. We are fortunate because there is a theorem which allows us to determine these conjugate directions.

THEOREM. If $\mathbf{p}_0$ is a maximum in the direction $\boldsymbol{\xi}_k$, and $\mathbf{p}_n$ is also a maximum in the direction $\boldsymbol{\xi}_k$, then the direction $(\mathbf{p}_k - \mathbf{p}_0)$ is conjugate to $\boldsymbol{\xi}_k$.

† It is easy to show that the set of n conjugate directions $\boldsymbol{\xi}_k$ is not unique, because if $\mathbf{Q} = \mathbf{I}$ (the identity matrix), then any set of mutually orthogonal directions would be conjugate directions.

Proof. Since $\mathbf{p}_0$ is a maximum, then

$$\frac{d}{d\alpha_k} I\left(\mathbf{p}_0 + \sum_{i=1}^{n} \alpha_i \boldsymbol{\xi}_i\right) = 0 \qquad \text{at} \quad \alpha_k = 0$$

or

$$\boldsymbol{\xi}_k^T = (\mathbf{Q}\mathbf{p}_0 + \mathbf{b}) = 0 \tag{3.3.3}$$

Also, since $\mathbf{p}_n$ is a maximum,

$$\frac{d}{d\alpha_k}\left[I\left(\mathbf{p}_n + \sum_{i=1}^{n} \alpha_i \boldsymbol{\xi}_i\right)\right] = 0 \qquad \text{at} \quad \alpha_k = 0$$

or

$$\boldsymbol{\xi}_k^T(\mathbf{Q}\mathbf{p}_n + \mathbf{b}) = 0 \tag{3.3.4}$$

Subtraction of Eq. 3.3.3 from 3.3.4 gives

$$\boldsymbol{\xi}_k^T\mathbf{Q}(\mathbf{p}_n - \mathbf{p}_0) = 0 \tag{3.3.5}$$

the definition of conjugacy.

Thus by taking the difference between maxima in a direction $\boldsymbol{\xi}_k$, we can determine a new direction conjugate to $\boldsymbol{\xi}_k$ and in this way construct a set of n conjugate directions. Powell suggests the following iteration procedure:

1. Select the coordinate axes as the initial set of directions

$$\boldsymbol{\xi}_1 = \mathbf{e}_1, \quad \boldsymbol{\xi}_2 = \mathbf{e}_2, \ldots, \boldsymbol{\xi}_n = \mathbf{e}_n$$

2. From an initial point $\mathbf{p}_0$, find the maximum in each of the n directions $\boldsymbol{\xi}_1, \ldots, \boldsymbol{\xi}_n$, that is, find $\alpha_k = \alpha_k^*$ which satisfies

$$\max_{\alpha_k} I(\mathbf{p}_{k-1} + \alpha_k \boldsymbol{\xi}_k) \qquad \text{for} \quad k = 1, 2, \ldots, n$$

where $\mathbf{p}_k = \mathbf{p}_{k-1} + \alpha_k^* \boldsymbol{\xi}_k$ is updated each time. This requires n one-dimensional searches.

3. Now remove $\boldsymbol{\xi}_1$ from the set of directions and let

$$\boldsymbol{\xi}_k = \boldsymbol{\xi}_{k+1} \qquad k = 1, \ldots, n-1$$

Then let $\boldsymbol{\xi}_n = \mathbf{p}_n - \mathbf{p}_0$.

4. Now find $\lambda = \lambda^*$ which maximizes $I(\mathbf{p}_n + \lambda\boldsymbol{\xi}_n)$ and let

$$\mathbf{p}_0 = \mathbf{p}_n + \lambda^* \boldsymbol{\xi}_n$$

5. Go back to step 2 and continue iterating.

Powell has shown by induction that after m iterations, the directions $\boldsymbol{\xi}_{n-m+1}, \boldsymbol{\xi}_{n-m+2}, \ldots, \boldsymbol{\xi}_n$ are all mutually conjugate. Thus after n iterations,

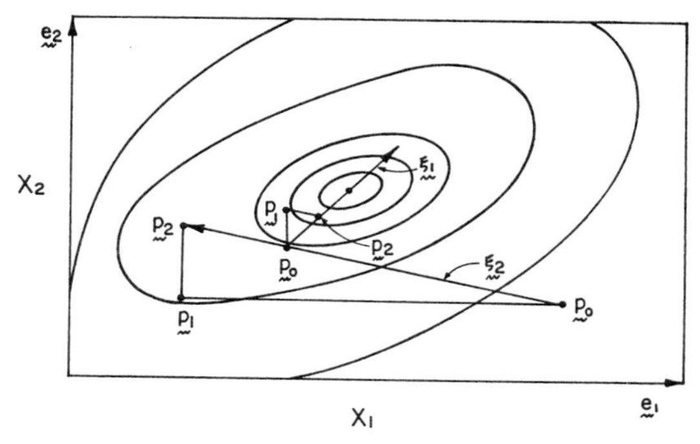

Fig. 3.14 Method of conjugate directions.

all the ξ_k will be conjugate and we will have reached the maximum because we have searched each of the directions once in the last iteration.

For an illustration of the technique in two dimensions, Fig. 3.14 shows the steps involved for a quadratic. As can be seen the method converges to the optimum after two iterations.

Powell mentions that for large n, there is a problem of including directions in the basis which are very nearly parallel to a coordinate axis which is still in the basis. To prevent this, he gives a criterion which can be used at each stage to decide which direction should be removed and whether a new direction should be added. He reports very good results with this modification for functions having as many as 20 variables.

EXAMPLE 3.3.1. As a more practical example, let us consider the optimization of part of the hydrometallurgical extraction plant, sketched in Fig. 3.15.

It is seen that ore of average size d_o is fed through a series of crushers and ball mills; the ore leaves this cascade with an average size d_p, which is a function of the residence time within the size reduction system, t_c.

The ore is then introduced into the leaching unit, at flow rate F, where a soluble ingredient is leached. The rate of recovery in the leaching unit will

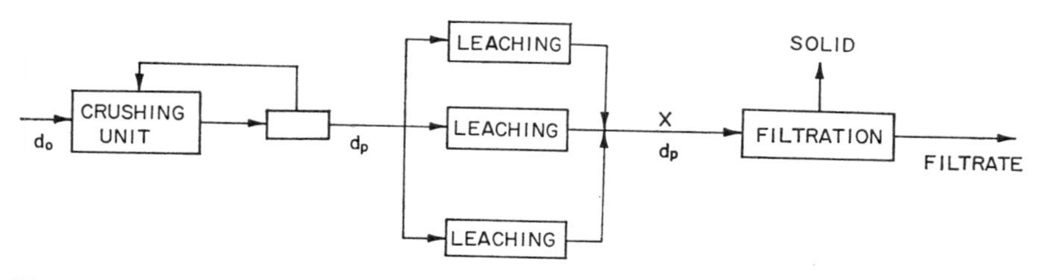

Fig. 3.15 Schematic representation of the hydrometallurgical plant, in Example 3.3.1.

depend on the average particle size (d_p), the operating temperature of the unit, the power used for stirring, and the residence time t_R.

From the leaching unit the liquid-solid suspension is passed through a filter, the solid from which is discarded or collected for later recycling, and the liquid is sent for further processing. The rate of filtration achieved per unit filter area depends on the particle size (d_p).

The objective is to maximize the profit in the operation of this system.

The following relationships are available.

Comminution depends on the residence time in the crushing-grinding cascade as follows:

$$d_p(t_c) = \frac{1}{d_0^{-1} + e_1 t_c} + d_\infty \tag{3.3.6}$$

where t_c is the residence time (hours) of the crushing unit.

The following relationship exists between $t_R(1)$, residence time required for *complete conversion*, and the operating parameters in the leaching unit

$$t_R(1) = \frac{10^5 d_p^{\,2}\left[1 + \dfrac{1}{e_2(P+1)}\right]}{1 + e_3 T} \tag{3.3.7}$$

For less than 100% conversion the dependence of $t_r(X)$ on X is given in the following tabulated form:

$$t_R(X) = f(x) t_R(1) \tag{3.3.8}$$

TABLE 3.3.1

X	$f(X)$
1	1
0.95	0.5
0.9	0.4
0.8	0.3
0.6	0.2
0.36	0.1

where $f(X) < 1$ is an effectiveness factor for incomplete conversion.

Finally the surface area of the filtration unit is related to the particle size as follows:

$$A = \frac{5}{1 + 100 d_p} \tag{3.3.9}$$

where A = ft² per lb/hr feed processed and d_p = particle diameter in cm.

The economic parameters needed for consideration are

Value of product	$\alpha_0 FX$
Raw material cost	$\alpha_1 F$
Waste disposal cost	$\alpha_2 F(1 - X)$
Cost of power	$\alpha_3 PF$
Cost of heating	$\alpha_4 T^2 F$
Cost of filter area	$\alpha_5 AF$
Cost of crusher size	$\alpha_6 t_c F$
Cost of extractor size	$\alpha_7 t_R F$

By recognizing that we have a fixed throughput rate F, we can represent our objective as

$$\text{Max}\ (I = c_0 X - c_1 P - c_2 T^2 - c_3 A - c_3 t_c - c_5 t_R) \tag{3.3.10}$$

We see that there are four independent variables (i.e., t_c, t_R, T, P) and three dependent variables (i.e., X, d_p, A) in this problem. Equations 3.3.6 to 3.3.9 can be used to represent the dependent variables in terms of the independent variables. Thus we wish to search on t_c, t_R, T, P in order to maximize the objective I in Eq. 3.3.10.

Let us assume that the parameters are

$d_0 = 0.65$ cm	$c_0 = 4000$
$d_\infty = 5 \times 10^{-4}$ cm	$c_1 = 10$
$e_1 = 25\ \text{hr}^{-1}$	$c_2 = 0.001$
$e_2 = 0.20(\text{Hp})^{-1}$	$c_3 = 200$
$e_3 = 0.05(°\text{F})^{-1}$	$c_4 = 10$
	$c_5 = 10$

Solution. We shall apply the conjugate direction method of Powell in the form of the standard subroutine BOTM available from the IBM share system (cf. list in Appendix B). Let us begin with an initial guess (which might represent the current operating conditions): $t_c = t_R = 3$ hrs, $T = 100°\text{F}$, $P = 3$ Hp which requires $A = 2.12\ \text{ft}^2/\text{lb/hr}$ and produces $d_p = 0.0136$ cm from the crusher and a conversion of 72.86% in the extractor. The profit for this initial guess is \$2390/day.

The first search along each of the coordinate axes in turn yields the results labeled iteration 1 in Table 3.3.2. The search proceeds by replacing each of the coordinate axes by a new direction. The results of the iterations in Table 3.3.2 show that a very good approximation to the optimal operating conditions is obtained after about 15 iterations.

The optimal operating conditions are a residence time of 1.54 hr in the crusher which produces an average particle size of 0.0255 cm. The optimal conditions for the extractor are a residence time of 10.94 hr, a temperature

TABLE 3.3.2 RESULTS OF POWELL METHOD APPLIED TO THE CRUSHING-EXTRACTION-FILTERING SEQUENCE

Iteration	t_c	t_r	T	P	$I \times 10^{-3}$	Number of evaluations of I
0	3.00	3.00	100.0	3.00	2.3899	0
1	4.74	3.09	100.2	2.92	3.3472	18
2	4.35	3.78	102.6	2.31	3.3618	30
3	3.77	4.43	135.9	1.62	3.3891	54
4	2.91	5.22	171.1	2.08	3.4283	77
5	2.20	6.56	230.0	2.41	3.4681	100
10	1.83	6.87	257.5	4.39	3.4838	205
15	1.57	10.24	227.3	4.37	3.4968	335
20	1.54	10.94	224.0	4.31	3.4974	397

of 224°F, and a power input for agitation of 4.32 Hp—producing a leaching conversion of 99.93%. The filter requirements then become 1.41 ft²/lb/hr. The optimal profit then is $3497/day, a saving of more than $350,000/yr over the initial operating policy.

We must caution the reader at the conclusion of this example that Eqs. 3.3.6 to 3.3.10 were "empirical relationships"; in applying this technique to the optimization of cascades involving crushing, leaching, and filtration, appropriate modeling equations must be developed for the system. It is quite likely that these modeling equations would be rather more complex than the simple expressions selected here.

Method of Nelder and Mead [9]

This method, while not at all sophisticated, gives remarkably good results when the number of variables is not large. The technique forms a simplex from $n + 1$ points (e.g., a triangle for two variables) and reflects this away from the low point in an iterative fashion.

The basic idea behind the method of Nelder and Mead may be illustrated on a two-dimensional example. Let us consider a hill (in two dimensions) and let us place a triangular plane against this hill. In principle the hill may be climbed by repeatedly overturning the triangle about suitably chosen pivotal points.

The procedure will be given and an example worked in two variables. The steps are as follows:

(a) Select $(n + 1)$ initial points $\mathbf{x}_k$, $I(\mathbf{x}_k)$ $\quad k = 1, 2, \ldots, n + 1$
(b) Find the highest and lowest of these points, $\mathbf{x}^h$, $\mathbf{x}^l$

(c) Find the centroid in the $(n - 1)$ dimensional space made up of all points except $\mathbf{x}^l$ and call this $\bar{\mathbf{x}}$.

(d) Reflect $\mathbf{x}^l$ through the centroid $\bar{\mathbf{x}}$ a distance α to

$$\mathbf{x}^* = \mathbf{x}^l + (1 + \alpha)(\bar{\mathbf{x}} - \mathbf{x}^l)$$

If this produces $I(\mathbf{x}^*) > I(\mathbf{x}^h)$, then we extend our reflection a further distance γ to

$$\mathbf{x}^{**} = \bar{\mathbf{x}} + \gamma(\mathbf{x}^* - \bar{\mathbf{x}})$$

If $I(\mathbf{x}^l) < I(\mathbf{x}^*) < I(\mathbf{x}^h)$, then just replace $\mathbf{x}^l$ by $\mathbf{x}^*$

If $I(\mathbf{x}^*) < I(\mathbf{x}^l)$, then we have a failure and we must contract a distance β to

$$\mathbf{x}^{***} = \bar{\mathbf{x}} - \beta(\bar{\mathbf{x}} - \mathbf{x}^l)$$

If this contraction produces $I(\mathbf{x}^{***}) < I(\mathbf{x}^l)$, then we must shrink our simplex about our best point, $\mathbf{x}^h$, and try again.

(e) Go back to step (b) and continue iterating until the method has converged.

In order to demonstrate the application of this technique, we will work a two-dimensional example. Initially we select 3 points $\mathbf{x}_1$, $\mathbf{x}_2$, $\mathbf{x}_3$ in Fig. 3.16 and determine that $\mathbf{x}_1 = \mathbf{x}^l$ and $\mathbf{x}_3 = \mathbf{x}^h$. Then the centroid $\bar{\mathbf{x}}$ is determined and the first reflection to $\mathbf{x}_4 = \mathbf{x}^*$ is carried out. Since $I(\mathbf{x}_4) > I(\mathbf{x}^h)$, then we move the further distance γ to $\mathbf{x}^{**} = \mathbf{x}_5$.

Now we are ready to begin a second iteration by forming the simplex made up of $\mathbf{x}_2$, $\mathbf{x}_3$, $\mathbf{x}_5$ where now $\mathbf{x}_5 = \mathbf{x}^h$, $\mathbf{x}_2 = \mathbf{x}^l$. Again the centroid $\bar{\mathbf{x}}$ is found and reflection a distance α to $\mathbf{x}_6$ is done. Since $I(\mathbf{x}^l) < I(\mathbf{x}_6) < I(\mathbf{x}^h)$ then we just replace $\mathbf{x}^l$ by $\mathbf{x}_6$ and form a new simplex of $\mathbf{x}_3$, $\mathbf{x}_5$, $\mathbf{x}_6$ where now $\mathbf{x}_3 = \mathbf{x}^l$, $\mathbf{x}_5 = \mathbf{x}^h$. Reflection of $\mathbf{x}_3$ through $\bar{\mathbf{x}}$ produces $\mathbf{x}_7$, and the new simplex of $\mathbf{x}_5$, $\mathbf{x}_6$, $\mathbf{x}_7$ now has the maximum contained within it.

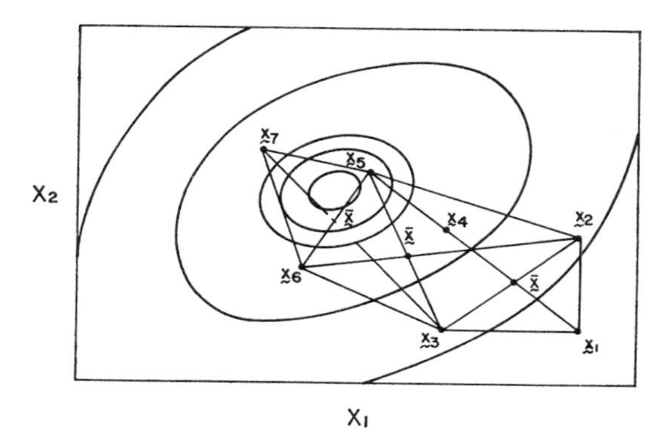

Fig. 3.16 The method of Nelder and Mead.

We can now continue the iteration procedure until our simplex has shrunk to the desired approximation of the optimum.

The method has been tested on several very ill-conditioned functions and was found to perform as well as or better than the conjugate gradient method for functions of two, three, and four variables. However, as we shall see, the performance deteriorates for more than four variables.

3.4 QUADRATIC CONVERGENCE

Because many very complicated objective functions $I(\mathbf{x})$ can be approximated by a second-order Taylor series sufficiently close to the optimum

$$I(\mathbf{x}) = I(\mathbf{x}^*) + \left(\frac{\partial I}{\partial \mathbf{x}}\right)_{\mathbf{x}^*}^T (\mathbf{x} - \mathbf{x}^*) + \tfrac{1}{2}(\mathbf{x} - \mathbf{x}^*)^T \left(\frac{\partial^2 I}{\partial \mathbf{x}^2}\right)_{\mathbf{x}^*} (\mathbf{x} - \mathbf{x}^*) + \text{H.O.T.} \tag{3.4.1}$$

and it is close to the optimum that the different peak-seeking methods distinguish themselves, then it is of some value to have a technique work well on quadratic functions. Thus if a method can be shown to converge always to the optimum of a quadratic in a given, finite number of steps, then the method is said to possess *quadratic convergence*.† By using this definition, we can see that the Fletcher-Powell method, the second-order method, and the conjugate direction method all possess quadratic convergence while the steepest ascent-gradient methods and the method of Nelder and Mead do not. As we shall demonstrate in the next section, methods having quadratic convergence do indeed behave better than methods that do not have this property.

3.5 A COMPARISON OF UNCONSTRAINED PEAK-SEEKING METHODS

Although it is generally recognized that some procedures work best on some functions and other procedures work best on others, there is some merit in running time and efficiency trials on a number of functions to obtain an idea of the expected behavior of some optimizing procedures. M. J. Box [10] has run tests on functions having up to 20 variables using Fletcher and Powell's method (I), Powell's conjugate direction method (II), and Nelder

† Some authors restrict the term "quadratic convergence" to mean convergence in n steps, but we shall broaden the definition.

and Mead's simplex method (III), as well as several other methods. The functions tested were

(a) Two-dimensional exponential function

$$I(x_1, x_2) = -\sum_{k=1}^{10} [(e^{-k/10x_1} - e^{-k/10x_2}) - (e^{-k/10} - e^{-k})]^2$$

which has an asymmetric curved valley with a maximum at $x_1 = 1$, $x_2 = 10$, $I = 0$.

(b) Three-dimensional exponential function

$$I(x_1, x_2, x_3) = -\sum_{k=1}^{10} [(e^{-k/10x_1} - e^{-k/10x_2}) - x_3(e^{-k/10} - e^{-k})]^2$$

which has maximum $I = 0$ at $x_1 = 1$, $x_2 = 10$, $x_3 = 1$.

(c) n-dimensional trigonometric function

$$I(x_1, \ldots, x_n) = -\sum_{i=1}^{n} \left[E_i - \sum_{j=1}^{n} (A_{ij} \sin x_j + B_{ij} \cos x_j) \right]^2$$

where E_i, A_{ij}, B_{ij} were generated as random numbers.

Each method was tried from a number of different starting points and the number of equivalent function evaluations were tabulated. In order to compare gradient and nongradient methods, the set of partial derivatives in n directions needed to form a gradient was assumed the equivalent of $n + 1$ function evaluations. This is reasonable, since this is the amount of computation normally required whether derivatives are computed analytically or by differences. The results are presented in Table 3.3.3 for each of the test functions. As can be seen, the conjugate gradient method of Fletcher and Powell is the most consistently successful for all problems tested. This conclusion was reached independently in a similar study by Leon [11]. Box also concludes that Powell's conjugate direction method is nearly as efficient as conjugate gradients and has the added advantage that no derivatives are required. The failures noted for function (b) Box attributes to the fact that the program he used found the minimum in each direction too precisely, and this led to a spurious saddle point for function (b). We note that the simplex method of Nelder and Mead is very good for two or even three dimensions, but behaves poorly in problems involving higher dimensions.

Box concludes from his study that the Fletcher and Powell algorithm and the conjugate direction method seem to be the best (as might be expected, since they both possess quadratic convergence) and their efficiency becomes more apparent as the number of variables increases.

Recently a number of extended conjugate gradient methods (e.g., the memory gradient [12] and super-memory gradient [13] procedures) have

TABLE 3.3.3[a] A COMPARISON OF VARIOUS OPTIMIZATION METHODS

	Function a, $n = 2$, Starting point					Function b, $n = 3$, Starting point									Function c, $n = 5$, Starting point		Function c, $n = 10$, Starting point		Function c, $n = 20$, Starting point	
Optimization method	1	2	3	4	5	1	2	3	4	5	6	7	8	9	1	2	1	2	1	2
Fletcher and Powell	51	48	45	63	27	92	68	144	24	148	140	96	148	140	114	138	396	319	1,764	1,428
Powell's conjugate direction	64	51	84	77	23	78	F	F	56	F	F	19	F	F	104	103	239	269	1,519	2,206
Nelder and Mead's simplex method	41	43	39	49	40	119	128	112	73	110	307	79	164	315	229	195	752	962	6,970	12,100

[a] n = number of dimensions; F = failure to converge to correct answer.

been developed which perform several dimensional searches in the step sizes at each iteration. Only limited computational results are available, but these methods seem to be superior to ordinary conjugate gradient techniques in some instances. They have the disadvantage, however, that they require second derivatives of the objective function.

Another class of variable metric methods which allows more freedom in the choice of step-size and direction of search, has been analyzed by Murtagh and Sargent [14, 15]. They have shown that in some problems, these methods can be more efficient than the standard conjugate gradient algorithms. A very recent monograph [16] gives more detailed comparisons of the various unconstrained search methods.

PROBLEMS

1. Find the maximum of

$$I = x^3 - \tfrac{9}{2}x^2 + 6x + 20$$

in the interval $0 < x \leq 2$ with accuracy in x_{opt} of at least 0.05 using
(1) Fibonacci search.
(2) Golden section search.
(3) Dichotomous search.

2. We wish to find the optimum temperature for a continuous stirred reactor in which a first-order, reversible, exothermic reaction is being carried out [17]. The following information is available: The reaction is

$$A \rightleftarrows S$$

The inlet concentration of A and S is

$$C_{A0} = 1, \quad C_{S0} = 0, \quad (\text{g mole/l})$$

The fractional conversion at equilibrium X_{Ae} is given by

$$X_{Ae} = \frac{K}{1 + K} \tag{i}$$

where K, the equilibrium constant is given by

$$\ln K = \frac{9060}{T} - 27.46 \tag{ii}$$

with T in °K.
The temperature dependence of the forward rate constant is given as

$$k = 3 \times 10^7 e^{-11,600/RT} (\text{min}^{-1})$$

and the conversion, X_A, is related to the holding time, τ, by

$$X_A = \frac{X_{Ae} k_1 \tau}{X_{Ae} + k_1 \tau} \tag{iii}$$

Our objective is then to find the optimum temperature T, which maximizes X_A, for a fixed holding time of $\tau = 10$ min. Use a single variable elimination technique to determine the optimal temperature within 5°C. Plot your experiments on a X_A, T plot.

3. Find the minimum of

$$I = x_1^2 + x_1 x_2 + x_2^2 - 3x_1 - 3x_2$$

using
(1) Steepest descent method.
(2) Gradient method.
(3) Second-order method.
(4) Fletcher-Powell conjugate gradient method.
Start from $x_1 = 1$, $x_2 = 0$ for each method. Plot your progress on graph paper for each case so that you can see clearly what is happening.

4. Find the maximum of

$$I = 3x_1 + 2x_2 - 3x_1 x_2 - 4x_1^2 - 5x_2^2$$

using
(1) Powell's conjugate direction method, and
(2) Nelder and Mead's simplex algorithm.
Start from $x_1 = 1$, $x_2 = 0$ for each method. Plot your progress on graph paper for each case so that you can see clearly what is happening.

5. It is desired to develop an optimal design for the leaching of ores with sulfuric acid, in a set of steam-jacketed autoclaves that operate in a batchwise manner. The following empirical expression is available, relating the rate of extraction to batch time, temperature, power input, and to the acid concentration:

$$X = X_0 \left\{ \exp\left[- \frac{0.02t(T - T_0)(1 + 0.05C_{ac})}{1 + \dfrac{1}{P^2 + 0.2}} \right] \right\}$$

where X_0 = the initial metal content of the ore
X = the residue
t = batch time, in hours
C_{ac} = the sulfuric acid concentration, in g/liter
P = the power input, in horsepower

$T =$ the temperature,

$T_0 =$ reference temperature.

We wish to maximize the profit from the operation of the unit of fixed volume V.

The rate of production is given as

$$\frac{V(X_0 - X)}{t + t_c}$$

where t_c is the time required to empty and clean the vessel after each operation. Then our objective function may take the form:

$$I(t, P, T, C_{ac}) = \alpha_1 \frac{V(X_0 - X)}{t + t_c} - \alpha_2(C_{ac} + 5C_{ac}^{\ 2}) - \alpha_3(T - T_0)^2 - \alpha_4 P^2 - \alpha_5 VX$$

Here $\alpha_1, \alpha_2 \cdots \alpha_5$ are the appropriate cost factors. Let us assign the following numerical values:

t_c	1 hr
$\alpha_1 V$	10^5
α_2	5.0
α_3	30
T_0	80°F
α_4	10
$\alpha_5 V$	0.01
X_0	0.10

Use one of the techniques of Chapter 3 to solve this problem.

REFERENCES

1. J. Szekely and G. P. Martins, *Chem. Eng. Sci.* **26,** 147 (1971).
2. J. Szekely and G. P. Martins, *Trans. Met. Soc. AIME* **245,** 629 (1969).
3. J. Szekely and G. P. Martins, *Met. Trans.*, to be published.
4. J. Szekely, *Int. J. Heat Mass Transfer* **6,** 417 (1963).
5. G. E. P. Box and N. R. Draper, *Evolutionary Operation*, Wiley, 1969.
6. R. Fletcher and M. J. D. Powell, *Comput. J.* **6,** 163 (1963).
7. B. A. Murtagh and R. W. H. Sargent, *Comput. J.* **13,** 185 (1970).
8. M. J. D. Powell, *Comput. J.* **7,** 155 (1964).
9. J. A. Nelder and R. Mead, *Comput. J.* **8,** 308 (1965).
10. M. J. Box, *Comput. J.* **9,** 67 (1966).
11. A. Leon, "A Comparison among Eight Known Optimization Procedures," in *Recent Advances in Optimization Techniques*, A. Lavi and T. P. Vogel, Eds., Wiley, 1966.

12. A. Miele and J. W. Cantrell, *J. Opt. Theory Appl.* **3,** 459 (1968).
13. E. E. Cragg and A. V. Levy, *J. Opt. Theory Appl.* **4,** 191 (1969).
14. B. A. Murtagh and R. W. H. Sargent, *A constrained minimization Method with Quadratic Convergence*, paper presented at IMA-BCS Joint Conference on Optimization, University of Keele, March 1968.
15. B. A. Murtagh and R. W. H. Sargent, *Comput. J.* **13,** 185 (1970).
16. David M. Himmelblau, *Applied Nonlinear Programming*, McGraw-Hill, 1972.
17. O. Levenspiel, *Chemical Reaction Engineering*, Wiley, 1962, Chap. 8.

4 Constrained Optimization Techniques

4.1 INTRODUCTION

In Chapter 3 we considered various numerical techniques for the solution of unconstrained optimization problems. Basically these techniques involved "hill-climbing" procedures, that is, we sought to find the peak of a function made up of several independent control variables.

As we recall from our discussion in Section 2.4, problems involving "loose" inequality constraints may also be tackled by this method since these do not exclude the region where the "unconstrained" optimum would be located. Figure 4.1*a* shows the problem of maximizing $I(x_1, x_2)$ subject to $g_j(x_1, x_2) \leq 0$, $j = 1, 2, 3$, but there the unconstrained maximum (at A) falls within the feasible region.

However, in many real problems, the optimum occurs on the boundary of at least one of the constraints, and this may cause complications. This is illustrated in Fig. 4.1*b* where one additional constraint, $g_4(x_1, x_2) \leq 0$, removes the unconstrained optimum from the feasible region, and the optimum is constrained at the boundary by $g_4(x_1, x_2) = 0$. In this case, the methods of Chapter 3 break down, and new techniques are needed to

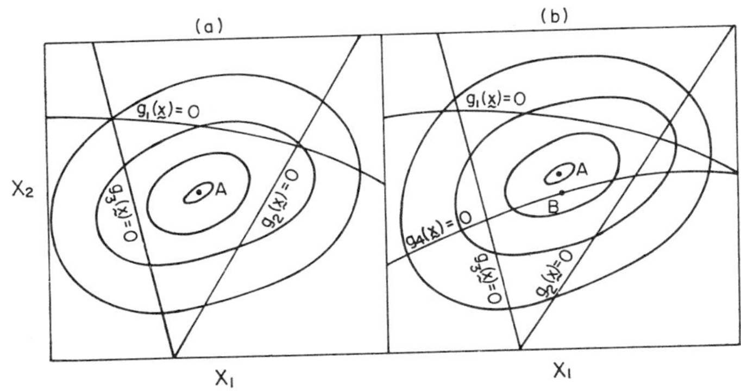

Fig. 4.1 Constrained optimization problems: (*a*) location of optimum unaffected by the constraints and (*b*) optimum located on a constraint.

search within the constraints to find the new optimum at B. The present chapter is devoted to a discussion of these techniques.

Regarding the organization of this chapter, Section 2 deals with the transformation of constrained optimization problems to unconstrained problems by a change of the variables. The technique of handling constraints through the use of penalty functions is discussed in Section 3. Section 4 is devoted to linear programming. Quadratic programming is discussed in Section 5, and Section 6 is concerned with separable programming, whereas geometric programming is discussed in Section 7. A general discussion of nonlinear programming techniques is given in Section 8, and a comparison of these is presented in Section 9. A short discussion of integer programming can be found in Section 10.

4.2 TRANSFORMATION OF CONSTRAINED OPTIMIZATION PROBLEMS BY THE CHANGE OF VARIABLE

Under certain conditions it is possible to transform a constrained optimization problem to an unconstrained problem by the appropriate change of the control variables. When this technique is applicable, we may readily apply the methods discussed in the previous chapter.

A general statement of the nonlinear programming problem is

$$\operatorname*{Max}_{x_i} I(\mathbf{x}) \qquad i = 1, 2, \ldots, n \tag{4.2.1}$$

subject to

$$g_j(\mathbf{x}) \leq 0 \qquad j = 1, 2, \ldots, m \tag{4.2.2}$$

Let us consider some types of constraints $g_j(\mathbf{x})$ which can be eliminated by transformation.

(i) If we require some or all of the variables x_i be positive, then the following change of variables

$$x_j = e^{y_j} \qquad j = 1, 2, \ldots, m \tag{4.2.3}$$

or

$$x_j = (y_j)^2 \qquad j = 1, 2, \ldots, m \tag{4.2.4}$$

will ensure that the constraints are automatically satisfied. For example, the problem

$$\operatorname*{Max}_{x_i} I(\mathbf{x}) = x_1 - x_1^2 - x_1x_2 + x_2 - x_2^2 \tag{4.2.5}$$

subject to

$$g_j(\mathbf{x}) = -x_j \leq 0 \qquad j = 1, 2 \tag{4.2.6}$$

can be transformed by Eq. 4.2.4 to the unconstrained problem

$$\operatorname*{Max}_{y_i} I(\mathbf{y}) = (y_1)^2 - (y_1)^4 - (y_1y_2)^2 + (y_2)^2 - (y_2)^4 \tag{4.2.7}$$

the solution of which is

$$x_1 = y_1^2 = \tfrac{1}{3}, \qquad x_2 = (y_2)^2 = \tfrac{1}{3}$$

(ii) If we require some or all of the variables to lie within the hypercube

$$x_{i*} \leq x_i \leq x_i^* \tag{4.2.8}$$

then the transformation

$$x_i = x_{i*} + (x_i^* - x_{i*})\{\sin y_i\}^2 \tag{4.2.9}$$

will insure that Eq. 4.2.8 is always satisfied. For example, the problem given by Eq. 4.2.5 and

$$\begin{aligned} 0 &\leq x_1 \leq 0.2 \\ 0.2 &\leq x_2 \leq 0.5 \end{aligned} \tag{4.2.10}$$

becomes

$$\begin{aligned} \operatorname*{Max}_{y_i} I(\mathbf{y}) = {} & 0.2 \sin^2 y_1 - [0.2 \sin^2 y_1]^2 \\ & - [0.2 \sin^2 y_1][0.2 + 0.3 \sin^2 y_2] \\ & + [0.2 + 0.3 \sin^2 y_2] - [0.2 + 0.3 \sin^2 y_2]^2 \end{aligned} \tag{4.2.11}$$

the solution of which is

$$y_1 = \frac{\pi}{2} + n\pi, \quad y_2 = n\pi + \sin^{-1}\left[\left(\frac{2}{3}\right)^{1/2}\right]$$

or

$$x_1 = 0.2, \qquad x_2 = 0.4$$

(iii) There are many other transformations that can be used in certain cases, for example, if $0 \leq x_1 \leq x_2 \leq x_3 \cdots \leq x_n$ is required, then

$$\begin{aligned} x_1 &= (y_1)^2 \\ x_2 &= (y_1)^2 + (y_2)^2 \\ &\cdot \\ &\cdot \\ &\cdot \\ x_n &= \sum_{i=1}^{n} (y_i)^2 \end{aligned} \tag{4.2.12}$$

will render the problem unconstrained. These are only a few of the transformations possible; further examples can be found in the excellent paper by Box [1].

The reader may consider the inequality constraints that can be handled in this manner so simple as to be trivial. However, if we wish to carry out a multivariable optimization problem with one of the techniques described in Chapter 3, we shall be relying on standard computer subroutines developed for unconstrained problems. Unless we are careful, the optimum found might specify negative particle diameters, or negative absolute temperatures. The techniques described in this section allow us to avoid these pitfalls.

The reader is advised to refer to Example 3.4.2 and question how this can be regarded as an unconstrained optimization problem, since it contains parameters such as particle diameter, void fraction, and absolute temperature. The answer is simply that the nonnegativity constraints are there, but they have not been invoked because inspection of the objective function shows that the selection of negative values for the temperature, particle diameter, or the void fraction would have invariably reduced the profitability of the operation.

Clearly, if these constraints on the problem had been binding, then the methods described in this section would have been useful for guaranteeing that they were indeed satisfied.

4.3 PENALTY FUNCTIONS

An alternative method of converting constrained optimization problems to unconstrained problems is through the use of *penalty functions*. The penalty function technique requires that we adjoin the constraints to the objective function in such a way that the equations representing the constraints are multiplied by a factor ("a penalty"); thus any deviations from the constraints will tend to reduce the value of the objective function. In the extreme, when the constraints are fulfilled exactly, the "penalty" will reduce to zero. In the following, a general statement is given of certain types of penalty functions; their use will then be illustrated by a simple analytical example, although in the majority of cases the practical use of penalty functions is associated with numerical, rather than analytical optimization techniques.

Let us consider the general statement of the mathematical programming problem given in Chapter 1

$$\operatorname*{Max}_{x_i} I(\mathbf{x}) \qquad i = 1, 2, \ldots, n \tag{1.2.1}$$

subject to

$$h_k(\mathbf{x}) = 0 \qquad k = 1, 2, \ldots, m_1 \tag{1.2.2}$$

$$g_j(\mathbf{x}) \leq 0 \qquad j = 1, 2, \ldots, m_2 \tag{1.2.3}$$

The system of Eqs. 1.2.1 to 1.2.3 may be converted to the unconstrained problem

$$\operatorname*{Max}_{x_i} \left\{ \hat{I}(\mathbf{x}) = I(\mathbf{x}) - \sum_{k=1}^{m_1} \mu_k [h_k(\mathbf{x})]^2 - \sum_{j=1}^{m_2} \eta_j H[g_j(\mathbf{x})] \right\} \tag{4.3.1}$$

where

$$H(a) = \begin{cases} 1 & a > 0 \\ 0 & a < 0 \end{cases}$$

is the Heaviside function. Here μ_k and η_j appearing in Eq. 4.3.1 are the "costs" or penalties associated with violating the constraints of the problem. It is readily seen that if μ_k and η_j are made large enough, then for the maximization of $\hat{I}(\mathbf{x})$, $h_k(\mathbf{x})$ has to tend to zero and $g_j(\mathbf{x})$ will have to tend to values less than zero; in other words, the constraints (Eqs. 1.2.1 to 1.2.2) will be satisfied.

In the following we shall present an illustration of the use of penalty functions, using a very simple problem, which could, in fact, be tackled by other means.

EXAMPLE 4.3.1. Find the minimum of the function

$$I(\mathbf{x}) = 2x_1^2 + 3x_2^2 \tag{4.3.2}$$

subject to

$$x_1 + 2x_2 = 5 \tag{4.3.3}$$

using the method of penalty functions.

Solution. Equation 4.3.3 is an equality constraint; thus the third term on the right-hand side of Eq. 4.3.1 is nonexistent. Furthermore there exists only one equality constraint; thus the index k is unity and the function $[h_1(\mathbf{x})]^2$ is given as

$$[h_1(\mathbf{x})]^2 = (x_1 + 2x_2 - 5)^2$$

It follows that the modified objective function is now written as

$$\hat{I}(\mathbf{x}) = 2x_1^2 + 3x_2^2 + \mu_1(x_1 + 2x_2 - 5)^2 \tag{4.3.4}$$

We note that, since we wish to *minimize* the function $\hat{I}(\mathbf{x})$, the sign in front of the penalty function will be *positive*; if we wished to *maximize* $I(\mathbf{x})$ subject to the same constraint, the sign of μ_1 would have been negative, as shown in the general form of Eq. 4.3.1.

We shall now proceed by finding the stationary values of Eq. 4.3.4; thus, we have

$$\frac{\partial \hat{I}}{\partial x_1} = 4x_1 + 2\mu_1(x_1 + 2x_2 - 5) = 0 \tag{4.3.5}$$

and

$$\frac{\partial \hat{I}}{\partial x_2} = 6x_2 + 4\mu_1(x_1 + 2x_2 - 5) = 0 \tag{4.3.6}$$

On combining Eqs. 4.3.5 with 4.3.6, we obtain

$$x_1 = \tfrac{3}{4}x_2$$

then on substituting for x_1 in Eq. 4.3.5 we have that

$$3x_2 + 2\mu_1(\tfrac{3}{4}x_2 + 2x_2 - 5) = 0$$

that is,

$$x_2 = \frac{20}{\dfrac{6}{\mu_1} + 11} \tag{4.3.7}$$

When $\mu_1 \to \infty$, x_2 tends to 20/11 and x_1 tends to 15/11; thus $\hat{I}(\mathbf{x})$ has its stationary value at $x_1 = 20/11$, $x_2 = 15/11$.

It is left to the reader to show, through the use of the *Hessian* (see Chapter 2), that the above location is indeed a minimum.

It is noted, that the problem tackled in Example 4.2.1 through the use of the penalty function could have been easily solved either by direct *substitution*, or by using the *Lagrange multipliers*, as was done in Chapter 2.

The principal application of penalty function methods is in conjunction with numerical methods, where the "penalties", that is, the numerical values of μ_k and η_j, are gradually raised as the optimum is approached.

Optimization subroutines involving penalty functions are available in the libraries of computer centers (cf. Appendix C).

One problem with penalty-function techniques is that the function $\hat{I}(\mathbf{x})$ may be relatively flat within the feasible region, but at the same time may have a very steep hill just at the constraint. This type of behavior makes most unconstrained optimization techniques perform very badly and causes the maximization of $\hat{I}(\mathbf{x})$ to need many more iterations than would have been required to find the maximum of $I(\mathbf{x})$, that is, the unconstrained function.

Rosenbrock [2] suggested that this difficulty may be surmounted by defining a new objective function such that

$$\hat{I}(x) = I(x) \prod_{j=1}^{m_2} \phi_j \tag{4.3.8}$$

where

$$\phi_j = \begin{cases} 1 & \text{if} \quad g_j \leq -\epsilon \\ f(g_j) & \text{if} \quad -\epsilon \leq g_j \leq 0 \qquad \text{for} \quad \epsilon > 0 \\ 0 & \text{if} \quad g_j > 0 \end{cases}$$

Here $f(g_j)$ is a parabolic function with $f(-\epsilon) = 1, f(0) = 0$. On using this modified objective function, there is some narrow range, of width ϵ, over which the function $\hat{I}(\mathbf{x})$ goes from 0 to $I(\mathbf{x})$. This method seems to work well for some problems, but probably is not the most efficient when following a constraint.

Another difficulty with penalty-function methods is the choice of the costs μ_k, η_j associated with the constraints. If these costs are made too small, the objective $\hat{I}$ will be unrealistically high because of "profitable" violations of the system constraints. This results in a nonfeasible optimum. On the other hand, if these constraint costs are chosen too large, the optimization routine ignores the value of $I(\mathbf{x})$ and concentrates on satisfying equality constraints to within a far greater degree of accuracy than the capability of the computer allows. This then results in a highly feasible, but nonoptimal solution.

Carroll [3] attempted to obviate this problem by using a "created response surface" and solving a sequence of problems with slowly increasing cost of constraint violation. This approach, called a "sequentially unconstrained maximization technique," was extended and given mathematical respectability by Fiacco and McCormick [4–6]. The method forms the modified objective function

$$\hat{I}(\mathbf{x}, r_k) = I(\mathbf{x}) + r_k \sum_{j=1}^{m_2} [g_j(\mathbf{x})]^{-1} - \frac{1}{(r_k)^{1/2}} \sum_{i=1}^{m_1} [h_i(\mathbf{x})]^2 \tag{4.3.9}$$

where r_k is the "cost" of approaching the inequality constraints too closely and $(r_k)^{-1/2}$ is the cost of violating an equality constraint. This method requires a feasible starting point, and the unconstrained maximum is found for cost r_1; the value of r is then changed and the process is repeated. The procedure is as follows:

1. Find a feasible starting point and solve the unconstrained problem for a selected value of r_1.
2. Reduce the size of $r_k > 0$ so that $r_{k+1} < r_k$ and solve the unconstrained problem again.
3. Iterate until the method converges.

Toward the end of the problem r_k becomes close to zero as does $g_l(\mathbf{x})$ if the l-th constraint is binding, and the location of the maximum moves closer to the binding constraint.

Fiacco and McCormick have shown that this method has a number of interesting properties. They have proved that the method will converge to the constrained maximum if $I(\mathbf{x})$ is concave, $g_j(\mathbf{x})$ are convex, and $[h_i(\mathbf{x})]^2$ are convex. In addition, they have shown that there is a dual problem that is solved at each step and which provides an upper bound on the optimum $I(\mathbf{x})$ at each stage of the optimization. This provides a good criterion for stopping the iteration procedure.

There is very limited computational experience with this method. Although Fiacco and McCormick reported good results with large dimensional problems, other workers [7, 8] have found difficulties when there are nonlinear equality constraints. This problem is discussed in greater detail in Section 4.9.

Although other versions of penalty-function methods have been proposed (e.g. [9, 10]), Fiacco and McCormick's technique seems to be one of the more popular at present. As noted earlier, the objective of the techniques involving penalty functions is to avoid facing the problem of constraints.

We shall now consider the more general case of constrained optimization problems. We shall show that under certain conditions rather than eliminating or circumventing the constraints, the constraints may be used to good advantage for solving the problem. Such a special case of problems is represented by linear programming, which is the subject matter of the next section.

4.4 LINEAR PROGRAMMING

Linear programming is probably the best known and most fully explored area of optimization. Linear programming is concerned with finding the optimum of a linear function, subject to linear constraints, which may be of both equality and inequality types.

As in most areas of linear mathematics, the theory is very tidy; furthermore, the algorithms for solution are very efficient and a large number of "canned programs" are available for dealing with linear programming problems (cf. Appendix C).

In view of the extensive literature on linear programming, we shall confine ourselves to the statement of some basic principles, and will stress applications illustrated by examples.

The linear programming problem can be written as

$$\operatorname*{Max}_{x_i} \left[I(\mathbf{x}) = \sum_{i=1}^{n} c_i x_i \right] \tag{4.4.1}$$

Subject to:

$$h_k(\mathbf{x}) = \sum_{i=1}^{n} \hat{A}_{ki} x_i - \hat{b}_k = 0; \qquad k = 1, 2, \ldots, m_1 \tag{4.4.2}$$

$$g_j(\mathbf{x}) = \sum_{i=1}^{n} \bar{A}_{ji} x_i - \bar{b}_j \leq 0; \qquad j = 1, 2, \ldots, m_2 \tag{4.4.3}$$

In order to illustrate the important features of this problem, let us consider the following example:

EXAMPLE 4.4.1. Find the maximum of $I(x_1, x_2) = 6x_1 + 2x_2$
Subject to the constraints

$$\begin{aligned} 2x_1 + x_2 &\leq 8 \\ 4x_1 + x_2 &\leq 10 \\ x_1, x_2 &\geq 0 \end{aligned} \tag{4.4.4}$$

Solution. The solution of this problem is given in Fig. 4.2*a* in graphical form. There we see that the problem is reduced to finding, within the

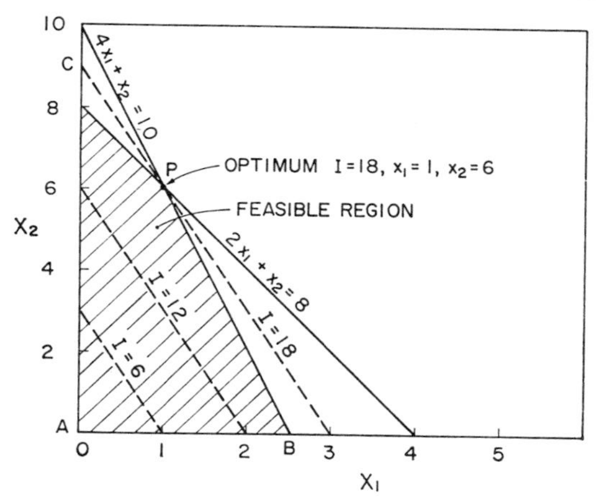

Fig. 4.2*a* Illustration of the linear programming Example 4.4.1 showing the *feasible region* designated by the shaded area and the optimum at the vertex *P*.

area below $2x_1 + x_2 = 8$ and $4x_1 + x_2 = 10$ (the feasible region), the point for which the line $6x_1 + 2x_2 = I$ has the highest value of I. We see that for this simple problem the solution is at point P where $x_1 = 1$, $x_2 = 6$, and $I = 18$. There are several comments that should be made here, and these are further illustrated in Fig. 4.2a:

(i) The feasible region (designated by the shaded area) is convex. (Readers who wish to refresh their memory concerning the concepts of the feasible region and convexity should refer to Chapter 2.) A little thought should make it clear that any two-dimensional, feasible region bounded by straight lines of the form of Eq. 4.4.3 will always be convex.

(ii) The optimum is located at the boundary of the feasible region. Intuition should suggest that for two-dimensional problems the optimum must be located on the boundary, because the feasible region is convex. As a counter example, let us imagine a case where the optimum is located at an interior point—clearly this is not possible.

(iii) There can be rare situations where the optimum does not lie only on a vertex, but rather lies along one of the constraints. This occurs when the profit line, $I = \sum c_i x_i$ is parallel to one of the constraints. In order to illustrate such a case, let us reconsider the procedure employed in solving Example 4.4.1.

We established the *feasible region* by plotting the constraints, Eq. 4.4.4 and then explored a number of constant values, K, for $I = 6x_1 + 2x_2 = K$, until we reached the optimum that provided the highest value of $I = K$ which still satisfied the constraints. (Point P in Fig. 4.2a.)

Let us now consider a similar problem, in which the constraints are given by Eq. 4.4.4, but let us assign the following form to the objective function

$$I(x_1, x_2) = 4x_1 + 2x_2$$

The graphical solution of this problem is illustrated in Fig. 4.2b. It is seen that the constraints and the feasible region are, of course, identical to those given in the previous example. However, in this case the form of the objective function causes the optimum to lie on the line $\overline{PC}$, rather than on a vertex. Clearly, both vertices P and C represent optima (as well as the line between them). Thus one can say that *the optimum is always located at a vertex of the feasible region.*

Since the optimum always occurs at a vertex, and there are only a finite number of these, then we have to test only the vertices in order to find the optimum. This means only a finite number of steps need be taken to reach the optimum, which is very helpful.

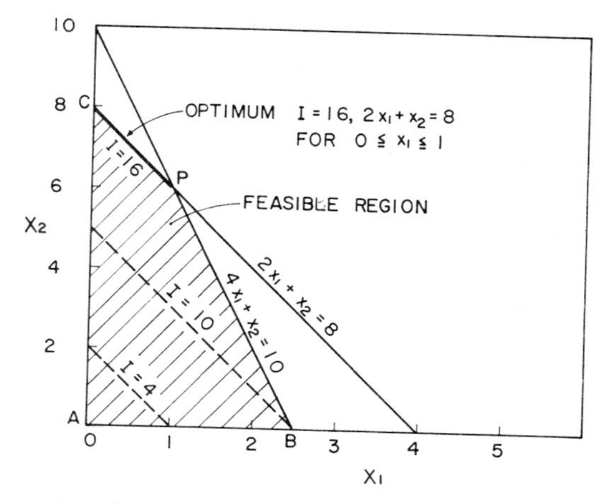

Fig. 4.2*b* Illustration of Example 4.4.1 for the case where the optimum coincides with one of the constraints; the optimum is the heavy line $\overline{PC}$.

The properties of two-dimensional linear programs which we have observed from the graphs are, indeed, general properties of n dimensional linear programs (cf. [11–13]). Thus any linear programming procedure should be a method of testing vertices until the optimum vertex is found. The most common procedures in use today are variations of the simplex method of Dantzig [9] which chooses the next vertex for testing on the basis that the profit is increased the most, by choice of that vertex. Thus the technique is a type of steepest ascent method with discrete variables.

We shall review the simplex procedure below, but first let us discuss the concept of *duality*.

The linear programming (LP) problem given in Eqs. 4.4.1 to 4.4.3 can be rewritten in vector notation as

$$\operatorname*{Max}_{\mathbf{x}} \{I = \mathbf{c}^T\mathbf{x}\} \tag{4.4.5}$$

subject to

$$\hat{\mathbf{A}}\mathbf{x} - \hat{\mathbf{b}} = \mathbf{0} \tag{4.4.6}$$

$$\bar{\mathbf{A}}\mathbf{x} - \bar{\mathbf{b}} \leq \mathbf{0} \tag{4.4.7}$$

or if one uses the partitioned matrices

$$\mathbf{A} = \begin{bmatrix} \hat{\mathbf{A}} \\ \text{---} \\ \bar{\mathbf{A}} \end{bmatrix}; \qquad \mathbf{b} = \begin{bmatrix} \hat{\mathbf{b}} \\ \text{---} \\ \bar{\mathbf{b}} \end{bmatrix} \tag{4.4.8}$$

the constraints become

$$\mathbf{A}\mathbf{x} - \mathbf{b}(\overline{\overline{\leq}})\mathbf{0} \tag{4.4.9}$$

The reader will note that Eq. 4.4.9 is but a convenient shorthand notation for Eqs. 4.4.6 to 4.4.7 through the use of partitioned matrices.

Let us use Lagrange multipliers $\boldsymbol{\lambda}$ to form the *Lagrangian* of this problem

$$L(\mathbf{x}, \boldsymbol{\lambda}) = \mathbf{c}^T\mathbf{x} + \boldsymbol{\lambda}^T(\mathbf{b} - \mathbf{A}\mathbf{x}) \tag{4.4.10}$$

Let us recall from Chapter 2 that the Kuhn-Tucker conditions require L to be a maximum with respect to $\mathbf{x}$ and a minimum with respect to $\boldsymbol{\lambda}$ for this convex problem.

Let us now rearrange Eq. 4.4.10 by taking the transpose of both sides; the left-hand side of the equation is a scalar so it will not be affected by the transposition, and thus we obtain the following:

$$L(\mathbf{x}, \boldsymbol{\lambda}) = \mathbf{b}^T\boldsymbol{\lambda} + \mathbf{x}^T(\mathbf{c} - \mathbf{A}^T\boldsymbol{\lambda}) \tag{4.4.11}$$

Inspection of Eq. 4.4.11 shows that this is the Lagrangian of the following problem

$$\underset{\lambda}{\text{Min}} \; \{\mathbf{b}^T\boldsymbol{\lambda}\}$$

subject to

$$\mathbf{A}^T\boldsymbol{\lambda} - \mathbf{c} \;(\geqq)\; \mathbf{0} \tag{4.4.12}$$

where now $\boldsymbol{\lambda}$ is a state variable and $\mathbf{x}$ is the Lagrange multiplier for Eq. 4.4.12.

Equation 4.4.12 is called the *dual* of the problem given by Eq. 4.4.5 and fits the general definition of a dual problem discussed in Section 2.5.

The dual of the linear programming problem has several attractive properties. One of these is that a solution of the dual provides a solution of the primal problem and vice versa. Thus we can decide to solve the easiest of the two problems. Normally, one chooses the problem with fewer *constraints* because it has been found that the computational algorithms are more efficient in this case.

There is a considerable advantage to having the solution of the dual problem because the optimal $\boldsymbol{\lambda}$, called *shadow prices*, have the economic meaning

$$\lambda_i = \left.\frac{\partial I}{\partial b_i}\right|_{\text{opt}} \tag{4.4.13}$$

and give the marginal increase in profit if the constraint b_i could be relaxed. This information is quite valuable because it provides a cost for a possibly ill-considered constraint, as well as giving the economic incentives for expanding capacity, and so on. An additional attraction is that these Lagrange multipliers may be obtained from some of the computational algorithms with no additional work.

There are numerous computational algorithms available for dealing with linear programming problems. Here we consider the *simplex technique*, which may be suitable for hand calculation in the case of a few variables, say 3 to 4, and for which "canned" programs are readily available.

The Simplex Method

To illustrate the essential features of the simplex method of linear programming, we shall consider the following problem

$$\underset{\mathbf{x}}{\text{Max}}\ \{\mathbf{c}^T\mathbf{x}\} \tag{4.4.14}$$

subject to

$$\mathbf{A}^*\mathbf{x} \le \mathbf{b}$$
$$\mathbf{x} \ge 0$$

Let us eliminate the inequality constraints by the introduction of a set of artificial variables, called *slack variables*, $z_1, z_2, \ldots, z_m$ which are added to each constraint. Thus we have

$$\begin{aligned}
a_{11}x_1 + a_{12}x_2 + \cdots + a_{1n}x_n + z_1 &= b_1 \\
a_{21}x_1 + a_{22}x_2 + \cdots + a_{2n}x_n + z_2 &= b_2 \\
\vdots \qquad\qquad\qquad\qquad \vdots \qquad &\quad \vdots \\
a_{m1}x_1 + a_{m2}x_2 + \cdots + a_{mn}x_n + z_m &= b_m
\end{aligned} \tag{4.4.15}$$

It follows that in this manner all the constraints become equality constraints. The problem can then be posed as

$$\underset{\mathbf{y}}{\text{Max}}\ \{\mathbf{p}^T\mathbf{y}\} \tag{4.4.16}$$

subject to

$$\mathbf{A}\mathbf{y} = \mathbf{b}; \qquad \mathbf{y} \ge 0$$

where

$$\mathbf{y} = \begin{bmatrix} x_1 \\ x_2 \\ \cdot \\ \cdot \\ x_n \\ \hline z_1 \\ \cdot \\ \cdot \\ z_m \end{bmatrix} = \begin{bmatrix} \mathbf{x} \\ \hline \mathbf{z} \end{bmatrix} \tag{4.4.17}$$

is made up of n real variables and m slack variables, and $\mathbf{p}$, $\mathbf{A}$ become

$$\mathbf{p} = \begin{bmatrix} c_1 \\ c_2 \\ \cdot \\ \cdot \\ \cdot \\ c_n \\ \hdashline 0 \\ \cdot \\ \cdot \\ \cdot \\ 0 \end{bmatrix} = \begin{bmatrix} \mathbf{c} \\ \hdashline \mathbf{0} \end{bmatrix};$$

$$\mathbf{A} = \begin{bmatrix} a_{11} & a_{12} & \cdots & a_{1n} & 1 & 0 & \cdots & 0 \\ a_{21} & a_{22} & \cdots & a_{2n} & 0 & 1 & & \\ \cdot & & & & \cdot & 0 & \cdot & \\ \cdot & & & & \cdot & \cdot & & \cdot \\ \cdot & & & & \cdot & \cdot & & \cdot \\ \cdot & & & & \cdot & \cdot & & \\ a_{m1} & a_{m2} & \cdots & a_{mn} & 0 & 0 & & 1 \end{bmatrix} = [\mathbf{A}^* \,\vdots\, \mathbf{I}] \quad (4.4.18)$$

Let us denote the columns of $\mathbf{A}$ by the vectors $\mathbf{f}_j$, that is,

$$\mathbf{A} = (\mathbf{f}_1 \mathbf{f}_2 \cdots \mathbf{f}_{n+m})$$

Then from the constraints in Eq. 4.4.16 we obtain

$$\mathbf{b} = \sum_{j=1}^{m+n} y_j \mathbf{f}_j \qquad (4.4.19)$$

However, since the rank of $\mathbf{A}$ is at most m, then only m of the total $m + n$ vectors $\mathbf{f}_j$ are needed to represent $\mathbf{b}$. That is

$$\mathbf{b} = \sum_{j=1}^{m} y_j^0 \mathbf{f}_j^0 \qquad (4.4.20)$$

where only m independent vectors $\mathbf{f}_j^0$ are required, and where the superscript ^{0} denotes the fact that the $\mathbf{f}_j^0$ have been selected from the $\mathbf{f}_j$ and thus the subscripts on the $\mathbf{f}_j^0$ may not correspond to the $\mathbf{f}_j$.

It can be readily seen that one such choice is the $\mathbf{f}_j^0$ composing the identity matrix associated with the slack variables. Thus an immediate feasible solution to Eq. 4.4.16 is

$$y_{n+1} = b_1$$
$$y_{n+2} = b_2$$
$$\vdots$$
$$y_{n+m} = b_m$$

so that

$$\mathbf{b} = \sum_{i=1}^{m} y_i^0 \mathbf{f}_i^0 = \sum_{j=n+1}^{n+m} y_j \mathbf{f}_j \tag{4.4.21}$$

However, since there is zero profit associated with this feasible solution, let us describe a procedure for moving from vertex to vertex until the optimum is found. Since a vertex is associated with each set of m independent vectors $\mathbf{f}_j^0$ which form *a basic feasible solution* (Eq. 4.4.20), then by searching among the various *basic feasible solutions* (i.e., the vertices) we can find the optimal solution. The simplex method is a technique for the stepwise removal of an $\mathbf{f}_j^0$ from the basic set of m and for its replacement with a more profitable one. To illustrate this, let us set up the following tableau where the labels on the first m rows are the m basis vectors.

	$\mathbf{f}_1\mathbf{f}_2 \cdots \cdots \mathbf{f}_n\mathbf{f}_{n+1} \cdots \cdots \mathbf{f}_{n+m}$	$\mathbf{y}^0$	$\mathbf{p}^0$
Basis vectors $\mathbf{f}_1^0$		y_1^0	0
$\mathbf{f}_2^0$		$\cdot$	$\cdot$
$\cdot$	$\boldsymbol{\alpha}$	$\cdot$	$\cdot$
$\cdot$		$\cdot$	$\cdot$
$\mathbf{f}_m^0$		y_m^0	0
p_j	$p_1 p_2 \quad p_n \; 0 \quad 0$		
μ_j		μ_y	
$p_j - \mu_j$			

The matrix made up of the first m rows and $n + m$ columns, $\boldsymbol{\alpha}$, gives the representation of all the $m + n$ vectors $\mathbf{f}_j$ in terms of the m basis vectors, that is,

$$\mathbf{f}_j = \sum_{i=1}^{m} \alpha_{ij} \mathbf{f}_i^0 \qquad j = 1, 2, \ldots, m + n \tag{4.4.22}$$

For the particular feasible solution given in Eq. 4.4.21 (i.e., made up of only slack variables), $\boldsymbol{\alpha} = \mathbf{A}$, and $\mathbf{f}_i^0 = \mathbf{f}_{n+i}$, $i = 1, 2, \ldots, m$ in the tableau.

The quantity μ_j is the incremental profit associated with each vector $\mathbf{f}_j$ and is given by

$$\mu_j = \sum_{i=1}^{m} \alpha_{ij} p_i^0 \tag{4.4.23}$$

The column $\mathbf{y}^0$ gives the elements of $\mathbf{y}^0$ that satisfy the constraints at each iteration. Thus if Eq. 4.4.23 is applied to this column

$$\mu_y = \sum_{i=1}^{m} y_i^0 p_i^0 \tag{4.4.24}$$

is the profit resulting from the choice of $\mathbf{y}^0$.

By examining the conditions for the largest profit increase when vector $\mathbf{f}_s$ is added to the basis in place of $\mathbf{f}_k^0$, Dantzig [11] established the following iterative procedure.

1. In the tableau, examine the row $p_j - \mu_j$ for the initial feasible solution. Select the vector $\mathbf{f}_s$ for which the value of $(p_s - \mu_s)$ is the largest. This vector is to be added to the basis.
2. For the column $\mathbf{f}_s$, compute

$$y_j^0/\alpha_{js} \qquad j = 1, 2, \ldots, m$$

and choose the smallest positive value y_k^0/α_{ks}. This determines the vector $\mathbf{f}_k^0$ to be removed from the basis.

3. Perform the following row and column operations to replace $\mathbf{f}_k^0$ by $\mathbf{f}_s$.
   (a) Divide the kth row, α_{kj}, by α_{ks} to produce a "one" in the α_{ks} element.
   (b) Obtain zeros in the remaining elements, α_{is} $(i \neq k)$, by multiplying the new kth row by α_{is} and subtracting it from the ith row. Do this for all $m - 1$ rows.
4. Compute μ_j, μ_y by Eqs. 4.4.23 to 4.4.24 and compute $p_j - \mu_j$ for each column.
5. If all $p_j - \mu_j < 0$, the optimum has been found. Otherwise, choose the largest $p_j - \mu_j$ so as to determine the new vector $\mathbf{f}_s$ and go back to step 2 and iterate.

Step 1 chooses the vector to be added that increases the profit the most whereas step 2 eliminates the vector that is most expensive in terms of taxing the constraints. The row and column operations in step 3 insure that each iteration is a feasible solution and step 5 stops the algorithm when no further addition of vectors will increase the profit.

In order to illustrate the technique, we shall work a simple example and then proceed to discuss complications, special cases, and so on.

EXAMPLE 4.4.2

$$\text{Max } (4x_1 + 2x_2 + \tfrac{1}{3}x_3 + x_4) \tag{4.4.25}$$

subject to

$$\begin{aligned} 4x_1 + 3x_2 + 2x_4 &\leq 8 \\ 8x_1 + x_2 \qquad\quad &\leq 6 \\ x_2 + \tfrac{8}{3}x_3 + 2x_4 &\leq 6 \\ x_i \geq 0 \qquad i &= 1, 2, 3, 4 \end{aligned}$$

On adding slack variables z_1, z_1, z_3 to each of the constraints, we obtain this statement of the problem:

$$\text{Maximize } (4x_1 + 2x_2 + \tfrac{1}{3}x_3 + x_4 + 0.z_1 + 0.z_2 + 0.z_3) \tag{4.4.26}$$

subject to

$$\begin{aligned} 4x_1 + 3x_2 + 2x_4 + z_1 &= 8 \\ 8x_1 + x_2 + z_2 \qquad &= 6 \\ x_2 + \tfrac{8}{3}x_3 + 2x_4 + z_3 &= 6 \\ x_i \geq 0, \quad z_j &\geq 0 \end{aligned}$$

or in terms of the variables:

$$\mathbf{y} = \begin{bmatrix} x_1 \\ x_2 \\ x_3 \\ x_4 \\ \hline z_1 \\ z_2 \\ z_3 \end{bmatrix}; \quad \mathbf{p} = \begin{bmatrix} 4 \\ 2 \\ \frac{1}{3} \\ 1 \\ \hline 0 \\ 0 \\ 0 \end{bmatrix}; \quad \mathbf{b} = \begin{bmatrix} 8 \\ 6 \\ 6 \end{bmatrix}; \quad \mathbf{A} = \left[\begin{array}{cccc|ccc} 4 & 3 & 0 & 2 & 1 & 0 & 0 \\ 8 & 1 & 0 & 0 & 0 & 1 & 0 \\ 0 & 1 & \frac{8}{3} & 2 & 0 & 0 & 1 \end{array}\right] \tag{4.4.27}$$

we have the problem given in Eq. 4.4.16.

The initial feasible tableau becomes

TABLEAU I

	$\mathbf{f}_1$	$\mathbf{f}_2$	$\mathbf{f}_3$	$\mathbf{f}_4$	$\mathbf{f}_5$	$\mathbf{f}_6$	$\mathbf{f}_7$	$\mathbf{y}^0$	$\mathbf{p}^0$
$\mathbf{f}_1^0 = \mathbf{f}_5$	4	3	0	2	1	0	0	8	0
$\mathbf{f}_2^0 = \mathbf{f}_6$	(8)	1	0	0	0	1	0	6	0
$\mathbf{f}_3^0 = \mathbf{f}_7$	0	1	$\frac{8}{3}$	2	0	0	1	6	0
p_j	4	2	$\frac{1}{3}$	1	0	0	0		
μ_j	0	0	0	0	0	0	0	0	
$p_j - \mu_j$	(4)	2	$\frac{1}{3}$	1	0	0	0		

By application of the procedure we see that $\mathbf{f}_s = \mathbf{f}_1$ is to be added to the basis, and since $y_1^0/\alpha_{11} = 2$; $y_2^0/\alpha_{21} = \frac{3}{4}$, $y_3^0/\alpha_{31} \to \infty$ $\mathbf{f}_2^0 = \mathbf{f}_6$ is to be removed from the basis. Thus we must convert α_{21} to unity and perform the row and column operations to get Tableau II:

TABLEAU II

	$\mathbf{f}_1$	$\mathbf{f}_2$	$\mathbf{f}_3$	$\mathbf{f}_4$	$\mathbf{f}_5$	$\mathbf{f}_6$	$\mathbf{f}_7$	$\mathbf{y}^0$	$\mathbf{p}^0$
$\mathbf{f}_1^0 = \mathbf{f}_5$	0	$\frac{5}{2}$	0	2	1	$-\frac{1}{2}$	0	5	0
$\mathbf{f}_2^0 = \mathbf{f}_1$	1	$\frac{1}{8}$	0	0	0	$\frac{1}{8}$	0	$\frac{3}{4}$	4
$\mathbf{f}_3^0 = \mathbf{f}_7$	0	1	$\frac{8}{3}$	2	0	0	1	6	0
p_j	4	2	$\frac{1}{3}$	1	0	0	0		
μ_j	4	$\frac{1}{2}$	0	0	0	$\frac{1}{2}$	0	3	
$p_j - \mu_j$	0	$\frac{3}{2}$	$\frac{1}{3}$	1	0	$-\frac{1}{2}$	0		

which shows a profit of 3.

Iterating again yields $\mathbf{f}_s = \mathbf{f}_2$ to be added and $\mathbf{f}_1^0 = \mathbf{f}_5$ to be removed from the basis. This produces the third tableau:

TABLEAU III

	$\mathbf{f}_1$	$\mathbf{f}_2$	$\mathbf{f}_3$	$\mathbf{f}_4$	$\mathbf{f}_5$	$\mathbf{f}_6$	$\mathbf{f}_7$	$\mathbf{y}^0$	$\mathbf{p}^0$
$\mathbf{f}_1^0 = \mathbf{f}_2$	0	1	0	$\frac{4}{5}$	$\frac{2}{5}$	$-\frac{1}{5}$	0	2	2
$\mathbf{f}_2^0 = \mathbf{f}_1$	1	0	0	$-\frac{1}{10}$	$-\frac{1}{20}$	$\frac{3}{20}$	0	$\frac{1}{2}$	4
$\mathbf{f}_3^0 = \mathbf{f}_7$	0	0	$\frac{8}{3}$	$\frac{6}{5}$	$-\frac{2}{5}$	$\frac{1}{5}$	1	4	0
p_j	4	2	$\frac{1}{3}$	1	0	0	0		
μ_j	4	2	0	$\frac{6}{5}$	$\frac{3}{5}$	$\frac{1}{5}$	0	6	
$p_j - \mu_j$	0	0	$\frac{1}{3}$	$-\frac{1}{5}$	$-\frac{3}{5}$	$-\frac{1}{5}$	0		

which shows a profit of 6.

Going through another iteration produces $\mathbf{f}_s = \mathbf{f}_3$ to be added and $\mathbf{f}_3^0 = \mathbf{f}_7$ to be removed from the basis. This results in the fourth tableau:

TABLEAU IV

	$\mathbf{f}_1$	$\mathbf{f}_2$	$\mathbf{f}_3$	$\mathbf{f}_4$	$\mathbf{f}_5$	$\mathbf{f}_6$	$\mathbf{f}_7$	$\mathbf{y}^0$	$\mathbf{p}^0$
$\mathbf{f}_1{}^0 = \mathbf{f}_2$	0	1	0	$\frac{4}{5}$	$\frac{2}{5}$	$-\frac{1}{5}$	0	2	2
$\mathbf{f}_2{}^0 = \mathbf{f}_1$	1	0	0	$-\frac{1}{10}$	$-\frac{1}{20}$	$\frac{3}{20}$	0	$\frac{1}{2}$	4
$\mathbf{f}_3{}^0 = \mathbf{f}_3$	0	0	1	$\frac{9}{20}$	$-\frac{3}{20}$	$\frac{3}{40}$	$\frac{3}{8}$	$\frac{3}{2}$	$\frac{1}{3}$
p_j	4	2	$\frac{1}{3}$	1	0	0	0		
μ_j	4	2	$\frac{1}{3}$	$\frac{27}{20}$	$\frac{11}{20}$	$\frac{9}{40}$	$\frac{1}{8}$	$\frac{13}{2}$	
$p_j - \mu_j$	0	0	$-\frac{8}{3}$	$-\frac{7}{20}$	$-\frac{11}{20}$	$-\frac{9}{40}$	$-\frac{1}{8}$		

which shows a profit of $\frac{13}{2}$.

This is the optimal solution, since $p_j - \mu_j < 0$ for all $\mathbf{f}_j$; the optimal values of $\mathbf{y}^0$ are given as

$$y_1{}^0 = y_2 = x_2 = 2$$
$$y_2{}^0 = y_1 = x_1 = \tfrac{1}{2}$$
$$y_3{}^0 = y_3 = x_3 = \tfrac{3}{2}$$

and since the other variables are not in the basis, then

$$x_4 = z_1 = z_2 = z_3 = 0$$

Example 4.4.2 was a simple numerical calculation to which no particular physical significance was attached. However, these types of problems, with appropriate values for the constants, do occur in a number of practical situations, as illustrated by the following example, which will be posed, but not solved.

EXAMPLE 4.4.3. Let us consider a processing operation sketched in Fig. 4.3, where raw material is shipped from a central source (mine, port, etc.) to

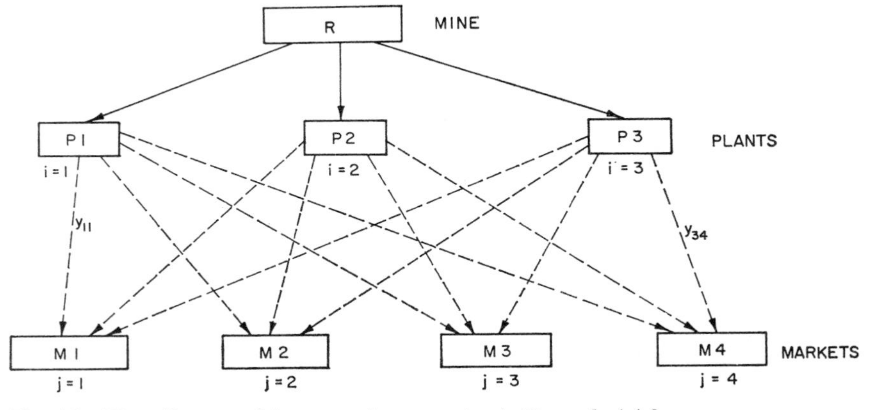

Fig. 4.3 Flow diagram of the processing operation in Example 4.4.3.

three processing plants, designated P_1, P_2, and P_3. From these plants the products are then shipped to four distinct markets, designated by M_1, M_2, M_3, and M_4.

The problem is then, given the requirements of the four markets and the capacities of the three plants, how to minimize the transportation costs.

Let us define the following parameters:

- e_i, $i = 1, 2, 3$ capacity of plant i
- b_j, $j = 1, 2, 3, 4$ market at point j
- y_{ij} the amount of material shipped from plant i to market (warehouse) j
- c_i the cost of shipping to and production cost at plant i (per unit material produced)
- d_{ij} the cost of shipping from plant i to market (warehouse) j (per unit material)

Then the problem may be stated as

$$\operatorname*{Min}_{y_{ij}} \left[\sum_{i=1}^{3} \sum_{j=1}^{4} (c_i + d_{ij}) y_{ij} \right] \tag{1}$$

Subject to the constraints

$$\sum_{j}^{4} y_{ij} \leq e_i, \qquad i = 1, 2, 3 \tag{2}$$

which expresses the fact that each plant has a limited capacity, and

$$\sum_{i=1}^{3} y_{ij} = b_j, \qquad j = 1, 2, 3, 4 \tag{3}$$

which defines the size of the four markets—to which material may be shipped from any one of the three plants.

Finally,

$$y_{ij} \geq 0 \tag{4}$$

which states that the amount shipped from the plant to the warehouse must be positive.

Equations 1 to 4 provide a complete definition of the problem, which may be solved for any set of known coefficients either by hand calculations as illustrated in Example 4.4.2 or by the use of a suitable computer algorithm. A practical problem of this type was treated by Bouman and Kwasnoski [14].

Some Properties of Linear Programming and the Simplex Method

Let us discuss briefly a number of important points concerning linear programming and the simplex method. The dual problem given in Eq. 4.4.12 has a solution that appears automatically in the simplex tableau. For our tableau, the solution can be found in the row μ_j under the vectors associated with the slack variables. For Example 4.4.2 the values $\lambda_1 = \frac{11}{20}$, $\lambda_2 = \frac{9}{40}$, $\lambda_3 = \frac{1}{8}$ given in Tableau IV are the solution to the dual of the example problem, that is,

$$\text{Min } [8\lambda_1 + 6\lambda_2 + 6\lambda_3] \tag{4.4.28}$$

subject to

$$\begin{aligned} 4\lambda_1 + 8\lambda_2 \qquad &\geq 4 \\ 3\lambda_1 + \lambda_2 + \lambda_3 &\geq 2 \\ 8\lambda_3 &\geq 1 \\ 2\lambda_1 + 2\lambda_3 &\geq 1 \\ \lambda_i &\geq 0 \end{aligned}$$

Thus the simplex method automatically determines the Lagrange multipliers (dual variables) associated with each constraint, which is often useful in practical situations.

There are a number of complications which can occur in linear programs of which we shall mention only a few. If the constraints (Eq. 4.4.9) have mixed positive and negative elements of $\mathbf{b}$, then there may be problems getting an initial feasible solution to begin the simplex method. This can usually be accomplished by solving a short linear program to get an initial feasible solution. Another possible difficulty is that of "degeneracy," which occurs if at some point one or more of the variables $\mathbf{y}^0$ takes the value zero. This can cause cycling and nonconvergence of the method. However, Hadley [12] claims that this never happens in practice and that all known degenerate problems are artificially contrived.

Finally, the problem may have insufficient constraints to keep it bounded, resulting in infinite profit, or there may exist conflicting constraints. Both this case and the degenerate difficulties are the results of improperly posed linear programs. Thus one must exercise care in the statement of the optimization problem.

The prospective user of linear programming is fortunate indeed, because there are a large number of general purpose "canned" programs available for treating linear programming problems (cf. Appendix C). A widely used program comes from the IBM-MPS system and can handle problems with thousands of variables and constraints while automatically determining an initial feasible solution and diagnosing possible inconsistencies in the problem formulation. Thus with a minimum of preparation, one can quickly obtain a solution to linear programming problems.

In the preceding examples we have considered relatively simple, rather sterile problems because our objective has been to illustrate the nature of linear programming problems. We shall conclude this section by turning our attention to a more realistic and, therefore, more complex example, involving the optimization of the primary end of an integrated steel plant—a problem which we will work in detail.

EXAMPLE 4.4.4. THE OPTIMIZATION OF THE PRIMARY END OF AN INTEGRATED STEEL PLANT Let us develop the optimal production policy for the "primary end" of an integrated steel plant for specified crude steel production levels. The system consists of a coke plant, three blast furnaces, a basic oxygen furnace (BOF) shop, an electric furnace shop, and an open-hearth shop as sketched in Fig. 4.4.

The blast furnaces are fed with sinter and possibly with prereduced pellets, and the hot metal is processed in the BOF shop, the operation of which may be supplemented through the activation of the open-hearth shop.

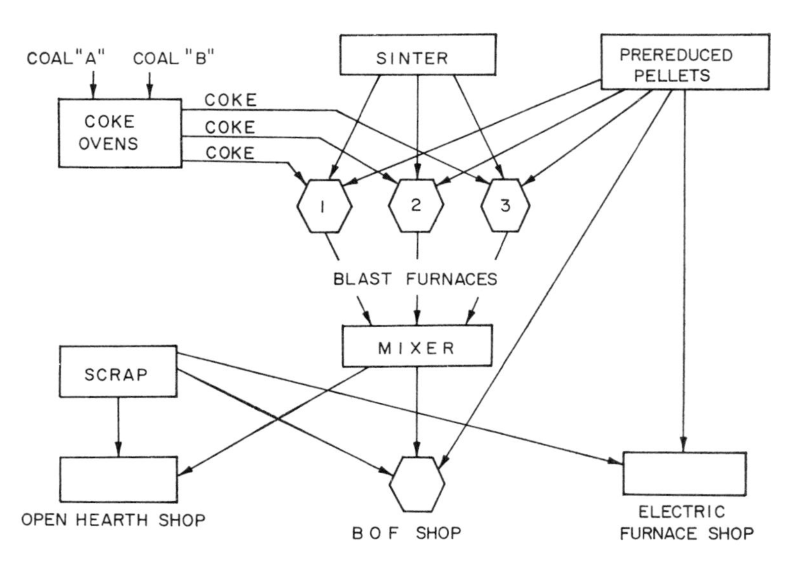

Fig. 4.4 Flow diagram for Example 4.4.4 illustrating material flow in the primary end of an integrated steel plant.

Scrap is used in both the BOF and in the open-hearth shops, and in addition there exists an electric furnace shop that may use either scrap or prereduced pellets.

The problem is, given the cost of the raw materials and the cost of production, determine the strategy which should be used to minimize the overall cost of producing a given quantity of crude steel. In any given practical situation a large number of variables (control variables) could enter the picture. Obvious parameters that one could choose to examine could include the optimal ore mix to be obtained from a variety of sources, as discussed by Kwasnoski and Bouman [14], the optimal scrap purchasing policy, again through the selection from a number of alternatives, and the like. These quantities, together with the cost factors used, will necessarily vary from plant to plant.

Since our primary objective is still to illustrate a technique rather than to solve a given specific problem, a number of deliberate choices have been made so as to render the problem readily tractable, while retaining some of the essential features of the operation of an integrated steel plant.

The assumptions made in defining the problem are stated in the following, where comments will also be made on their appropriateness.

1. In describing the operation of the coke plant, allowance is made for the fact that the feed may consist of a mixture of different grades of coal. In the example chosen only two such feed streams are considered, together with the associated constraints on capacity. In practical situations a larger number of possible feed combinations is likely to occur, and additional constraints may arise from the need to meet certain requirements of chemical composition.

2. In describing the blast furnace plants, we have considered three furnaces, each with a different range of production capacities; appropriate coke rates have been assigned to given production rates, so as to allow for the fact that an increased coke rate is required both when the furnace is "driven hard" and when the production rate is low.

The blast furnaces are allowed to operate with a feed which is composed in part of pellets, but no provision is made for a mixed ore or mixed sinter feed.

3. In the BOF shop we specify a maximum ratio of

$$(\text{scrap} + \text{pellets})/(\text{hot metal});$$

however, no such limitation was placed on the operation of the open-hearth shop, which if activated, was considered capable of handling a 100% scrap charge.

4. The electric furnace shop was assumed capable of handling either scrap or a mixture of scrap and pellets.

All the above assumptions are somewhat restrictive but it is thought that through their use the problem is reduced to manageable proportions, and at the same time we have retained the essential features of the real system. The complicating factors mentioned could be readily included in the formulation, without altering the basic nature of the computational procedure.

The following property values and cost factors were assumed for the quantitative statement of the problem.

Coke Ovens

Maximum capacity, when fed with coal "A"	2.5 M tons/annum
Capacity when fed with coal "B"	2.0 M tons/annum
Cost of coal "A"	\$26/ton of coke produced
Cost of coal "B"	\$21/ton of coke produced
Fixed costs	\$5 $\times$ 10^6 per annum

Blast Furnace No. 1

Operating range: 0.9 to 1.5 M tons/annum
1.4 tons of sinter produce 1 ton of hot metal
1.1 tons of prereduced pellets + 0.3 ton of coke produce 1 ton of iron

The use of prereduced pellets increases the maximum capacity to 1.7 M tons/annum. The maximum amount of prereduced pellets that may be fed: 0.3 M tons/annum.

Coke Rates (for Use with Sinter Loading)

At 1.2 M tons/annum metal production, 0.4 ton of coke/ton of metal produced.
At 1.5 M tons/annum metal production, 0.45 ton of coke/ton of metal produced.
At 0.9 M tons/annum metal production, 0.45 ton of coke/ton of metal produced.

Cost of sinter	\$21/ton
Cost of coke	\$$d_3$/ton
Cost of prereduced pellets	\$$f_1$/ton
Fixed costs:	\$7.2 $\times$ 10^6 per annum

It is noted that the costs denoted by letters rather than numerical quantities (i.e., d_3, f_1, etc.) are either derived terms, or parameters the values of which may be varied.

Blast Furnace No. 2

Operating range 0.7 to 1.3 M tons/annum, 1.4 if prereduced pellets are used
1.4 tons of sinter produce 1 ton of hot metal
1.1 tons of prereduced pellets +0.3 ton of coke produce 1 ton of hot metal

The amount of prereduced pellets that may be fed is limited to 0.2×10^6 tons/annum.

Coke Rates (for Use with Sinter Loading)

At 1.2 M tons/annum metal production, 0.45 ton of coke/ton of metal produced.
At 0.7 M tons/annum metal production, 0.55 ton of coke/ton of metal produced.
At 1.3 M tons/annum metal production, 0.55 ton of coke/ton of metal produced.

Cost of sinter	\$21/ton
Cost of coke	$\$d_3$/ton
Cost of prereduced pellets	$\$f_1$/ton
Fixed costs:	$\$5.0 \times 10^6$ per annum

Blast Furnace No. 3

Operating range: 1.7 to 2.3 M tons/annum, with prereduced pellets up to 2.6 M tons/annum
Maximum amount of prereduced pellets that may be charged: 0.4 M tons/annum

Coke Rates (for Use with Sinter Loading)

At 2 M tons/annum metal production, 0.35 ton of coke/ton of metal produced.
At 2.3 M tons/annum metal production, 0.4 ton of coke/ton of metal produced.

At 1.7 M tons/annum metal production, 0.4 ton of coke/ton of metal produced.

Cost of sinter	\$21/ton
Cost of coke	\$$d_3$/ton
Cost of prereduced pellets	\$$f_1$/ton
Fixed costs:	\$11.6 × 10^6 per annum

The BOF Shop

Maximum capacity 5 M tons of crude steel

Yield: 90% from hot metal, 90% from scrap, and 85% from prereduced pellets

The amount of hot metal charged must be larger than 4 × [(scrap charge) + 1.5 × (the amount of prereduced pellet charge)]

Fixed costs: \$10.5 × 10^6/annum

Cost of production (excluding fixed costs and cost of material) = \$17/ton of crude steel

The Electric Furnace

Operating range: 0.2 to 1.0 M tons/annum of charged material

Yield: 95% from scrap and 85% from pellets

Fixed costs: \$3.0 × 10^6/annum

Cost of production (excluding fixed cost and cost of material) = \$23/ton

Open-Hearth Shop

Operating capacity: 3 M tons/annum of charge if fed with 100% hot metal, and 2 M tons/annum of charge if fed with 100% scrap

Yield: 90% for both scrap and hot metal

Fixed costs: \$3.0 × 10^6/annum

Operating cost (excluding material cost and fixed cost) = \$24/ton

Let us define the principal symbols, as follows:

x_i material flow rate in tons/annum
c_i cost of a stream (\$/annum)
d_i unit cost of a stream (\$/ton)
f_i cost of raw material

More specifically, we have the following:

x_1 coke produced from coal A
x_2 coke produced from coal B

x_3 coke output from coke oven
x_4 sinter input to blast furnace no. 1
x_5 coke input to blast furnace no. 1
x_6 prereduced pellet input to blast furnace no. 1
x_7 hot metal output from blast furnace no. 1
x_8 sinter input to blast furnace no. 2
x_9 coke input to blast furnace no. 2
x_{10} prereduced pellet input to blast furnace no. 2
x_{11} hot metal output from blast furnace no. 2
x_{12} sinter input to blast furnace no. 3
x_{13} coke input to blast furnace no. 3
x_{14} prereduced pellet input to blast furnace no. 3
x_{15} hot metal output from blast furnace no. 3
x_{16} total hot metal output from three blast furnaces
x_{17} hot metal input to BOF
x_{18} scrap input to BOF
x_{19} prereduced pellets input to BOF
x_{20} crude steel output from BOF
x_{21} scrap fed into electric furnace
x_{22} prereduced pellets fed into electric furnace
x_{23} crude steel production from electric furnace
x_{24} hot metal input into open hearth
x_{25} scrap input into open hearth
x_{26} crude steel from open hearth
x_{27} total crude steel production
x_{28} total scrap input
x_{29} total pellet input
x_{30} total sinter input
c_3 cost of coke stream
d_3 unit cost of coke stream
c_7 cost of hot metal from blast furnace no. 1
d_7 unit cost of hot metal from blast furnace no. 1
f_1 cost of prereduced pellets (unit cost)
f_2 unit cost of scrap
c_{11} cost of hot metal stream from blast furnace no. 2
d_{11} unit cost of hot metal stream from blast furnace no. 2
c_{15} cost of hot metal stream from blast furnace no. 3
d_{15} unit cost of hot metal stream from blast furnace no. 3
c_{20} cost of crude steel from BOF
d_{20} unit cost of crude steel from BOF
c_{23} cost of crude steel stream from electric furnace
d_{23} unit cost of crude steel from electric furnace

c_{26} cost of crude steel from the open hearth
d_{26} unit cost of crude steel from the open hearth

Because of the nonlinearity in the relationship between production rate and coke requirements for the blast furnaces, we discretize this by letting

z_{i0} = fraction of time at zero throughput

z_{i1} = fraction of production time at minimum throughput

z_{i2} = fraction of production time at optimal throughput

z_{i3} = fraction of production time at maximal throughput

for blast furnace no. i. This is a convenient way to interpolate between the discrete values that were given. It is stressed that the use of the variables $z_{i0} \cdots z_{in}$ is dictated by computational convenience and that fractional values of z_i may be interpreted as operation at an intermediate throughput, maintained constant over the whole working period examined. A larger number of throughputs could be thus selected for interpolation, if desired. It is noted that this technique would be particularly suitable for incorporating nonlinear relationships between the operating parameters into a linear programming optimization procedure.

The system described on the preceding pages may now be represented by the following set of linear equations which express the constraints posed by the mass balances and other operational criteria.

Thus we have:

Coke Plant

$$x_3 = x_1 + x_2 \tag{1}$$

Total coke produced = coke produced from A + coke produced from B

$$x_1 + 1.25x_2 \leq 2.5 \times 10^6 \tag{2}$$

Capacity constraint

$$c_3 = 5.0 \times 10^6 + 26x_1 + 21x_2 \tag{3}$$

Total cost of coke stream

$$d_3 = \frac{5 \times 10^6 + 26x_2 + 21x_2}{x_3} \tag{4}$$

Unit cost of coke stream

Blast Furnace No. 1

$$x_7 = 0.715x_4 + 0.91x_6 \tag{5}$$

Hot metal output = 0.715 (sinter input) + 0.91 (prereduced pellet input)

$$x_4 = 1.4(0.9 \times 10^6 z_{11} + 1.2 \times 10^6 z_{12} + 1.5 \times 10^6 z_{13}) \tag{6}$$

$$\text{Sinter input} = \begin{bmatrix}\text{Fractional input}\\ \text{at lowest level}\end{bmatrix} + \begin{bmatrix}\text{Fractional input}\\ \text{at optimum level}\end{bmatrix} + \begin{bmatrix}\text{Fractional input}\\ \text{at maximum level}\end{bmatrix}$$

$$x_5 = (0.405z_{11} + 0.480z_{12} + 0.675z_{13}) \times 10^6 + 0.275x_6 \tag{7}$$

Relationship between coke rates and production rates

$$z_{10} + z_{11} + z_{12} + z_{13} = 1 \tag{8}$$

Sum of fractions equals unity.

$$0 \leq x_6 \leq 0.3 \times 10^6 \tag{9}$$

Limits on the inputs of prereduced pellets

$$c_7 = 21x_4 + d_3x_5 + f_1x_6 + 7.2 \times 10^6 \tag{10}$$

Total cost of the hot metal stream

$$d_7 = \frac{c_7}{x_7} \tag{11}$$

Unit cost of the hot metal stream

Identical considerations apply to the equations describing the operation of blast furnaces nos. 2 and 3, and therefore these expressions will be given without further comment.

Blast Furnace No. 2

$$x_9 = 0.715x_8 + 0.91x_{10} \tag{12}$$

$$x_8 = 1.4(0.7 \times 10^6 z_{21} + 1.2 \times 10^6 z_{22} + 1.3 \times 10^6 z_{23}) \tag{13}$$

$$x_{11} = 0.275x_{10} + 0.385 \times 10^6 z_{21} + 0.540 \times 10^6 z_{22} + 0.715 \times 10^6 z_{23} \tag{14}$$

$$z_{20} + z_{21} + z_{22} + z_{23} = 1 \tag{15}$$

$$0 \leq x_{10} \leq 0.2 \times 10^6 \tag{16}$$

$$c_{11} = 21x_8 + d_3x_9 + f_1x_{10} + 5.0 \times 10^6 \tag{17}$$

$$d_{11} = \frac{c_{11}}{x_{11}} \tag{18}$$

Blast Furnace No. 3

$$x_{13} = 0.715x_{12} + 0.91x_{14} \tag{19}$$

$$x_{12} = 1.4(1.7z_{31} + 2.0z_{32} + 2.3z_{33}) \times 10^6 \tag{20}$$

$$x_{15} = 0.275x_{14} + (0.68z_{31} + 0.70z_{32} + 0.92z_{33}) \times 10^6 \tag{21}$$

$$z_{30} + z_{31} + z_{32} + z_{33} = 1 \tag{22}$$

$$0 \le x_{14} \le 0.4 \times 10^6 \tag{23}$$

$$c_{15} = 21x_{12} + d_3x_{13} + x_{14}f_1 + 11.6 \times 10^6 \tag{24}$$

$$d_{15} = \frac{c_{15}}{x_{15}} \tag{25}$$

Mixer

$$\underset{\begin{bmatrix}\text{Hot metal from the}\\ \text{three blast furnaces}\end{bmatrix}}{x_7 + x_{11} + x_{15}} = \underset{\begin{bmatrix}\text{Total hot metal}\\ \text{production}\end{bmatrix}}{x_{16}} = \underset{\begin{bmatrix}\text{Hot metal}\\ \text{to BOF}\end{bmatrix}}{x_{17}} + \underset{\begin{bmatrix}\text{Hot metal}\\ \text{to open hearths}\end{bmatrix}}{x_{24}} \tag{26}$$

$$\underset{\begin{bmatrix}\text{Total coke}\\ \text{output}\end{bmatrix}}{x_3} = \underset{\begin{bmatrix}\text{Coke used in the three}\\ \text{blast furnaces}\end{bmatrix}}{x_5 + x_9 + x_{13}} \tag{27}$$

$$d_{16} = \frac{c_7 + c_{11} + c_{15}}{x_{16}} \tag{28}$$

Unit cost of hot metal

BOF Shop

$$\underset{\begin{bmatrix}\text{Crude steel}\\ \text{output}\end{bmatrix}}{x_{20}} = \underset{\begin{bmatrix}\text{Yield from}\\ \text{hot metal}\end{bmatrix}}{0.9x_{17}} + \underset{\begin{bmatrix}\text{Yield from}\\ \text{scrap}\end{bmatrix}}{0.9x_{18}} + \underset{\begin{bmatrix}\text{Yield from}\\ \text{pellets}\end{bmatrix}}{0.85x_{19}} \tag{29}$$

$$x_{17} \ge 4(x_{18} + 1.5x_{19}) \tag{30}$$

Constraint on scrap and pellet content of the charge

$$x_{20} \le 5.0 \times 10^6 \tag{31}$$

Overall capacity constraint

$$c_{20} = 10.5 \times 10^6 + 17x_{20} + d_{16}x_{17} + f_1x_{19} + f_2x_{18} \tag{32}$$

Total cost of crude steel from BOF

$$d_{20} = \frac{c_{20}}{x_{20}} \tag{33}$$

Unit cost of crude steel from BOF

Electric Furnace

$$x_{23} = 0.95x_{21} + 0.85x_{22} \tag{34}$$

Crude steel output — Yield from scrap — Yield from pellets

$$0 \leq x_{21} + x_{22} \leq 1.0 \times 10^6 \tag{35}$$

Capacity constraint

$$c_{23} = f_2 x_{21} + f_1 x_{22} + 23.0x_{23} + 3.0 \times 10^6 \tag{36}$$

Cost of crude steel from the electric furnace

$$d_{23} = \frac{c_{23}}{x_{23}} \tag{37}$$

Unit cost of crude steel from electric furnace

Open-Hearth Shop

$$x_{26} = 0.9x_{24} + 0.9x_{24} \tag{38}$$

$$\begin{bmatrix}\text{Crude steel from}\\ \text{open hearth}\end{bmatrix} = \begin{bmatrix}\text{Yield from}\\ \text{hot metal}\end{bmatrix} + \begin{bmatrix}\text{Yield from}\\ \text{scrap}\end{bmatrix}$$

$$x_{24} + 1.5x_{25} \leq 3.0 \times 10^6 \tag{39}$$

Capacity constraint

$$c_{26} = 24x_{26} + f_2 x_{25} + d_{16} x_{24} + 3.0 \times 10^6 \tag{40}$$

Cost of crude steel from open hearth

$$d_{26} = \frac{c_{26}}{x_{26}} \tag{41}$$

Unit cost of crude steel from open hearth

In addition we have the following overall balances and constraints:

$$x_{27} = x_{20} + x_{23} + x_{26} \tag{42}$$

$$\begin{bmatrix}\text{Total crude}\\ \text{steel production}\end{bmatrix} = \begin{bmatrix}\text{Crude steel}\\ \text{from BOF}\end{bmatrix} + \begin{bmatrix}\text{Crude steel}\\ \text{from electric}\\ \text{furnace}\end{bmatrix} + \begin{bmatrix}\text{Steel from}\\ \text{open hearth}\end{bmatrix}$$

$$c_{27} = d_{20}x_{20} + d_{23}x_{23} + d_{26}x_{26} \tag{43}$$

Cost of crude steel production

$$0 \leq x_{19} + x_{22} + x_6 + x_{10} + x_{14} = x_{29} \leq \alpha_1 \tag{44}$$

Total pellet input may be constrained

$$0 \leq x_{28} = x_{18} + x_{21} + x_{25} \leq \alpha_2 \tag{45}$$

Total scrap input may be constrained

$$0 \leq x_4 + x_8 + x_{12} = x_{30} \leq \alpha_3 \tag{46}$$

Total sinter input may be constrained

The use of Eqs. 44 to 46 provides us with the desirable option to constrain the supply of pellets, scrap, and sinter, respectively. Clearly, if high enough values are chosen for α_1, α_2, and α_3, these constraints will not become operative.

Our objective is to minimize the total cost of crude steel production:

$$c_T = c_{26} + c_{23} + c_{20} \tag{47}$$

$$c_T = 45.3 \times 10^6 + 26x_1 + 21x_2 + f_1x_{29} + f_2x_{28} + 21x_{30} \times 17x_{20} + 23x_{23} + 24x_{26} \tag{48}$$

for a given value of x_{27}, that is, the total crude steel production required.

The Solution by Linear Programming

The essential equality and inequality constraints from Eqs. 1 to 46 form a constraint equation

$$\mathbf{Ax}(\lesseqqgtr)\mathbf{b} \tag{49}$$

where $\mathbf{A}$ is an $m \times n$ matrix of coefficients, $\mathbf{x}$ is an n vector of variables, and $\mathbf{b}$ is an m vector of right-hand sides. The objective to be maximized is given in Eq. 48. For this problem $\mathbf{A}$ was a 36×43 matrix, $\mathbf{x}$ a 43 vector, and $\mathbf{b}$ a 36 vector.

This information was fed to a standard linear programming algorithm, using a modification of the previously described simplex procedure for converging to the optimal policy.† The computed results, a selection of which is given below, required approximately 3 sec and cost 50 cents for each optimal policy found.

Computed Results

A selected set of the computed results is shown in Tables 4.4.1 to 4.4.8. These tables contain the optimal operating levels of the individual units, corresponding to the input cost, production costs, and the constraints specified.

In addition, we also show the unit costs, and where applicable, *the shadow prices*, that is, the unit production costs for an incremental output, if it were possible to relax the limiting constraint.

† A listing of standard linear programming subroutines is given in Appendix C.

We discuss Table 4.4.1 in detail to demonstrate the type of information that is being generated through the use of this procedure. The optimal production rates of each of the units in M tons is shown along with the unit cost ($/ton of product) of processing the material in each unit. These costs are cumulative in the sense, for example, that the unit cost of the BOF includes earlier processing costs. These unit costs are obtained by dividing the total processing and fixed costs by the production of that unit.

The shadow prices (dual variables) in the last row of the table give the marginal value of each raw material and product. For example, in this particular case the steel plant would like to use more pellets than are available (2 M tons); thus the shadow price associated with constraint (44b) shows that relaxing that constraint would be worth $0.765 for one ton of additional pellets in cost reduction. This is also reflected in the shadow price associated with constraint (44a)—this number tells us that we could profitably afford to purchase pellets for up to $30.765/ton, a very valuable piece of information.

The numbers in this last row under "Hot metal" and "Steel produced" are the shadow prices associated with constraints (26) and (42). These tell us the cost of producing the last ton of the material made. For example, the unit cost of hot metal produced was $43 to $44/ton depending on the blast furnace used. However, the cost of the last ton was only $40.154. Similarly, the unit cost of steel was $65.18/ton, but the cost of the last ton was only $62.89. Thus if a competitor offered to sell us steel to fill orders with, he would have to sell below the shadow price figure, $62.89, rather than the unit cost figure, in order to make it profitable for us to reduce production and buy steel "from outside."

It should be emphasized that these shadow prices are only local rates of change of cost close to the optimum found. One must re-solve the linear programming problem to determine the actual cost reduction when a constraint is relaxed.

The remaining Tables 4.4.2–4.4.8 show the effects of production rates, price changes, unplanned shutdown of a unit, and the like on the optimal operation. As can be seen, the use of optimization reduces the costs associated with these eventualities to a minimum.

Inspection of these tables shows certain trends which are consistent with expectations, such as:

1. The relatively inefficient open-hearth furnaces would be used only when excess capacity is required.
2. The overall production costs are increased significantly when the plant is operated much below its nominal capacity.
3. The production costs also increase if the plant is "driven very hard," that is, operated significantly above the nominal capacity.

Other less obvious optimal operating policies that may be worthwhile enumerating are listed in the following:

1. As seen in Table 4.4.2, when there is a limited availability of sinter, and the production requirement is 5.5 M tons, the optimal policy is to shut down the largest, and the most efficient blast furnace.
2. As seen in Tables 4.4.4 to 4.4.5, for a production rate of 5.5 M tons, when one of the smaller blast furnaces is shut down, the price structure assumed will lead to a policy of operating the two remaining blast furnaces at their optimum (and not maximum) output level, and to using the open-hearth furnaces to make up for the extra capacity required.
3. A comparison of Tables 4.4.1 and 4.4.3 shows that the increase in output from 7.5 to 8.0 M tons will lead to an increase in the unit cost, because one of the blast furnaces will have to be operated above its optimum level.
4. On comparing Tables 4.4.6 and 4.4.7, it is seen that an increase in the production cost of the open hearth would lead to a change in the operating policy, that is, to a reduction of the fraction of material processed in the open hearth.

In some instances these decisions could have been made intuitively, but in many cases the large number of interrelated factors make it unlikely that such intuitive decisions would, indeed, correspond to the optimum.

In applying the considerations given here to actual practical operations, the reader must be aware that the "optimal policies" will depend critically on the price structure and on the functional relationships between the individual parameters, that is, on the appropriateness of the process model. Thus the example given here should be regarded as an illustration of the methodology, rather than as a set of specific guidelines, applicable to a particular operation.

It is quite likely that the development of an optimal policy for an actual integrated steel plant will involve a much larger number of parameters than used in this simplified treatment. It is thought, however, that the methodology outlined here should be applicable to these more complex situations. This approach should have a considerable appeal because the balance type modeling equations could be readily generated and their solution is an essentially routine procedure.

In concluding this example we note that the problem discussed is representative of a broad class of linear programming problems, usually termed blending. A large class of practical problems fall within this category; examples are:

1. To find the combination of nutrients to provide a satisfactory diet at a minimum cost.

TABLE 4.4.1 OPTIMAL POLICY FOR HIGH STEEL PRODUCTION RATES

	Coal									
	A	B	Sinter	Pellet	Scrap	Hot metal	Steel produced	Unit cost	Shadow price	Eq. no.
Coke oven		1.818						23.747		
BF-1			1.68	0.3		1.47		44.107		
BF-2			1.68	0.0577		1.25		44.172		
BF-3			2.8	0		2.002		43.468		
BOF				0.6423	0.219	4.73	5.0	65.9	1.031[b] 2.419[a]	31 29
E.F.				1.0	0		0.85	61.824	3.14[a]	35
O.H.					1.833	0	1.65	64.70		
Total		1.818	6.16	2.0	2.052	4.73	7.5			
Shadow price			46a: 21[c]	44b: 0.765[a] 44a: 30.765[c]	45a: 35.0[c]	26: 40.154[c]	42: 62.889[c]			

Steel production: 7.5 megatons.
Pellet cost (f_1): \$30/ton.
Scrap cost (f_2): \$35/ton.
Pellets available (α_1): 2 megatons.
Scrap available (α_2): 10 megatons.
Sinter available (α_3): 10 megatons.
Total cost: \$4.888 $\times 10^8$.
Unit cost: \$65.18/ton of steel produced.

[a] Capacity constraint.
[b] Mixture constraint.
[c] Outside purchase option.

TABLE 4.4.2 OPTIMAL POLICY WHEN PELLETS ARE LIMITED

	Coal									
	A	B	Sinter	Pellet	Scrap	Hot metal	Steel produced	Unit cost	Shadow price	Eq. no.
Coke oven		1.72						23.907		
BF-1			1.68	0		1.201		44.925		
BF-2			1.68	0		1.201		44.288		
BF-3			2.8	0		2.002		43.524		
BOF				0	1.101	4.404	4.955	66.105	0.889[b]	31
E.F.				0.5	0.126		0.545	64.119		
O.H.					0	0	0			
Total		1.72	6.16	0.5	1.227	4.404	5.5			
Shadow price			46a: 21[c]	44b: 1.316[a] 44a: 31.316[c]	45a: 35[c]	26: 39.447[c]	42: 59.842[c]			

Steel production: 5.5 megatons.
Pellet cost (f_1): \$30/ton.
Scrap cost (f_2): \$35/ton.
Pellets available (α_1): 0.5 megatons.
Scrap available (α_2): 10 megatons.
Sinter available (α_3): 10 megatons.
Total cost: \$3.655 $\times 10^8$.
Unit cost: \$66.45/ton of steel produced.

[a] Capacity constraint.
[b] Mixture constraint.
[c] Outside purchase option.

TABLE 4.4.3 OPTIMAL POLICY FOR VERY HIGH STEEL PRODUCTION RATES

	Coal									
	A	B	Sinter	Pellet	Scrap	Hot metal	Steel produced	Unit cost	Shadow price	Eq. no.
Coke oven	0.014	1.989						23.531	20[a]	2
BF-1			1.756	0.3		1.529		43.916		
BF-2			1.68	0.2		1.383		43.587		
BF-3			2.8	0.4		2.366		42.883		
BOF				0	1.111	4.444	5.0	65.421	4.848[b] 66.255[a]	31 29
E.F.				0	1.0		0.95	63.00	79.657[a]	35
O.H.					1.444	0.833	2.05	67.739		
Total	0.014	1.989	6.236	0.9	3.555	5.278	8.0	65.73		
Shadow price			46a: 21[c]	44a: 30[c]	45a: 35[c]	26: 59.241[c]	42: 143.691[c]			

Steel production: 8 megatons.
Pellet cost (f_1): \$30/ton.
Scrap cost (f_2): \$35/ton.
Pellets available (α_1): 2 megatons.
Scrap available (α_2): 10 megatons.
Sinter available (α_3): 10 megatons.
Total cost: \$5.259 $\times 10^8$.
Unit cost: \$65.73/ton of steel produced.

[a] Capacity constraint.
[b] Mixture constraint.
[c] Outside purchase option.

TABLE 4.4.4 OPTIMAL POLICY WHEN SINTER IS LIMITED

	Coal									
	A	B	Sinter	Pellet	Scrap	Hot metal	Steel produced	Unit cost	Shadow price	Eq. no.
Coke oven		1.078						25.638		
BF-1			1.68	0.3		1.474		44.701		
BF-2			1.32	0.2		1.126		46.113		
BF-3			0	0		0				
BOF				0	0.65	2.6	2.925	72.611	1.575[b]	31
E.F.				1.0	0		0.85	61.824	3.906[a]	35
O.H.					1.917	0	1.725	64.635		
Total		1.078	3.0	1.5	2.567	2.6	5.5			
Shadow price			46b: 5.006[a] 46a: 26[c]	44a: 30[c]	45a: 35[c]	26: 42.88[c]	42: 62.889[c]			

Steel production: 5.5 megatons.
Pellet cost (f_1): \$30/ton.
Scrap cost (f_2): \$35/ton.
Pellets available (α_1): 2 megatons.

Scrap available (α_2): 10 megatons.
Sinter available (α_3): 3 megatons.
Total cost: \$3.751 $\times 10^8$.
Unit cost: \$68.447/ton of steel produced.

[a] Capacity constraint.
[b] Mixture constraint.
[c] Outside purchase option.

TABLE 4.4.5 OPTIMAL POLICY WHEN BLAST FURNACE NO. 1 IS DOWN

	Coal									
	A	B	Sinter	Pellet	Scrap	Hot metal	Steel produced	Unit cost	Shadow price	Eq. no.
Coke oven		1.405						24.559		
BF-1			0	0		0				
BF-2			1.68	0.2		1.383		44.029		
BF-3			2.8	0.4		2.366		43.234		
BOF				0	0.937	3.749	4.218	67.659	1.575[b]	31
E.F.				1.0	0		0.85	61.824	3.906[a]	35
O.H.					0.48	0	0.432	69.833		
Total		1.405	4.48	1.6	1.417	3.749	5.5			
Shadow price			46a: 21[c]	44a: 30[c]	45a: 35[c]	26: 42.875[c]	42: 62.889[c]			

Steel production: 5.5 megatons.
Pellet cost (f_1): \$30/ton.
Scrap cost (f_2): \$35/ton.
Pellets available (α_1): 2 megatons.
Scrap available (α_2): 10 megatons.
Sinter available (α_3): 10 megatons.
Total cost: \$3.681 $\times 10^8$.
Unit cost: \$66.927/ton of steel produced.

[a] Capacity constraint.
[b] Mixture constraint.
[c] Outside purchase option.

TABLE 4.4.6 OPTIMAL POLICY WHEN BLAST FURNACE NO. 3 IS DOWN

	Coal									
	A	B	Sinter	Pellet	Scrap	Hot metal	Steel produced	Unit cost	Shadow price	Eq. no.
Coke oven		1.157						25.332		
BF-1			1.68	0.3		1.474		44.58		
BF-2			1.68	0.2		1.383		44.357		
BF-3			0	0		0				
BOF				0	0.714	2.857	3.215	71.167	1.575[b]	31
E.F.				1.0	0		0.85	61.824	3.906[a]	35
O.H.					1.595	0	1.435	64.993		
Total		1.157	3.36	1.5	2.309	2.857	5.5			
Shadow price			46a: 21[c]	44a: 30[c]	45a: 35[c]	26: 42.875[c]	42: 62.889[c]			

Steel production: 5.5 megatons.
Pellet cost (f_1): \$30/ton.
Scrap cost (f_2): \$35/ton.
Pellets available (α_1): 2 megatons.

Scrap available (α_2): 10 megatons.
Sinter available (α_3): 10 megatons.
Total cost: \$3.746 $\times 10^8$.
Unit cost: \$68.112/ton of steel produced.

[a] Capacity constraint.
[b] Mixture constraint.
[c] Outside purchase option.

TABLE 4.4.7 OPTIMAL POLICY WHEN BLAST FURNACE NO. 3 IS DOWN AND THE OPEN HEARTH OPERATING COSTS ARE INCREASED TO $30/TON

	Coal									
	A	B	Sinter	Pellet	Scrap	Hot metal	Steel produced	Unit cost	Shadow price	Eq. no.
Coke oven		1.353						24.695		
BF-1			2.1	0.3		1.775		44.504		
BF-2			1.68	0.2		1.383		44.088		
BF-3			0	0		0				
BOF				0	0.789	3.158	3.552	70.402	2.925[b]	31
E.F.				1.0	0		0.85	61.824	9.006[a]	35
O.H.					1.219	0	1.097	71.627		
Total		1.353	3.78	1.5	2.009	3.158	5.5			
Shadow price			46a: 21[c]	44a: 30[c]	45a: 35[c]	26: 49.625[c]	42: 68.889[c]			

Steel production: 5.5 megatons.
Pellet cost (f_1): $30/ton.
Scrap cost (f_2): $35/ton.
Pellets available (α_1): 2 megatons.
Scrap available (α_2): 10 megatons.
Sinter available (α_3): 10 megatons.
Total cost: $\$3.813 \times 10^8$.
Unit cost: $69.32/ton of steel produced.

[a] Capacity constraint.
[b] Mixture constraint.
[c] Outside purchase option.

TABLE 4.4.8 OPTIMAL POLICY FOR VERY LOW STEEL PRODUCTION

	Coal									
	A	B	Sinter	Pellet	Scrap	Hot metal	Steel produced	Unit cost	Shadow price	Eq. no.
Coke oven		0.678						28.375		
BF-1			0	0		0				
BF-2			0	0		0				
BF-3			2.685	0		1.92		45.429		
BOF				0.32	0	1.92	2.0	85.431	0.435[b]	31
E.F.				0	0					
O.H.					0	0				
Total		0.678	2.685	0.32	0	1.92	2.0			
Shadow price			46a: 21[c]	44a: 30[c]	45a: 35[c]	26: 34.965[c]	42: 55.366[c]			

Steel production: 2 megatons.
Pellet cost (f_1): \$30/ton.
Scrap cost (f_2): \$35/ton.
Pellets available (α_1): 2 megatons.
Scrap available (α_2): 10 megatons.
Sinter available (α_3): 10 megatons.
Total cost: \$1.601 $\times$ 10^8.
Unit cost: \$79.76/ton of steel produced.

[a] Capacity constraint.
[b] Mixture constraint.
[c] Outside purchase option.

2. To provide fertilizers of given N, P, and K content at a minimum cost.
3. To provide gasoline blends of given octane number and given impurity specifications, at a minimum cost, and so on.

Furthermore, the methodology outlined in the discussion of this example could be applied to many large-scale chemical, petroleum, or metallurgical operations [11–15]).

4.5 QUADRATIC PROGRAMMING

Quadratic programming, as the name might suggest, denotes algorithms that are designed to optimize quadratic functions subject to linear constraints. The general problem is

$$\underset{\mathbf{x}}{\text{Max}}\ [I(\mathbf{x}) = \mathbf{c}^T\mathbf{x} + \tfrac{1}{2}\mathbf{x}^T\mathbf{H}\mathbf{x}]$$

subject to

$$\mathbf{h}(\mathbf{x}) = \hat{\mathbf{A}}\mathbf{x} - \hat{\mathbf{b}} = \mathbf{0}$$

$$\mathbf{g}(\mathbf{x}) = \bar{\mathbf{A}}\mathbf{x} - \bar{\mathbf{b}} \leq \mathbf{0} \tag{4.5.1}$$

For quadratic programming maximization problems, it is assumed that $I(\mathbf{x})$ is a concave function, so that the Kuhn-Tucker conditions are both necessary and sufficient for a global maximum (since we already know that the linear constraints form a convex feasible region).

The theory here is not quite as tidy as for linear programming because we may have our maximum at a vertex, on an edge, or in the interior of the feasible region as demonstrated in Fig. 4.5 for two dimensions.

However, the derivatives of a quadratic function are linear, and therefore the Kuhn-Tucker conditions are linear and linear programming methods can be used. If one collects the constraints into the form of Eq. 4.4.9, that is, through the use of a partitioned matrix, then the Lagrangian may be written as

$$L(\mathbf{x}, \boldsymbol{\lambda}) = \mathbf{c}^T\mathbf{x} + \tfrac{1}{2}\mathbf{x}^T\mathbf{H}\mathbf{x} + \boldsymbol{\lambda}^T(\mathbf{b} - \mathbf{A}\mathbf{x}) \tag{4.5.2}$$

where $\mathbf{A}$, $\mathbf{b}$ are defined by Eq. 4.4.8. The condition that L be a minimum with respect to $\boldsymbol{\lambda}$ results in

$$\boldsymbol{\lambda}^T \frac{\partial L}{\partial \boldsymbol{\lambda}} = 0, \quad \frac{\partial L}{\partial \boldsymbol{\lambda}} = (\mathbf{b} - \mathbf{A}\mathbf{x}) \geq \mathbf{0}$$

or

$$\boldsymbol{\lambda}^T(\mathbf{A}\mathbf{x} - \mathbf{b}) = 0 \tag{4.5.3}$$

$$\mathbf{A}\mathbf{x} - \mathbf{b} \leq \mathbf{0} \tag{4.5.4}$$

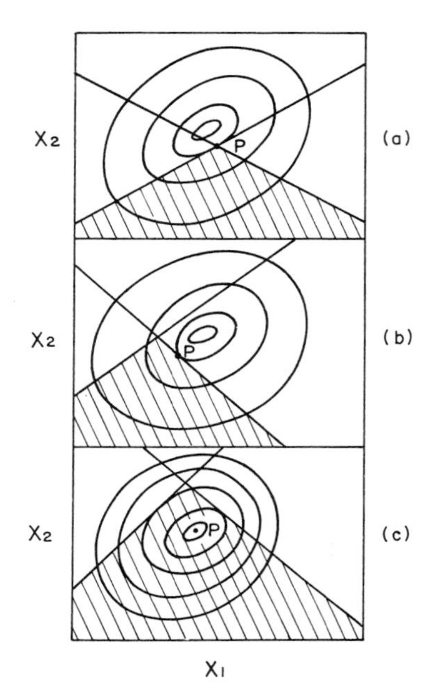

Fig. 4.5 The location of the optimum in quadratic programming problems. The feasible region is denoted by the shaded area: (*a*) optimum at a vertex, (*b*) optimum at the boundary of the feasible region, and (*c*) optimum within the feasible region—location unaffected by the constraints.

The condition that L be a maximum with respect to $\mathbf{x}$ gives

$$\mathbf{x}^T \frac{\partial L}{\partial \mathbf{x}} = 0; \quad \frac{\partial L}{\partial \mathbf{x}} \leq \mathbf{0}$$

or

$$\mathbf{x}^T(\mathbf{c} + \mathbf{Hx} - \boldsymbol{\lambda}^T\mathbf{A}) = 0 \tag{4.5.5}$$

$$\mathbf{c} + \mathbf{Hx} - \boldsymbol{\lambda}^T\mathbf{A} \leq \mathbf{0} \tag{4.5.6}$$

Since we know from Eq. 4.5.3 that

$$\lambda_j = 0 \qquad \text{if} \quad \sum_{i=1}^{n} A_{ji}x_i - b_j < 0$$

$$\lambda_j > 0 \qquad \text{if} \quad \sum_{i=1}^{n} A_{ji}x_i - b_j = 0$$

we can introduce the slack variables w_j so that Eqs. 4.5.3 and 4.5.4 become

$$\begin{gathered} \mathbf{Ax} - \mathbf{b} + \mathbf{w} = \mathbf{0} \\ \lambda_j w_j = 0 \\ \boldsymbol{\lambda} \geq 0, \qquad \mathbf{w} \geq 0 \end{gathered} \tag{4.5.7}$$

Similarly we know from Eq. 4.5.5 that Eq. 4.5.6 is an equality when $x_i > 0$, and an inequality when $x_i = 0$, so that we can introduce the slack variables z_i so that Eqs. 4.5.5 and 4.5.6 become

$$\begin{aligned} \mathbf{c} + \mathbf{Hx} - \boldsymbol{\lambda}^T\mathbf{A} + \mathbf{z} &= \mathbf{0} \\ \mathbf{x} &\geq 0 \\ x_i z_i &= 0 \end{aligned} \tag{4.5.8}$$

Now if we begin with the initial nonfeasible point

$$x_i = 0, \quad \lambda_j = 0, \quad w_j = b_j, \quad z_i = -c_i < 0$$

(which violates Eq. 4.5.6) and try to maximize $\sum_{i=1}^{n} z_i$ subject to Eqs. 4.5.7 and 4.5.8, then ultimately we will reach the position where either all z_i are zero and x_i positive, or some z_i are positive and x_i are zero. This produces a feasible solution to Eqs. 4.5.7 and 4.5.8 and an optimal solution to the quadratic programming problem.

This problem is the same as an ordinary linear programming problem except that if $\lambda_j \neq 0$, then we must retain $w_j = 0$ in the basis and vice versa. Similarly, if $x_i \neq 0$, then we retain $z_i = 0$ in the basis. Wolfe [16], Beale [17], and others have developed quadratic programming algorithms based on these ideas and report that the computing time required for solution of the quadratic program is about the same as for linear programming problems of comparable size. Both the Wolfe and Beale algorithms are available as standard subroutines (cf. Appendix C).

There are other approaches to quadratic programming which have been developed. For details, the reader is referred to the discussions in [18, 19].

4.6 SEPARABLE PROGRAMMING

The term separable programming is used to describe techniques for optimizing nonlinear but separable functions subject to nonlinear, but separable constraints [cf. 18, Chapter 4]. A function is separable if it can be written as

$$f(\mathbf{x}) = \sum_{i=1}^{n} f_i(x_i) \tag{4.6.1}$$

that is, each term is a function of only a single variable x_i. For example, the function

$$f(\mathbf{x}) = x_1^2 + x_2 e^{-x_2} + (x_3^3 - x_3)^2 \tag{4.6.2}$$

is highly nonlinear, but separable.

The technique normally employed for such functions is to transform the standard problem

$$\underset{x_i}{\text{Max}}\left[I(\mathbf{x}) = \sum_{i=1}^{n} f_i(x_i)\right]$$

subject to

$$h_k(\mathbf{x}) = \sum_{i=1}^{n} h_{ki}(x_i) = 0 \qquad k = 1, 2, \ldots, m_1 \tag{4.6.3}$$

$$g_j(\mathbf{x}) = \sum_{i=1}^{n} g_{ji}(x_i) \leq 0 \qquad j = 1, 2, \ldots, m_2$$

into a piecewise linear problem, and solve by a modified linear programming technique. Each of the nonlinear functions, say $f_i(x_i)$, is fitted with a piecewise linear approximation as shown in Fig. 4.6. Each function is broken up

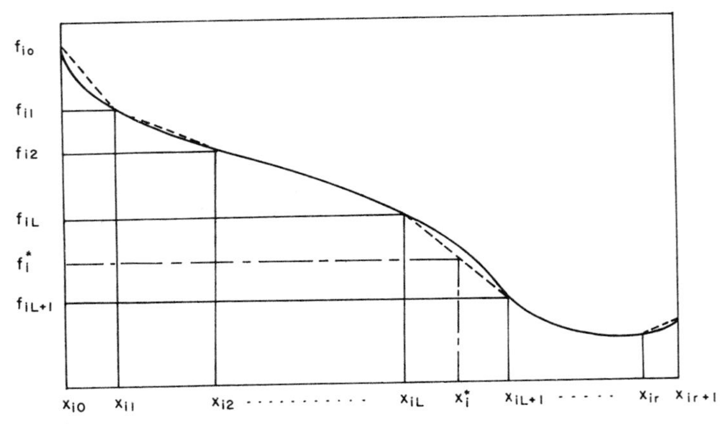

Fig. 4.6 Piecewise linear approximation of a separable function.

into $r + 1$ linear parts so that at any point x_i^* between x_{iL} and x_{iL+1} we have the linear interpolation,

$$f_i(x_i^*) = f_{iL} + \left(\frac{f_{iL+1} - f_{iL}}{x_{iL+1} - x_{iL}}\right)(x_i^* - x_{iL}) \tag{4.6.4}$$

or if we let

$$x_i^* = \lambda_{iL}x_{iL} + \lambda_{iL+1}x_{iL+1}$$
$$\lambda_{iL} + \lambda_{iL+1} = 1 \tag{4.6.5}$$

then

$$f_i(x_i^*) = \lambda_{iL}f_{iL} + \lambda_{iL+1}f_{iL+1} \tag{4.6.6}$$

Now if we specify that $0 \leq \lambda_{iL} \leq 1.0$ for all L, and $\lambda_{iL} = 0$ for points other than those bordering x_i^*, then

$$\begin{aligned} x_i^* &= \sum_{L=0}^{r+1} \lambda_{iL} x_{iL} \\ f_i(x_i^*) &= \sum_{L=0}^{r+1} \lambda_{iL} f_{iL} \\ h_{ki}(x_i^*) &= \sum_{L=0}^{r+1} \lambda_{iL} h_{kiL} \\ g_{ji}(x_i^*) &= \sum_{L=0}^{r+1} \lambda_{iL} g_{jiL} \\ \sum_{L=0}^{r+1} \lambda_{iL} &= 1 \end{aligned} \tag{4.6.7}$$

so that our problem (Eq. 4.6.3) becomes

$$\operatorname*{Max}_{\lambda_{iL}} \left[I(\mathbf{x}) = \sum_{i=1}^{n} \sum_{L=0}^{r+1} \lambda_{iL} f_{iL} \right]$$

subject to

$$\begin{aligned} h_k(\mathbf{x}) &= \sum_{i=1}^{n} \sum_{L=0}^{r+1} \lambda_{iL} h_{kiL} = 0 \qquad k = 1, 2, \ldots, m_1 \\ g_j(\mathbf{x}) &= \sum_{i=1}^{n} \sum_{L=0}^{r+1} \lambda_{iL} g_{jiL} \leq 0 \qquad j = 1, 2, \ldots, m_2 \end{aligned} \tag{4.6.8}$$

which is a linear programming problem in the $n(r + 2)$ variables λ_{iL}. The only difference between this problem and the standard linear programming problem is that one does not allow λ_{iL} to enter the basis unless either λ_{iL+1} or λ_{iL-1} are already in the basis. This assures that the linear interpolation always occurs between adjacent grid points x_{iL}, x_{iL+1}. The fact that there are $n(r + 2)$ physical variables and m_2 slack variables in the system as opposed to only $n + m_2$ total variables in the standard linear programming problem, means that there will be increased computation time; however, linear programming algorithms are much more efficient than most alternative nonlinear programming techniques so that the additional dimensionality is usually fully justified by the overall savings in computer time. A general purpose algorithm for solving separable programming problems is available from the IBM Share Library, as program SCM3 (1962) (cf. Appendix C).

4.7 GEOMETRIC PROGRAMMING

This technique, which is entirely different in nature from any of the others we have treated, is based on the arithmetic mean—geometric mean inequality

$$\sum_{i=1}^{p} w_i y_i \geq \prod_{i=1}^{p} (y_i)^{w_i} \qquad w_i \geq 0 \tag{4.7.1}$$

which is a special case of the more general Bernoulli inequality [20]. The positive weights w_i must sum to unity

$$\sum_{i=1}^{p} w_i = 1 \tag{4.7.2}$$

Also, the equality sign in Eq. 4.7.1 holds only when $y_1 = y_2 = \cdots y_p$. Zener and Duffin [21] have shown that if the y_i are all positive and are posynomials (a posynomial is a polynomial with positive coefficients) that is, $y_i = c_i/w_i \prod_{i=1}^{q} x_j^{\alpha_{ij}}$, $c_i \geq 0$, then the left-hand side of Eqs. 4.7.1 attains its minimum when the equality holds, so that the right-hand side is the infinum (greatest lower bound) of the left-hand side there. In this case, Eq. 4.7.1 becomes

$$\sum_{i=1}^{p} c_i \prod_{j=1}^{q} x_j^{\alpha_{ij}} \geq \prod_{i=1}^{p} \left[\frac{c_i}{w_i} \prod_{j=1}^{q} (x_j)^{\alpha_{ij}} \right]^{w_j} = \prod_{i=1}^{p} \prod_{j=1}^{q} \left(\frac{c_i}{w_i} \right)^{w_i} x_j^{\alpha_{ij} w_i} \tag{4.7.3}$$

and the equality is valid only if the w_i are selected so that the right-hand side is independent of the x_j [21]. This requires

$$\sum_{i=1}^{p} \alpha_{ij} w_i = 0 \qquad j = 1, 2, \ldots, q \tag{4.7.4}$$

Thus if we wish to minimize the left-hand side of Eq. 4.7.3, then the dual problem would be the right-hand side—which has to be maximized.

In order to illustrate the use of this technique let us consider the following problem:

EXAMPLE 4.7.1. We wish to maximize the profits resulting from the operation of a furnace in which a slag-metal reaction is being carried out. The following relationships are used for describing the behavior of the system in the region of interest

$$R = a_0 L^3 T^2 \tag{4.7.5}$$

where R is the rate of production, L is the characteristic dimension of the furnace, T is the absolute temperature in the furnace, and a_0 is a constant. We note that the temperature dependence of the production rate is a quadratic approximation to the Arrhenius type relationship which should be valid over a limited temperature range.

The expression

$$Q = a_1 L^2 T^4 \tag{4.7.6}$$

describes the heat loss from the system, where Q is the total rate of heat loss, and a_1 is a coefficient which incorporates the Boltzmann constant and the emissivity. The capital charges on the installation are assumed to be given by

$$C_c = a_2 L^2 \tag{4.7.7}$$

where C_c represents the capital charges and a_2 is a coefficient.

It is desired to operate close to a standard production rate R^* while minimizing the capital costs and heat losses. Thus we choose the objective to be minimized as

$$I = \frac{a_3 R^*}{R} + a_4 a_2 L^2 + a_5 a_1 L^2 T^4 \tag{4.7.8}$$

where the first term emphasizes in a nonlinear fashion, the relative value of production, the second term the relative value of capital costs, and the third term the relative value of heat losses. If we let $x_1 = L$, $x_2 = T$, $a_3 R^*/a_0 = 10^{13}$, $a_4 a_2 = 10^2$, $a_5 a_1 = 5 \times 10^{-11}$, then our objective to be minimized is

$$I = \frac{10^{13}}{x_1^3 x_2^2} + 10^2 x_1^2 + 5 \times 10^{-11} x_1^2 x_2^4 \tag{4.7.9}$$

Solution. The y_i for this problem then becomes

$$\mathbf{y} = \begin{bmatrix} \dfrac{c_1}{w_1 x_1^3 x_2^2} \\[2ex] \dfrac{c_2 x_1^2}{w_2} \\[2ex] \dfrac{c_3 x_1^2 x_2^4}{w_3} \end{bmatrix} \tag{4.7.10}$$

where the costs c_i are

$$\mathbf{c} = \begin{bmatrix} 10^{13} \\ 10^2 \\ 5 \times 10^{-11} \end{bmatrix} \tag{4.7.11}$$

and the α_{ij} are given by

$$\boldsymbol{\alpha} = \begin{bmatrix} -3 & -2 \\ 2 & 0 \\ 2 & 4 \end{bmatrix} \tag{4.7.12}$$

Equations 4.7.2 and 4.7.4, which define the weights, w_i, are

$$\begin{aligned} w_1 + w_2 + w_3 &= 1 \\ -3w_1 + 2w_2 + 2w_3 &= 0 \\ -2w_1 \qquad\quad + 4w_3 &= 0 \end{aligned} \tag{4.7.13}$$

On solving Eq. 4.7.13 for the w_i which causes the right-hand side of Eq. 4.7.3 to be independent of the x_i, the equality in Eq. 4.7.1 holds. Thus we shall seek the solution to the dual problem to Eq. 4.7.9.

Equation 4.7.13 yields the solution

$$w_1 = \tfrac{2}{5}, \quad w_2 = \tfrac{2}{5}, \quad w_3 = \tfrac{1}{5} \tag{4.7.14}$$

and the right-hand side of Eq. 4.7.1 becomes

$$I = \prod_{i=1}^{3} \left(\frac{c_i}{w_i}\right)^{w_i} = (\tfrac{5}{2} \times 10^{13})^{0.4}(\tfrac{5}{2} \times 10^{2})^{0.4}(25 \times 10^{-11})^{0.2} = 25{,}000 \tag{4.7.15}$$

Thus we now know the optimal value of the objective is 25,000. In addition, from the weights we have found that at the optimum the first and second terms of Eq. 4.7.9 will each contribute 40% to this cost and the third term will contribute 20%.

From the fact that each of the y_i must be equal we have

$$\frac{5 \times 10^{13}}{2x_1^{\,3}x_2^{\,2}} = \frac{5}{2} \times 10^2 x_1^{\,2} = 25 \times 10^{-11} x_1^{\,2} x_2^{\,4} \tag{4.7.16}$$

which yields the optimal values

$$x_1 = L = 10 \text{ ft}, \quad x_2 = T = 1000°\text{K} \tag{4.7.17}$$

Thus the general procedure is:

(i) Form the dual problem.

(ii) Solve Eqs. 4.7.2 and 4.7.4 for the weights w_i and use the right-hand side of Eq. 4.7.1 to compute the optimal objective.

(iii) Equate the y_i to determine the x_j.

There are several special problems with this technique that should be mentioned. First, it is only when $p = q + 1$ (the number of variables is one less than the number of terms in the objective) that Eq. 4.7.4 produces a unique solution for the w_i. In other cases, a minimization must be performed to determine the extra variables w_i. A second difficulty arises when all terms in the objective are not positive. This requires a special computational approach.

Geometric Programming can be considered a special case of the more general area of "inequality optimization" discussed in some detail by Ferron [23] and Eben and Ferron [20, 24]. Their approach, which is somewhat more general than the one given here, appears attractive computationally. There have been a large number of papers reporting the applications [e.g., 25–27] of these methods. Thus "geometric" or "inequality" programming seems to be a reasonable method for problems that have a polynomial structure. The extension to problems having posynomial constraints is straightforward [20–27] and involves renormalizing the weights w_i.

A practical disadvantage of this approach to optimization is that at present it requires somewhat more sophistication on the part of the user than most other approaches to nonlinear programming. This situation may change in the future as standard programs are developed and more experience is obtained with large practical problems.

4.8 GENERAL NONLINEAR PROGRAMMING

If we extend the nonlinear programming problem to the most general form

$$\operatorname*{Max}_{\mathbf{x}} I(\mathbf{x}) \tag{4.8.1}$$

subject to

$$h_k(\mathbf{x}) = 0 \qquad k = 1, 2, \ldots, m_1 \tag{4.8.2}$$

$$g_j(\mathbf{x}) \leq 0 \qquad j = 1, 2, \ldots, m_2 \tag{4.8.3}$$

much of the useful theory disappears and about all we can say is that:

(i) The Kuhn-Tucker stationarity conditions are necessary and sufficient for a global maximum if $I(\mathbf{x})$ is concave and $g_j(\mathbf{x})$ convex.
(ii) These are only necessary conditions for nonconcave $I(\mathbf{x})$ and nonconvex $g_j(\mathbf{x})$.

Most of the methods developed for this general form of the problem are designed by assuming $I(\mathbf{x})$ to be concave and the $g_j(\mathbf{x})$ to be convex; these are then applied to nonconvex problems in the hope that the global optimum will be found.

In addition to the penalty function methods discussed earlier, a number of other techniques have been used for treating the general nonlinear programming problem. Space does not allow an exhaustive description of these algorithms; however, a number of review articles are available for further reading [28–31]. In this section, we shall only describe a number of representative techniques that seem to enjoy some popularity at present. Rosen [32] developed a technique for nonlinear $I(\mathbf{x})$ and linear constraints that projects the gradient direction onto linear constraints and moves in this space of reduced dimensionality. Subsequently, the method was extended to nonlinear constraints [33]. Recently Goldfarb and Lapidus [34, 35] modified this technique to project the more efficient conjugate gradient method of Fletcher and Powell onto linear constraints.

Good convergence properties have been observed for both of these methods which will be illustrated in the following example.

EXAMPLE 4.8.1. We wish to minimize the cost of reducing the chromium content of carbon steel, which is brought into contact with an oxidizing slag in an electric furnace.

The slag-metal reaction responsible for the chromium transfer may be written as

$$(FeO)_{slag} + (Cr)_{metal} \leftrightarrows (Fe)_{metal} + (Cr)_{slag} \tag{4.8.4}$$

which takes place at the slag-metal interface.

By combining the available thermodynamic information with a postulate for the kinetic expression, Frohberg et al. [36] proposed the following expression, which relates W_t, the fraction of chromium removed, to the other process parameters:

$$W_t = \left\{[X]_0 + 10^5 \exp\left(-\frac{18{,}750}{T}\right) + Y_0\phi_0 - \left(\left[[X]_0 - 10^5 \exp\left(-\frac{18{,}750}{T}\right) - Y_0\phi_0\right]^2 + 4 \cdot 10^5 \cdot [X]_0 \exp\left(-\frac{18{,}750}{T}\right)\right)^{1/2}\right\} \times \left\{1 - \exp\left[-\frac{3.2 \times 10^6 t}{h\,[X]_0} \exp\left(-\frac{25{,}300}{T}\right)\right]\right\} \tag{4.8.5}$$

here

$$W_t = \frac{[Cr]_{initial} - [Cr]_{at\ time\ t}}{[Cr]_{initial}} \tag{4.8.6}$$

X_0 = initial chrome content of the metal (wt %)
T = absolute temperature (°K)
Y_0 = the ratio: weight of slag/weight of metal
ϕ_0 = iron content of the slag (wt %)
h = depth of the metal bath (cm)
t = time, minutes

The objective function, describing the total cost of the operation may be written as:

I ($/ton of steel processed)

$= 0.05\,(T - 1675)$ (Cost of raising the temperature above 1675°K) $+ 0.25t$ (Cost of heat losses)

$+ 0.5t$ (Fixed capital cost of the furnace) $+ 1.4Y_0\phi_0$ (Cost of slag)

Thus the objective function is given as

$$I = 0.05(T - 1675) + 0.75t + 1.4Y_0\rho_0 \tag{4.8.7}$$

The equality constraint is provided by Eq. 4.8.5, and the following inequality constraints are dictated by the physical and chemical nature of the system:

$$1675 \leq T(°\text{K}) \leq 1900 \tag{4.8.8}$$

$$4 \leq (\phi_0 Y_0) \leq 15 \tag{4.8.9}$$

Furthermore, let us set the initial chromium content, $X_0 = 0.5$ wt %, $h = 60$ cm, and then apply the additional constraint that $W_t = 0.70$.

Thus the problem may be written as

$$\text{Min } I(t, T, \lambda) = 0.05(T - 1675) + 0.75t + 1.4\lambda$$

subject to the constraints posed by Eqs. 4.8.5, 4.8.8, and 4.8.9 where $\lambda = \phi_0 Y_0$.

Solution. The problem was solved by the gradient projection method of Rosen [32], which is available as a general purpose Subroutine OPTM (see Appendix C). Table 4.8.1 gives the rates of convergence from two

TABLE 4.8.1 OPTIMIZATION FOR EXAMPLE 4.8.1

	Starting point 1				Starting point 2			
Iteration[a]	t(min)	T(°K)	λ(wt %)	I($/ton)	t(min)	T(°K)	λ(wt %)	I($/ton)
0	10.00	1750	10.00	—	5.00	1800	6.00	—
1	28.95	1752	15.00	46.55	22.13	1801	14.87	43.70
5	29.13	1752	14.73	46.31	22.34	1801	14.60	43.48
10	35.02	1754	9.85	43.99	29.93	1802	9.93	42.70
25	32.97	1781	9.42	43.21	25.51	1803	11.72	41.94
35	27.50	1796	11.00	42.06	23.89	1818	11.98	41.86
50	23.97	1819	11.92	41.86	24.00	1818	11.92	41.86
100	24.45	1815	11.78	41.85	24.45	1815	11.78	41.85

Computing time for 100 iterations is ~6 sec CDC 6400.

starting points. The optimum of

$$t = 24.45 \text{ min}, \quad T = 1815°\text{K}, \quad Y_0\phi_0 = 11.78 \text{ wt } \%$$

was found within 100 iterations from both starting points with a computing time of ~6 sec (computing cost ~80 cents). The equality constraints were satisfied with a tolerance of $\sim 10^{-13}$

Kelley [37] developed a "cutting plane" method that applies to this most general problem, and which solves a sequence of linear programming problems converging to the optimum of the nonlinear programming problem. The constraints on the problem are linearized about another point. The method iterates until convergence is attained. Griffith and Stewart [38] have developed a simple version of this technique which was used by Dibella and Stevens [39] to optimize a chemical plant. A standard program using an iterative LP approach, Process Optimization Program (POP) has been developed through the IBM Share system [40], but experience indicates that the algorithm is not the most efficient [7, 41].

The performance of some of these algorithms will be discussed and compared in the next section.

4.9 A COMPARISON OF NONLINEAR PROGRAMMING ALGORITHMS

The performance of NLP algorithms can be determined only by testing a wide variety of problems. Unfortunately, not enough testing has been reported to draw positive conclusions about the various algorithms proposed. Several comparative studies have been described recently [7, 8, 29, 41, 42], but the majority of these seem rather inconclusive. The only trend that is clear from these investigations is that many techniques will perform reasonably well in unconstrained problems or in problems with a few inequality constraints. However, nearly all of the techniques tested experienced difficulty with problems having nonlinear equality constraints.

To illustrate these difficulties, the results of one such investigation will be treated in some detail [8].† To test some of the optimization procedures discussed above, a mathematical model of a chemical plant, originally developed by Williams and Otto [43] for the study of computer control, was selected. This model was also studied by Dibella and Stevens [39] and again by Stevens and Adelman [44] in a slightly modified form, and more recently by Vinturella and Law [45, 46] under another modification. This model contains many characteristics of chemical process optimization problems, and in view of the individual components (namely, CSTR, heat exchanger, and distillation column) is thought realistic enough to illustrate the concepts under discussion. Also the model was used in several previous optimization studies, and these provide a basis for comparison.

† This material is used with the permission of The Canadian Society of Chemical Engineering.

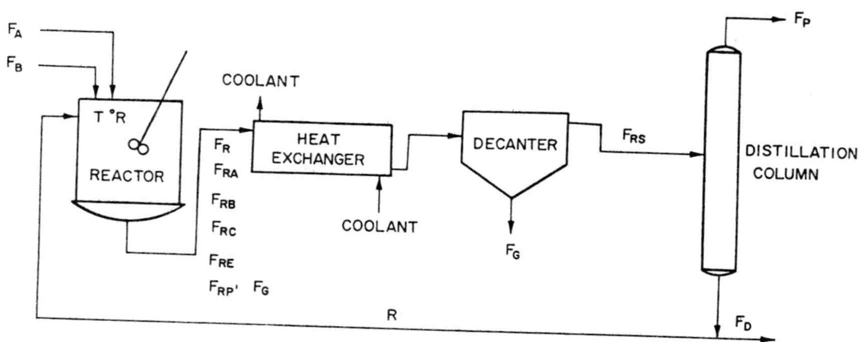

Fig. 4.7 Flow diagram of the chemical plant.

The plant under consideration is to manufacture a certain chemical, P. This plant, shown in Figure 4.7, consists of a perfectly mixed stirred-tank reactor, a heat exchanger to cool the reactor outflow, a decanter to separate a heavy, oily by-product, G, and a distillation column to separate the product P, from other components as the column top product. A portion of the bottom product of the distillation column is recycled to the reactor, while the remainder is used as plant fuel. The by-product G must be discarded after a waste treatment process which incurs additional cost.

The chemical reactions involved in manufacturing P are assumed to be irreversible and of second order, and the reaction mechanism is considered well known. The following three reactions are involved:

1. $A + B \xrightarrow{K_1} C$
2. $C + B \xrightarrow{K_2} P + E$
3. $P + C \xrightarrow{K_3} G$

where $k_i = A_i \text{ Exp } (-B_i/T)$ and T is the reactor temperature in °R.

The values of A and B are summarized in Table 4.9.1

TABLE 4.9.1 DATA ON CHEMICAL REACTIONS

i	A_i (hr) (wt fraction)	B_i(°R)
1	5.9755×10^9	12000
2	2.5962×10^{12}	15000
3	9.6283×10^{15}	20000

Reactants A and B are fed from an outside source in pure form, and it is assumed that they are available in unlimited quantities. Components C and E are intermediate by-products; neither of them has any market value as an independent commodity, but both can be used as the plant fuel. Their price as fuel is valued at \$0.0068 per pound based on their calorific values. It is also known that the reaction rate is negligible below 120°F, and undesirable decompositions take place over 220°F. It is necessary, therefore, to maintain the reactor temperature between 120°F and 220°F.

The outflow from the reactor is passed to the heat exchanger and cooled to a sufficiently low temperature to allow complete separation of G in the decanter. The remaining products are then sent to the distillation column where P is recovered as the column overhead product. However, it was found that P forms an azeotrope with E. The composition of P at the azeotropic point is 10% by weight and this contributes an additional constraint on the operation of the distillation column. The bottom product from the column is split into two streams: one is sent to the utilities plant as fuel, and the other is recycled to the reactor.

Some additional data for the mathematical model are listed in Table 4.9.2.

TABLE 4.9.2 PHYSICAL AND ECONOMIC DATA

Components	Molecular weights	Price (\$/lb)
A	100	0.02
B	100	0.03
C	200	(0.0068)[a]
E	200	(0.0068)[a]
G	300	−0.01[b]
P	100	0.30

[a] Price evaluated as the plant fuel.

[b] Waste treatment cost.

The objective function for this model is taken to be the annual rate of return on the investment, which may be expressed as follows:

I. Sales Volume:	$(0.3F_P + 0.0068F_D)/\text{hr}$
II. Raw Material Cost:	$(0.02F_A + 0.03F_B)/\text{hr}$
III. Waste Treatment Cost:	$0.01F_G/\text{hr}$
IV. Utilities Cost:	$2.22F_R/\text{yr}$
V. SARE†:	12.4% of Sales
VI. Plant Fixed Charge:	10% of the Plant Investment per yr
VII. Plant Investment:	$\text{PIN} = 600V\rho$

† Sales, Administration, and Research Expenses.

Total operating hours per year are assumed to be 8400 hr. By using the above information, the annual rate of return on the fixed investment can be expressed as:

$$\begin{aligned}\text{Max } I(x) = {} & 100 \times [8400(0.3F_P + 0.0068F_D - 0.02F_A \\ & - 0.03F_B - 0.01F_G) - (0.124)(8400) \\ & \times (0.3F_P + 0.0068F_D) - 180000 - 2.22F_R]/1800000 \end{aligned} \tag{4.9.1}$$

The equality constraint equations for the problem are the mass balances, that is, overall material balances:

$$h_1 = F_A + F_B - F_G - F_P - F_D = 0 \tag{4.9.2}$$

constraint on the separation efficiency of the distillation column:

$$h_2 = F_{RP} - 0.1F_{RE} - F_P = 0 \tag{4.9.3}$$

material balance on component E:

$$h_3 = (M_E/M_B)K_2F_{RB}F_{RC}V\rho/F_R^2 - F_D[F_{RE}/(F_R - F_G - F_P)] = 0 \tag{4.9.4}$$

material balance on component P:

$$\begin{aligned} h_4 = {} & [K_2F_{RB}F_{RC} - (M_P/M_C)K_3F_{RC}F_{RP}]V\rho/F_R^2 \\ & - F_D(F_{RP} - F_P)/(F_R - F_G - F_P) - F_P = 0 \end{aligned} \tag{4.9.5}$$

material balance on component A:

$$h_5 = (-K_1F_{RA}F_{RB})V\rho/F_R^2 - F_D[F_{RA}/(F_R - F_G - F_P)] + F_A = 0 \tag{4.9.6}$$

material balance on component B:

$$\begin{aligned} h_6 = {} & (-K_1F_{RA}F_{RB} - K_2F_{RB}F_{RC})V\rho/F_R^2 \\ & - F_D[F_{RB}/(F_R - F_G - F_P)] + F_B = 0 \end{aligned} \tag{4.9.7}$$

material balance on component C:

$$\begin{aligned} h_7 = {} & [(M_C/M_B)K_1F_{RA}F_{RB} - (M_C/M_B)K_2F_{RB}F_{RC} - K_3F_{RC}F_{RP}] \\ & \times V\rho/F_R^2 - F_DF_{RC}/(F_R - F_G - F_P) = 0 \end{aligned} \tag{4.9.8}$$

material balance on component G:

$$h_8 = (M_G/M_C)K_3F_{RC}F_{RP}V\rho/F_R^2 - F_G = 0 \tag{4.9.9}$$

definition of F_R:

$$h_9 = F_{RA} + F_{RB} + F_{RC} + F_{RP} + F_G - F_R + F_{RE} = 0 \tag{4.9.10}$$

The inequality constraints are the upper and lower bounds on the reactor temperature and production rate

$$580°R \leq T \leq 680°R \tag{4.9.11}$$

$$0 \leq F_P \leq 4763 \tag{4.9.12}$$

together with the non-negativity conditions on each of the variables. Thus there are 12 variables, 9 equality constraints (one of them is redundant), and 14 inequality constraints in this problem.

The optimization algorithms to be tested are the orthogonal search

TABLE 4.9.3 NLP ALGORITHM CHARACTERISTICS

Program name	Method of treating constraints	Information required	Method of ascent
CLIMB [2]	Penalty functions	No derivatives	Orthogonal direct search
SUMT [5]	Sequence of penalty functions	First and second derivatives	Gradient and Newton-Raphson ascent
OPTM [33]	Gradient projection	First derivatives	Modified gradient ascent

procedure of Rosenbrock [2] (Program CLIMB), the sequential unconstrained maximization technique of Fiacco and McCormick [5] (Program SUMT), and the gradient projection technique of Rosen [33] (Program OPTM). Table 4.9.3 outlines the characteristics of the algorithms tested.

The optimization of the chemical plant model, Eqs. 4.9.1 to 4.9.12 was begun from three different starting points given in Table 4.9.4. Starting point 1 is identical with Dibella and Stevens [39] initial guess, starting point 2 is the same as their "improved" initial guess, and starting point 3 is the same as the one used by Vinturella and Law [45, 46] and is almost the same as Dibella and Stevens "optimal" solution. As can be noted from the last row in Table 4.9.4, all starting points are distinctly nonfeasible with respect to the equality constraints, although starting point 3 is only "out" by about $\sim 0.5\%$ on the average.

TABLE 4.9.4 STARTING POINTS FOR THE W-O MODEL

Variable	Starting point 1	Starting point 2	Starting point 3
F_A	10,000	11,540	13,546
F_B	40,000	31,230	31,523
F_R	100,000	92,640	157,301
F_{RA}	6,000	8,820	18,187
F_{RB}	35,000	39,910	60,815
F_{RC}	10,000	2,360	3,331
F_{RE}	15,000	31,660	60,542
F_{RP}	6,000	7,890	10,817
F_G	5,000	2,010	3,609
F_P	4,763	4,763	4,763
F_D	20,000	36,010	36,697
T	610	610	656
$I(\%)$	−59.27	108.50	72.15
$\sum_j h_j^2$	1.9×10^9	2.6×10^8	7.7×10^3

The results of the optimizations, given in Table 4.9.5, show quite clearly that only OPTM converged to the same solution from all three starting points. SUMT and CLIMB failed both to find the optimal solution and to satisfy the equality constraints in all cases. This poor performance may be explained in light of their penalty function treatment of the large number of equality constraints (nine of them) which in the case of CLIMB clouds the true nature of the response surface (possibly even introducing multiple maxima) to the extent that the solution will not converge. In the case of SUMT, it was observed that in the early stages of the optimization, when the penalties for constraint violation are small, the search moved into a very profitable but highly nonfeasible region of the response surface. As the penalties for constraint violation were increased in the latter stages of the optimization, the procedure was unable to give up its highly profitable position in order to satisfy the equality constraints. It appeared that no amount of adjustment of the r_k parameters would alleviate this problem. The results from starting point 3 were curious in that neither CLIMB nor SUMT would move from this point. This result, when combined with the fact that Dibella and Stevens [39] (who used an initial penalty function approach for their equality constraints) found starting point 3 as their optimum, suggests that there may be a local optima at starting point 3 in the artificial response surface generated by the penalty function approach. OPTM, which does not use the penalty function technique, had no difficulty moving to the true optimum from starting point 3.

TABLE 4.9.5 RESULTS OF THE OPTIMIZATION

	Starting point 1			Starting point 2			Starting point 3		
Variable	OPTM	SUMT	CLIMB	OPTM	SUMT	CLIMB	OPTM	SUMT	CLIMB
F_A	12,537	14,300	14,624	12,521	14,300	12,854	12,518	13,544	13,546
F_B	28,379	9,531	31,402	28,360	9,532	26,931	28,365	31,520	31,520
F_R	324,065	5,378	102,109	326,786	5,482	91,960	326,418	157,302	157,300
F_{RA}	38,251	434	19,612	38,493	458	18,264	38,420	18,187	18,187
F_{RB}	120,818	288	49,608	121,833	332	42,518	121,745	69,815	60,814
F_{RC}	7,266	6	4,574	7,311	5	4,325	7,296	3,331	3,331
F_{RE}	136,416	0.05	22,175	137,708	0.01	21,541	137,541	60,542	60,541
F_{RP}	18,405	4,616	5,682	18,534	4,590	5,320	18,517	10,817	10,817
F_G	2,911	3	1,000	2,908	0.05	1,000	2,908	3,608	3,609
F_P	4,763	4,763	3,440	4,763	4,763	3,167	4,763	4,763	4,763
F_D	33,242	20,000	41,246	33,209	20,002	35,628	33,212	36,697	36,697
T	654	599	617	654	607	616	654	656	656
$I(\%)$	98.68	362.15	67.00	98.68	373.79	−35.37	98.68	72.22	72.22
$\sum h_j^2$	0.24	9.2×10^5	3.6×10^5	0.24	4.9×10^6	1.5×10^6	0.25	8.2×10^3	7.7×10^3
Number of iterations	349[a]	198[b]	102[b]	323[a]	181[b]	127[b]	374[a]	5[b]	1[b]
Execution time (sec)	52.2	1,200.0	157.1	50.6	1,200	199.5	42.6	44.4	3.0

[a] Number of objective function evaluations.
[b] Number of reductions of the penalty functions.

All of these results together with the studies [7, 29, 41, 42] described earlier indicate that the penalty function approach is not recommended for the treatment of large numbers of equality constraints. In addition, our results demonstrate quite clearly the ability of projection methods to handle these problems reliably, and quite efficiently. In all the cases the solution was found by OPTM in less than 1 min on the IBM 360/75 and the equality constraints were satisfied within 0.002% on the average—which is much better than the penalty function method.

Unfortunately, not enough techniques have been tested on a wide range of problems to provide conclusive information on the performance of nonlinear programming techniques. However, this example should indicate some of the difficulties that can be encountered, and should point out some disadvantages of the penalty function approach.

4.10 INTEGER PROGRAMMING

In a large number of practical mathematical programming problems some or all of the variables may be required to take only integer values. The cutting stock problem of determining the optimal way of cutting billets to minimize the waste would be a typical example. Suppose that we have m pieces of steel stock of 50 ft in length, and that we wish to cut these into three products of (1) 12 ft in length, (2) 16 ft in length, and (3) 24 ft in length. Let us require b_1 pieces of product (1), b_2 pieces of product (2), and b_3 pieces of product (3), and let x_{ij} denote the number of pieces of product i cut from the jth billet; then the waste to be minimized is given by

$$I = \sum_{j=1}^{m} [50 - (12x_{1j} + 16x_{2j} + 24x_{3j})] \tag{4.10.1}$$

subject to a length constraint

$$12x_{1j} + 16x_{2j} + 24x_{3j} \leq 50j = 1, 2, \ldots, m \tag{4.10.2}$$

and production requirements

$$\sum_{j=1}^{m} x_{1j} \geq b_1 \tag{4.10.3}$$

$$\sum_{j=1}^{m} x_{2j} \geq b_2 \tag{4.10.4}$$

$$\sum_{j=1}^{m} x_{3j} \geq b_3 \tag{4.10.5}$$

as well as a nonnegativity condition

$$x_{ij} \geq 0 \tag{4.10.6}$$

Since it is clear that all the x_{ij} must be nonnegative integers (in this problem 0, 1, 2, 3, or 4), the problem is rather more complicated than the standard linear programming problem discussed in Section 4.4.

Integer programming problems (even linear ones) are so resistant to solution, that there is no general purpose algorithm for them. A standard technique in integer linear programming is to consider the variables initially as continuous, and solve the LP problem. If the resulting solution has integer values, the problem is solved; if it does not, then artificial constraints are added to "cajole" the solution into an integer form. Hadley [12, 18], Dantzig [11], and Hu [47] discuss various approaches to these problems. As this is a rather specialized area, we shall not pursue it further here.

4.11 CONCLUSIONS

From the wide range of approaches discussed in this chapter, it should be clear that one must exercise some care in the choice of a mathematical programming algorithm. It is perhaps useful to list some of the considerations involved in this choice.

1. *Special form of the problem.* Since the techniques for linear programming, quadratic programming, separable programming, and geometric programming are specially designed for a particular type of problem, one should first examine the problem at hand to see if it can be put into one of these formats. Since these techniques are usually the most efficient for the problems they can treat, it is best to explore these before going on to more general, but possibly less efficient methods.
2. *Information available from the model.* The choice of a search procedure may be determined by the type of information available from the model. For example, if gradients and second derivatives are easily available, then a conjugate gradient or second-order method would probably be the best choice. However, if only the objective function could be obtained from the model, then one must choose between a nongradient search procedure and a gradient procedure which uses a finite difference calculation for the partial derivatives. Neither alternative is completely satisfactory, so that much current research is concerned with the development of efficient algorithms, not requiring derivatives, which are able to handle constraints well.
3. *Type of constraints.* The type of constraints on the problem can be important in the choice of a method. Many methods work on unconstrained

problems or on problems with a few inequality constraints; however as discussed in Section 4.9, nonlinear equality constraints can frustrate a large number of procedures. Thus one should be very selective in choosing an optimization procedure for problems with equality constraints.

4. *Problem size and cost of computing.* The size of the optimization problem and the cost involved in solving it may influence the choice of a method. Some techniques which work well for small problems are known to become much less effective with increasing problem size. Thus for large problems, one should choose an algorithm which has shown good behavior with increasing size. The cost of computing also influences the care with which algorithms are selected. If the cost of the optimization calculation is negligible, then any readily available algorithm can be selected. However, if the cost of the computation is significant then one is usually justified in performing preliminary research and computational studies before selecting the algorithm to be used.

Another consideration—the structure of the physical and mathematical problem—will be discussed in the next chapter on problem decomposition.

PROBLEMS

1. Apply the transformation technique discussed in Section 4.2 to solve the problem discussed in Example 3.2.4 so as to prevent the spurious "optima."
2. A manufacturer of heavy chemicals has plants at four locations, P_1, P_2, P_3, and P_4, and has to supply markets at six locations, M_1, M_2, M_3, M_4, M_5, M_6. Given the production costs and maximum production capabilities at $P_1 \cdots P_4$, the market demands of $M_1 \cdots M_6$, and the transportation costs,

   (i) Calculate the optimal production schedules and transportation arrangements.

   (ii) Assume each production facility could be expanded (at minor cost) by as much as 10%, to meet an increased demand of 20% increase at markets M_1, M_3, and M_6. Which plant(s) should be expanded?

   Hint: It is suggested that a standard LP computer program be used for this problem.

The data are as follows:

Plant	Maximum capacity (tons/year)	Unit production cost ($/ton)
P_1	3×10^5	18
P_2	4×10^5	15.5
P_3	2×10^5	19.5
P_4	6×10^5	14.0

Market	Size (tons/year)
M_1	2×10^5
M_2	2.5×10^5
M_3	5×10^5
M_4	1×10^5
M_5	1.5×10^5
M_6	1.5×10^5

Cost of Transportation ($/ton)

Market	Plants			
	P_1	P_2	P_3	P_4
M_1	2.0	1.9	2.6	1.9
M_2	2.2	2.5	2.1	1.8
M_3	2.5	3.0	2.4	4.0
M_4	1.2	2.1	1.9	2.2
M_5	3.0	1.2	2.8	2.6
M_6	1.9	2.0	2.3	3.0

3. We wish to minimize the total cost of one BOF heat which is made up of hot metal (from a blast furnace), scrap and possibly a supplemental SiC charge, which allows the unit to melt more scrap. In ordinary operation the ratio (hot metal charged/scrap charged) is fixed by thermodynamic considerations, but this ratio may be decreased by preheating the scrap charge. Let us define the following symbols:

x_1 hot metal charge
x_2 scrap charge
T scrap preheat temperature °F
x_3 SiC charge
C_h cost of preheating one ton of scrap to temperature T

The following empirical relationships are available:

$$x_1\left[1 + 2.0\frac{T}{2700} + 3.5\left(\frac{T}{2700}\right)^2\right] - 3(x_2 - 11x_3) \geq 0$$

$$C_h = 0, \quad T = 0$$

when $T > 0$

$$C_h = 2.0 + \frac{5T}{2500} + 30\left(\frac{T}{2000}\right)^3 \text{ in \$/ton}$$

$$x_1 + x_2 = 210 \text{ tons}$$

$$x_3 < 10 \text{ tons}$$

cost of SiC = \$200/ton
$0 \leq T \leq 1300°F$
cost of hot metal: \$60/ton, cost of scrap: \$35/ton
Fixed charges: \$15/ton of charge

Use a nonlinear programming algorithm to determine the optimal values of T, x_1, x_2, x_3. Appendix C lists some standard programs available.

REFERENCES

1. M. J. Box, *Comput. J.* **9,** 67 (1966).
2. H. H. Rosenbrock, *Comput. J.* **3,** 175 (1960).
3. C. W. Carroll, *Oper. Res.* **9,** 169 (1961).
4. A. V. Fiacco and G. P. McCormick, *Manage. Sci.* **10,** 360 (1964).
5. Ibid., **12,** 816 (1966).
6. A. V. Fiacco and G. P. McCormick, *Non-linear Programming*, Wiley, 1968.
7. D. C. Stocker and D. M. Himmelblau, Paper 31a, 67th. Nat'l. AIChE Meeting, Atlanta, February (1970).
8. B. S. Jung, W. Mirosh, and W. H. Ray, *Can. J. Chem. Eng.* **49,** 844 (1971).
9. R. Hooke and T. A. Jeeves, *JACM* **8,** 212 (1961).
10. W. I. Zangwill, *Manage. Sci.* **13,** 344 (1967).
11. G. B. Dantzig, *Linear Programming and Extensions*, Princeton University Press, 1963.
12. G. Hadley, *Linear Programming*, Addison-Wesley, 1962.
13. N. R. Amundson, *Mathematical Methods in Chemical Engineering*, Prentice-Hall, 1966, Chap. 4.
14. D. Kwasnoski and R. W. Bouman, *Ironmaking Proceedings*, J. R. Dietz, Ed., The Metallurgical Society of A.I.M.E., (1967).
15. W. W. Garvin, *Linear Programming*, McGraw-Hill, 1960.
16. P. Wolfe, *Econometrica* **27,** 3 (1959).
17. E. M. L. Beale, *J. R. Stat. Soc. (B)* **17,** 173 (1955).

18. G. Hadley, *Non-Linear and Dynamic Programming*, Addison-Wesley, 1964.
19. H. P. Kunzi, W. Krelle, and W. Oettli, *Nonlinear Programming*, Blaisdell, 1966.
20. C. Eben and J. R. Ferron, *A. I. Ch. E. J.* **14,** 32 (1968).
21. R. J. Duffin, E. L. Peterson, and C. Zener, *Geometric Programming*, Wiley, 1967.
22. D. J. Wilde, *I and EC*, **57,** 18 (August 1965).
23. J. R. Ferron, *CEP Symp.* Ser. No. 50, **60,** 60 (1964).
24. C. Eben and J. R. Ferron, *I and EC Fund.* **8,** 749 (1969).
25. M. Avriel and D. J. Wilde, *I and EC Process Des. and Dev.* **6,** 256 (1967).
26. U. Passy and D. J. Wilde, *SIAM J. Appl. Math.* **16,** 363 (1968).
27. G. E. Blau and D. J. Wilde, *Can. J. Ch. E.* **47,** 317 (1969).
28. P. Wolfe, "Methods of Non-linear Programming," in *Recent Advances in Mathematical Programming*, R. L. Graves and P. Wolfe, Eds., McGraw-Hill, 1963, p. 67.
29. G. Zoutendijk, *SIAM J. Control* **4,** 194 (1966).
30. P. Wolfe, "Methods of Non-linear Programming," in *Non-linear Programming*, J. Abadie, Ed., Wiley, 1967, p. 99.
31. G. Zoutendijk, "Non-Linear Programming, Computational Methods," in *Integer and Non-linear Programming*, J. Abadie, Ed., Elsevier, 1970, p. 37.
32. J. B. Rosen, *SIAM J.* **8,** 181 (1960).
33. J. B. Rosen, *SIAM J.* **9,** 514 (1961).
34. D. Goldfarb and L. Lapidus, *I and EC Fund.* **7,** 142 (1968).
35. D. Goldfarb, *SIAM J. Appl. Math.* **17,** 739 (1969).
36. V. M. G. Frohberg, D. Papamantellos, and E. Hanert, *Arch. Eisenhuttenwes.* **38,** 91 (1967).
37. J. E. Kelley, *SIAM J.* **8,** 703 (1960).
38. R. E. Griffith and R. A. Stewart, *Manage. Sci.* **7,** 379 (1961).
39. C. Dibella and W. F. Stevens, *I and EC Process Des.* **4,** 16 (1965).
40. IBM Share Program No. 7090-H9 IBM 0021 (POP II).
41. G. K. Barnes and D. M. Himmelblau, paper presented at 1966 CIC Conference, Windsor, Ontario.
42. R. A. Barneson, N. F. Brannock, J. G. Moore, and C. J. Morris, Paper Presented at 64th Natl meeting A.I.Ch.E. New Orleans, March 1969.
43. T. J. Williams and R. E. Otto, *AIEE Trans.* **79,** 458 (1960).
44. W. F. Stevens and A. Adelman, Paper 31c, 67th Natl A.I.Ch.E. Meeting, Atlanta, February 1970.
45. J. B. Vinturella, Ph.D. Thesis, Tulane University, New Orleans, 1968.
46. J. B. Vinturella and V. J. Law, Paper 12a, 65th Natl A.I.Ch.E. Meeting, Cleveland, May 1969.
47. T. C. Hu, *Integer Programming and Network Flows*, Addison-Wesley, Reading, Massachusetts, 1969.

5 Optimization of Decomposable Structured Systems

5.1 INTRODUCTION

In this chapter we shall consider the constrained optimization of systems which have a decomposable structure. By decomposable structure we mean systems which may be represented as a series of interconnected units or have a mathematical structure amenable to decomposition. Decomposition may immediately suggest itself by the physical nature of the system, for example, a series of stirred reactors in a chemical process, a series of interconnected flotation cells, or the sequential operation of comminution equipment in mineral processing.

Alternatively, decomposition may be possible through the mathematical model used for the representation of the system, even if it consists of just one physical unit. As an example we may consider the ladle degassing operation discussed in Example 3.1.2. Here degassing occurs because of two distinct processes, namely, (1) mass transfer to the bubble during its initial growth and subsequent rise through the metal and (2) mass transfer induced by agitation of the free surface. The rates of degassing due to these two processes were shown in Fig. 3.8. We may consider this system as a decomposable structure consisting of the two different mass transfer mechanisms each with its own optimum ladle depth; however, the overall optimum is neither of these. We shall return to this problem later.

In the following three sections we deal with optimization procedures especially suited to problems with a *serial structure*. These techniques have considerable practical relevance because interconnected processing units usually have a fundamental serial structure.

Section 5.5 introduces the reader to the general concepts of hierarchial systems and multilevel optimization.

Finally, in the remaining sections we discuss the methods available for the decomposition of linear and nonlinear problems, such that the components of the decomposed structure are optimized individually and then coordinated to produce the optimum of the whole system.

5.2 SERIALLY STRUCTURED SYSTEMS

Consider the serial staged system shown in Fig. 5.1.

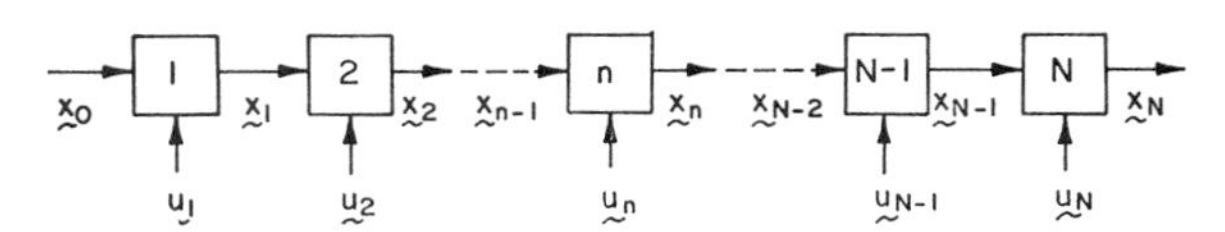

Fig. 5.1 A serial structure.

This could represent the actual flow of material from stage to stage with the state of the material at the exit of the nth stage given by the vector $\mathbf{x}_n$, $n = 1, 2, \ldots, N$ which might be composed of temperatures, concentrations, pressure, pH, and so forth. The variables which can be controlled at each stage are represented by $\mathbf{u}_n$, $n = 1, 2, \ldots, N$ and these could include holding times, heat input, reactant feed rates, agitation, and so on. An illustration of a serially structured metallurgical system with some branching is shown in Fig. 5.2, which depicts the processing of anodic slimes obtained in the electrorefining of copper [1].

Let us now define our staged optimization problem in general terms. The transformation equations

$$\mathbf{x}_n = \mathbf{f}_n(\mathbf{x}_{n-1}, \mathbf{u}_n) \qquad n = 1, 2, \ldots, N \tag{5.2.1}$$

are a set of equality constraints on the overall problem which allow the output of the nth stage, $\mathbf{x}_n$, to be represented in terms of the input, $\mathbf{x}_{n-1}$, and decision variables, $\mathbf{u}_n$, at that stage. In process optimization problems, Eq.

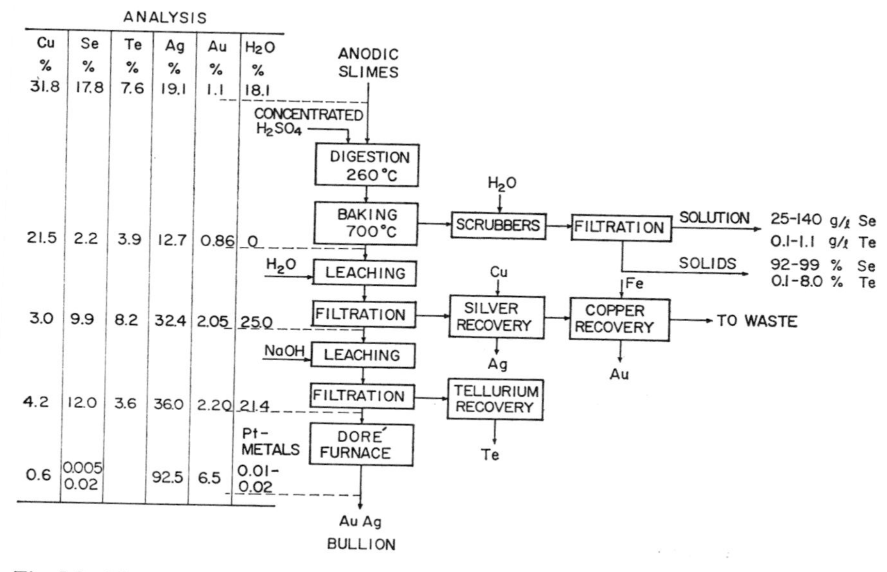

Fig. 5.2 Flow diagram for the processing of anodic slimes obtained in the electrorefining of copper, after Habashi ([1] F. Habashi, *Principles of Extractive Metallurgy*, Vol. 2, Gordon and Breach, 1970).

5.2.1 usually corresponds to the mass, energy, and momentum balances on the stage and thus can be considered a mathematical model of the process.

The constraints on the problem might be mixed, for example,

$$g_j(\mathbf{x}_1, \mathbf{x}_2, \ldots, \mathbf{x}_n, \mathbf{u}_1, \mathbf{u}_2, \ldots, \mathbf{u}_n) \leq 0 \tag{5.2.2}$$

which may present complications requiring special techniques to decompose the problem. However, for the moment let us assume, as is very often the case, that the constraints are decomposed according to stage, that is,

$$\mathbf{h}_n(\mathbf{x}_n, \mathbf{x}_{n-1}, \mathbf{u}_n) = \mathbf{0} \qquad n = 1, 2, \ldots, N \tag{5.2.3}$$

$$\mathbf{g}_n(\mathbf{x}_n, \mathbf{x}_{n-1}, \mathbf{u}_n) \leq \mathbf{0} \qquad n = 1, 2, \ldots, N \tag{5.2.4}$$

where we note that Eq. 5.2.1 could be included in the equality constraint (Eq. 5.2.3).

The objective function for staged problems will also be normally decomposable by repeated application of Eq. 5.2.1 to yield

$$I(\mathbf{x}_1, \mathbf{x}_2, \ldots, \mathbf{x}_n, \mathbf{u}_1, \mathbf{u}_2, \ldots, \mathbf{u}_n) = \sum_{n=1}^{N} F_n(\mathbf{x}_{n-1}, \mathbf{u}_n) + G(\mathbf{x}_N) \tag{5.2.5}$$

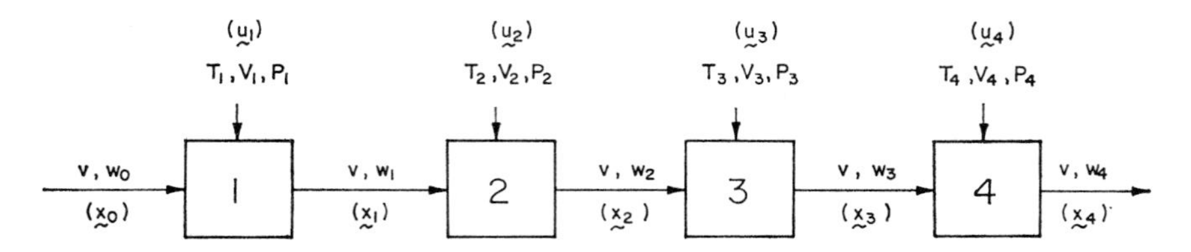

Fig. 5.3 The flow diagram for Example 5.2.1.

where F_n can be considered the objective or "profit" increase at each stage, and $G(\mathbf{x}_N)$ is the profit or cost associated with the stream leaving the nth stage.

EXAMPLE 5.2.1. As an illustration of a serially staged system let us consider a leaching operation in which zinc, present in the form of ZnS, is extracted from a solid concentrate, through contact with a dilute sulfuric acid solution, in a series of stirred tank units sketched in Fig. 5.3.

The concentrate enters the system at leaching unit 1, with a mass flow rate of v (lb/hr) and with a zinc mass fraction w_0. Upon passing through the cascade the concentrate stream leaves leaching unit 4, with a mass flow rate which is assumed to be the same as that at the entry, that is, v, with a zinc mass fraction, w_4.

The amount of material extracted in each of the units will depend on the temperature, T_n, on the power consumption, P_n, on the volume of the unit, V_n, and on the zinc content of the stream entering the unit.

Our objective is to maximize the profit from the operation of the system. In accordance with our earlier discussion, the fraction of zinc remaining in the concentrate, $x_n = w_n$, is the *state variable*, and temperature, T_n, power consumption, P_n, and the reactor volume, V_n are the *control variables* for this system, as indicated in Fig. 5.3.

Let us consider that the following relationships may be used for describing the system.

The extraction process is assumed to follow first-order kinetics; thus the inlet and exit concentration within any given unit is given by

$$v(x_{n-1} - x_n) = V_n k_n x_n \tag{5.2.6}$$

where x_n and x_{n-1} are the mass fractions of zinc in the streams leaving and entering unit n respectively, and k_n is the effective reaction rate constant.

Let us further assume that k is related to the temperature and the rate of agitation in the following manner:

$$k_n = k^\circ \frac{(T_n^{\,2} - T_r^{\,2})}{1 + \dfrac{1}{0.05 + \dfrac{P_n}{V_n v + 1}}}; \qquad T_r \leq T_n \leq T^* \tag{5.2.7}$$

where k° = a reference value of the rate constant
T_n = the temperature in unit n and
P_n = the rate of power consumption for agitation

The economic parameters needed for consideration are given as follows:

Value of product $\alpha_1 v(x_0 - x_4)$
Raw material cost $\alpha_2 v x_0$
Cost of power, including capital charges for stirrer, motor, etc. $\alpha_3 P_n$
Cost of heating, including capital charges for insulation, etc. $\alpha_4(T_n - T_r)$
Capital charges on reactor volume $\alpha_5 V_n^{2/3}$

If we consider a system with a fixed throughput, v, then the objective function may be written as:

$$\begin{aligned} \text{Max } [I = \alpha_1 v(x_0 - x_4) - \alpha_2 v x_0 \\ - \alpha_3(P_1 + P_2 + P_3 + P_4) - \alpha_4(T_1 + T_2 + T_3 + T_4) \\ - \alpha_5(V_1^{2/3} + V_2^{2/3} + V_3^{2/3} + V_4^{2/3})] \end{aligned} \tag{5.2.8}$$

Upon recasting Eqs. 5.2.6–8 in the form given by Eq. 5.2.1, we have:

$$x_n = w_n, \qquad n = 0, 1, 2, 3, 4 \tag{5.2.9}$$

and the decision (control variables):

$$\mathbf{u}_n = \begin{bmatrix} V_n v \\ T_n \\ P_n \end{bmatrix} = \begin{bmatrix} u_{1n} \\ u_{2n} \\ u_{3n} \end{bmatrix} \qquad n = 1, 2, 3, 4$$

and the transformation Eq. 5.2.1 becomes

$$x_n = \frac{x_{n-1}}{1 + k_n u_{1n}} \qquad n = 1, 2, \ldots, 4 \tag{5.2.10}$$

where

$$k_n = \frac{k^\circ[(u_{2n})^2 - T_r^{\,2}]}{1 + \left[0.05 + \dfrac{u_{3n}}{(u_{1n} + 1)}\right]^{-1}} \tag{5.2.11}$$

The objective function is then

$$I = a_1(x_0 - x_4) - a_2x_0 - a_3\sum_{n=1}^{4} u_{1n}^{2/3} - a_4\sum_{n=1}^{4}(u_{2n} - T_r) - a_5\sum_{n=1}^{4} u_{3n} \quad (5.2.12)$$

The constraints are

$$u_{1n} > 0, \quad u_{2n} > 0, \qquad \text{and} \quad T_r \leq u_{3n} \leq T^* \quad (5.2.13)$$

We shall form the profit at each stage as in Eq. 5.2.5

$$\begin{aligned} G(x_4) &= -a_1x_4 \\ F_n(x_{n-1}, \mathbf{u}_n) &= -a_3(u_{1n})^{2/3} - a_4u_{2n} - a_5u_{3n} \qquad n = 2, 3, 4 \\ F_1(x_0, \mathbf{u}_1) &= (a_1 - a_2)x_0 - a_3(u_{11})^{2/3} - a_4u_{21} - a_5u_{31} + a_4\sum_{n=1}^{4} T_r \end{aligned} \quad (5.2.14)$$

We shall return to this example in the next section.

5.3 DISCRETE DYNAMIC PROGRAMMING

Dynamic programming is a technique developed by Richard Bellman [2] in the mid-1950s and is well suited to the optimization of serial structures. The basis of dynamic programming is stated in Bellman's Principle of Optimality: "*An optimal policy has the property that whatever the initial state and the initial decision, the remaining decisions must constitute an optimal policy with regard to the state resulting from the first decision.*"

This principle may be illustrated by a simple practical example. Let us consider the operation of a three-stage system, such as sketched in Fig. 5.4. Here the vector $\mathbf{x}_0$ characterizes the feed into the first stage, and $\mathbf{x}_1$ defines the stream exiting the first stage, which is also the feed into the second stage, and so forth. The quantity $\mathbf{x}_n$ may denote concentrations, particle sizes, temperatures, and the like.

Our objective is to optimize the cascade as a whole, by finding some desired value for the vector $\mathbf{x}_3$, which characterizes the stream exiting the last processing unit in the cascade.

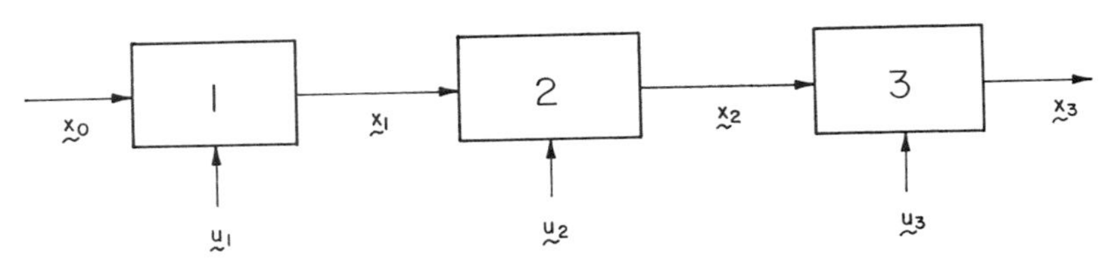

Fig. 5.4 A three-stage serial process.

According to Bellman's principle, this optimization may be carried out in a stagewise manner as follows:

(i) We consider the last stage, that is unit 3, in isolation and perform the optimization for various values of $\mathbf{x}_2$, that is, composition of feed into stage 3.

(ii) Then we proceed in a backward direction to unit 2, and perform the same operation, that is, optimize for various values of $\mathbf{x}_1$. The attractiveness of the procedure lies in the fact that at this stage we need not concern ourselves with reoptimizing unit 3, since we have already performed its optimization for the likely values of $\mathbf{x}_2$ resulting from unit 2.

(iii) We proceed "backwards" to unit 1, where the procedure is repeated.

Since we are concerned with discrete values of $\mathbf{x}_1$, $\mathbf{x}_2$ (rather than treating them as continuous functions), the technique is termed *discrete dynamic programming*. As we note, the characteristic of the technique is, that once we determine the optimum of a given stage for a variety of feeds, these values will remain fixed and unaltered by the optimization procedures performed on the upstream units.

A more colloquial expression of the guiding principle of dynamic programming is the exhortation, "Do not cry over spilt milk," or more appropriately, we may quote the Gilbertian definition proposed by Aris [3]: "If you do not do the best you can with what you happen to have got, you will never do the best you might have done with what you should have had."

Dynamic programming encompasses a large, growing field, with an extensive literature of its own. However, in this chapter, our attention will be confined to *discrete* dynamic programming, with application to staged systems.

Let us consider the serial system depicted in Fig. 5.1, the problem defined by Eqs. 5.2.1 to 5.2.5. In order to apply the principle of optimality, we must first find the optimum conditions for the last stage and then work our way backward until the entire sequence has been optimized.

In order to illustrate this procedure, let us define a number of useful parameters and work through a computational scheme.

Let us define

$$\mathbf{u}_n^*(\mathbf{x}_{n-1}) \equiv \text{the } \textit{optimal} \text{ value of the control variables } \mathbf{u}_n \text{ for feed } \mathbf{x}_{n-1}$$

$$P_n(\mathbf{x}_{n-1}) \equiv \text{the } \textit{optimal} \text{ profit for stages, } n, n+1, \ldots, N \text{ for feed } \mathbf{x}_{n-1} \text{ to the } n\text{th stage}$$

$$\equiv \underset{\mathbf{u}_n, \mathbf{u}_{n+1}, \ldots, \mathbf{u}_N}{\text{Max}} \left[\sum_{i=n}^{N} F_i(\mathbf{x}_{i-1}, \mathbf{u}_i) + G(\mathbf{x}_N) \right] \tag{5.3.1}$$

Now the optimization of the last stage for a discrete set (say m) of possible feeds (since we will not know the actual feed until the N stage problem is solved) requires the maximization with respect to $\mathbf{u}_N$ to yield m values of

$$P_N(\mathbf{x}_{N-1}) = \underset{\mathbf{u}_N}{\text{Max}} \, [F_N(\mathbf{x}_{N-1}, \mathbf{u}_N) + G(\mathbf{x}_N)] \tag{5.3.2}$$

as well as m values of $\mathbf{u}_N^*(\mathbf{x}_{N-1})$. These m quantities could be stored in tables or empirically fitted to interpolation formulae.

The optimization of the last two stages requires a maximization with respect to $\mathbf{u}_{N-1}$ for m discrete feeds $\mathbf{x}_{N-2}$ to yield

$$\begin{aligned} P_{N-1}(\mathbf{x}_{N-2}) &\equiv \underset{\mathbf{u}_{N-1},\mathbf{u}_N}{\text{Max}} \, [F_{N-1}(\mathbf{x}_{N-2}, \mathbf{u}_{N-1}) + F_N(\mathbf{x}_{N-1}, \mathbf{u}_N) + G(\mathbf{x}_N)] \\ &= \underset{\mathbf{u}_{N-1}}{\text{Max}} \, [F_{N-1}(\mathbf{x}_{N-2}, \mathbf{u}_{N-1}) + P_N(\mathbf{x}_{N-1})] \end{aligned} \tag{5.3.3}$$

where the second term is evaluated from the stored values of P_N and the transformation Eq. 5.2.1

$$\mathbf{x}_{N-1} = \mathbf{f}_{N-1}(\mathbf{x}_{N-2}, \mathbf{u}_{N-1}) \tag{5.3.4}$$

at each step of the maximization. The stored information in $P_N(\mathbf{x}_{N-1})$ allows us to reduce the two-stage optimization problem to a single problem in the variable $\mathbf{u}_{N-1}$.

Following this procedure from back to front, at any stage n, all one has to do is to solve the single-stage problem given by

$$\begin{aligned} P_n &\equiv \underset{\mathbf{u}_n,\mathbf{u}_{n+1},\ldots,\mathbf{u}_N}{\text{Max}} \left[\sum_{i=n}^{N} F_i(\mathbf{x}_{i-1}, \mathbf{u}_i) + G(\mathbf{x}_N) \right] \\ &= \underset{\mathbf{u}_n}{\text{Max}} \, [F_n(\mathbf{x}_{n-1}, \mathbf{u}_n) + P_{n+1}(\mathbf{x}_n)] \end{aligned} \tag{5.3.5}$$

because the influence of the decision $\mathbf{u}_n$ on the profit for stages $n + 1 \cdots N$ can be found from the previously calculated quantity, $P_{n+1}(\mathbf{x}_n)$ and the transformation Eq. 5.2.1. The optimal controls $\mathbf{u}_n^*(\mathbf{x}_{n-1})$ are stored at each stage for the m sets of feeds $\mathbf{x}_{n-1}$.

When one reaches stage 1, at the front of the sequence, the N stage optimal policy can be found for only one feed if that is desired, or for a discrete set of feeds with little additional effort.

Let us illustrate these ideas using a simple example.

EXAMPLE 5.3.1. Let us consider the leaching tank sequence developed in Example 5.2.1 and apply dynamic programming to determine the optimal operating policy for the system when the parameters are

$$\begin{aligned} k^\circ &= 2 \text{ hr}^{-1}/\text{ft}^3 & a_3 &= \$20/\text{ft}^2 \text{ hr} \\ T_r &= 40^\circ\text{C} & a_4 &= \$2.5/^\circ\text{C/hr} \\ T^* &= 80^\circ\text{C} & a_5 &= \$0.03/\text{hp-hr} \\ a_1 &= \$2.0 \times 10^4/\text{hr} & x_0 &= 0.2 \text{ wt fraction} \\ a_2 &= \$0.8 \times 10^4/\text{hr} \end{aligned}$$

TABLE 5.3.1 STAGE 4

x_3	$P_4(x_3)$ ($/hr)	u_{14}(hr)	u_{24}(°C)	u_{34}(hp)	x_4
0.0100	−115	0.175	41.72	12.24	0.000216
0.0095	−115	0.171	41.70	12.23	0.000213
0.0090	−114	0.167	41.67	12.01	0.000210
0.0085	−114	0.163	41.64	11.88	0.000206
0.0080	−114	0.159	41.61	11.75	0.000203
0.0075	−114	0.154	41.58	11.62	0.000199
0.0070	−113	0.150	41.55	11.47	0.000196
0.0065	−113	0.145	41.52	11.32	0.000192
0.0060	−113	0.140	41.48	11.16	0.000188
0.0055	−112	0.135	41.44	10.98	0.000183
0.0050	−112	0.129	41.40	10.79	0.000179
0.0045	−112	0.123	41.36	10.59	0.000174
0.0040	−111	0.117	41.31	10.37	0.000168
0.0035	−111	0.110	41.26	10.12	0.000162
0.0030	−110	0.103	41.20	9.85	0.000156
0.0025	−110	0.095	41.13	9.53	0.000149
0.0020	−109	0.085	41.06	9.15	0.000140
0.0015	−108	0.075	40.96	8.67	0.000130
0.0010	−107	0.061	40.84	8.03	0.000118
→					
0.0005	−106	0.042	40.65	6.96	0.000102

Solution. Let us assume a range of feed values x_{n-1} at each stage and determine the optimal operating conditions. At each stage this means the optimal selection of $\mathbf{u}_n$ (residence time, temperature, and power input) for each feed state x_{n-1}.

For the last stage, 20 values of the feed x_3 were chosen and the optimization performed for each using the Powell conjugate direction search technique of Chapter 3 (Program BOTM, cf. Appendix C). The results are shown in Table 5.3.1.

By going to stage 3 and optimizing to produce

$$P_3(x_2) = \underset{\mathbf{u}_3}{\text{Max}}\,[F_3(x_2, \mathbf{u}_3) + P_4(x_3)] \tag{5.3.6}$$

we obtain the results shown in Table 5.3.2. As at the previous stage, the optimization was performed using a multivariable search for each value of x_2. The value of $P_4(x_3)$ was found by interpolating in Table 5.3.1.

TABLE 5.3.2 STAGE 3

x_2	$P_3(x_2)$ (\$/hr)	u_{13}(hr)	u_{24}(°C)	u_{33}(hp)	x_3
0.1000	−220	0.119	41.32	10.43	0.004135
0.0950	−220	0.117	41.31	10.36	0.004032
0.0900	−219	0.116	41.30	10.32	0.003864
0.0850	−219	0.114	41.29	10.26	0.003750
0.0800	−219	0.112	41.27	10.18	0.003637
0.0750	−219	0.111	41.26	10.14	0.003470
0.0700	−219	0.109	41.24	10.06	0.003341
0.0650	−218	0.106	41.23	9.97	0.003216
0.0600	−218	0.104	41.20	9.87	0.003093
0.0550	−218	0.102	41.19	9.82	0.002897
0.0500	−218	0.099	41.17	9.71	0.002757
0.0450	−217	0.096	41.14	9.58	0.002620
0.0400	−217	0.094	41.13	9.51	0.002401
0.0350	−217	0.090	41.10	9.36	0.002244
0.0300	−216	0.086	41.06	9.17	0.002092
0.0250	−216	0.083	41.04	9.06	0.001827
0.0200	−215	0.077	40.99	8.79	0.001651
0.0150	−214	0.073	40.95	8.62	0.001304
0.0100	−213	0.063	40.86	8.13	0.001130
→					
0.0050	−212	0.054	40.77	7.64	0.000720

Repeating this process for stage 2, where we wish to maximize

$$P_2(x_1) = \underset{\mathbf{u}_2}{\text{Max}}\,[F_2(x_1, \mathbf{u}_2) + P_3(x_2)] \tag{5.3.7}$$

we obtain Table 5.3.3.

Because the feed to stage 1 is specified, the maximization

$$P_1(x_0) = \underset{\mathbf{u}_1}{\text{Max}}\,[F_1(x_0, \mathbf{u}_1) + P_2(x_1)] \tag{5.3.8}$$

need only be carried out for the known feed $x_0 = 0.2$. This produces the final optimal result shown in Table 5.3.4, where interpolation in Tables 5.3.1 to 5.3.3 (cf. arrow) was used to determine the optimal policies for stages 2 to 4.

The optimal policy is seen to produce a profit of \$2378/hr (excluding fixed costs) by using only ~3 min residence time in each tank and a uniform temperature of ~41°C with about 7.5 hp of agitation. This accomplishes better than 99.9% recovery of the available zinc.

TABLE 5.3.3 STAGE 2

x_2	$P_2(x_1)$ ($/hr)	u_{12}(hr)	u_{22}(°C)	u_{32}(hp)	x_2
0.2000	−321	0.0757	40.97	8.71	0.017184
0.1900	−321	0.0744	40.96	8.65	0.016758
0.1800	−321	0.0731	40.95	8.59	0.016339
0.1700	−320	0.0716	40.93	8.52	0.015922
0.1600	−320	0.0751	40.96	8.68	0.013912
0.1500	−320	0.0738	40.95	8.62	0.013418
0.1400	−320	0.0722	40.94	8.55	0.012945
0.1300	−320	0.0705	40.92	8.47	0.012486
0.1200	−320	0.0686	40.91	8.38	0.012037
0.1100	−319	0.0664	40.89	8.27	0.011591
0.1000	−319	0.0640	40.86	8.16	0.011146
0.0900	−319	0.0719	40.94	8.53	0.008386
0.0800	−319	0.0691	40.91	8.40	0.007922
0.0700	−318	0.0659	40.88	8.25	0.007469
0.0600	−318	0.0620	40.85	8.06	0.007017
0.0500	−317	0.0575	40.80	7.83	0.006577
0.0400	−317	0.0520	40.75	7.53	0.006077
→					
0.0300	−316	0.0453	40.68	7.14	0.005568
0.0200	−315	0.0361	40.59	6.56	0.005031
0.0100	−314	0.0199	40.39	5.24	0.004851

The reader is advised to consult Aris [3] for further examples of the application of this technique to chemical reactors, multistage compressors, and so forth.

From the computational algorithm and the example considered previously, one can see that dynamic programming allows one to decompose an N stage serially structured optimization problem into N single-stage optimization problems, which can be solved sequentially. This decomposition is done at the expense of having to *imbed* the desired optimization problem into

TABLE 5.3.4 OPTIMIZATION FOR $x_0 = 0.2$

Stage n	x_{n-1}	$P(x_{n-1})$ ($/hr)	u_{1n}(hr)	u_{2n}(°C)	u_{3n}(hp)	x_n
1	0.2000	2378	0.048	40.7	7.32	0.0339
2	0.0339	−317	0.048	40.7	7.32	0.0058
3	0.0058	−212	0.055	40.8	7.72	0.0008
4	0.0008	−107	0.054	40.8	7.66	0.0001

a wider class of problems, that is, the optimization of the system for a number of possible feeds.

In spite of these extra calculations, dynamic programming has some distinct computational advantages. If we were to decide naively to obtain all the maxima by an exhaustive search for 10 levels of each decision variable and to determine the optimum first-stage feed, by selecting 10 levels of each state variable, one can calculate the effort required to optimize an N stage problem with s state variables and r decision variables for each stage.

By using an exhaustive search on the entire N stage problem at once, it is seen that N^r decisions must be made for each selection of feed, so that 10^{Nr+s} calculations of the objective must be made to determine the optimum by our 10 level experimental design approach.

By only adding the strategy of dynamic programming and still using an exhaustive search for the single-stage optimizations, one sees that only $N10^{r+s}$ objective function evaluations are required. This means that dynamic programming is superior by a factor of $10^{(N-1)r}/N$. Aris [3] has calculated that for $N = 5$, $r = 3$, and $s = 1$, this comparison can result in a fraction of a cent computing cost for dynamic programming (using a 10 level exhaustive search to solve the single-stage problems) to solve the problem, versus a billion dollar expense associated with using a 10 level exhaustive search without the dynamic programming strategy.

Obviously, the more efficient multidimensional search algorithms discussed in Chapter 3 will perform much better than the exhaustive search. However, the dynamic programming strategy can often lead to a significant reduction of the amount of computation needed by these algorithms.

There is a very severe limitation to dynamic programming which arises when the number of state variables becomes large. At each stage in the dynamic programming optimization, the optimum must be found for a sufficient number of feed states to allow the subsequent interpolation. Thus the storage requirements must necessarily increase greatly with the number of state variables. If one wishes to use 10 levels of each state variable, then for an N stage problem with s state variables and r decisions per stage, the storage requirements are $Nr10^s$ for $\mathbf{u}_n^*(\mathbf{x}_{n-1})$ and 10^s for $P_n(\mathbf{x}_{n-1})$. Thus the total storage required just for these data corresponds to $(Nr + 1)10^s$ locations. To see what astronomical levels this requirement may reach, let us consider the case with $r = 2$, $N = 5$.

s	Memory locations
1	110
2	1100
3	11,000
4	110,000
5	1,100,000

This effect, which Bellman [2] called the "curse of dimensionality" is the chief disadvantage of dynamic programming. Perhaps equally troublesome is the s dimensional interpolation that must be done at each stage in order to use the tables of data $P_n(\mathbf{x}_{n-1})$, $\mathbf{u}_n^*(\mathbf{x}_{n-1})$.

Since the optimization is done stagewise, only the $P_n(\mathbf{x}_{n+1})$ need be kept in memory at any time and the other material may be put on disk. However, even this information increases exponentially with s.

Several approaches have been suggested for getting around the "curse of dimensionality," although none is completely successful. Curve fitting the data to interpolation formulae is a possibility; however, in s dimensions this can be difficult and the number of coefficients needed may well approach the number of data points in order to get the required accuracy. Other schemes are discussed by Aris [3] and by Larson [4].

Example 5.3.1 dealt with a rather simple problem, as our objective was to illustrate the application of dynamic programming. Let us close this section with an example taken from the "real world" dealing with the optimal scheduling of rolling mills.

EXAMPLE 5.3.2.

THE OPTIMAL SCHEDULING OF A REVERSING STRIP MILL†

We wish to devise an optimal schedule for the operation of a reversing hot strip mill, the physical description of which is given in the following.

Description of the Process

The essential features of the process are shown in Fig. 5.5, where it is seen that the ingot enters the *roughing mill* with an initial thickness of h_1 and upon passing through the mill this thickness is reduced to h_2.

The mill itself consists of two driven rolls with the top roll adjustable in the vertical plane to allow variation of the roll gap. After the slab has passed through the rolls, the mill is reversed and then the strip is driven through the rolls in the opposite direction.

Our objective is then to roll a slab from an initial thickness h_1 to a final thickness h_n in a discrete, odd number of passes, so that the total rolling time is minimized.

The need for the odd number of passes arises from the requirement that the strip must leave the system in the same direction as it first entered the mill.

In the following, this objective will be expressed in a quantitative form,

† This example is based on the M.S. thesis of R. Bayko, in the Dept. of Electrical Engineering, University of Toronto (1970). The authors wish to acknowledge the help and permission of Mr. Bayko and Prof. H. W. Smith for using this material.

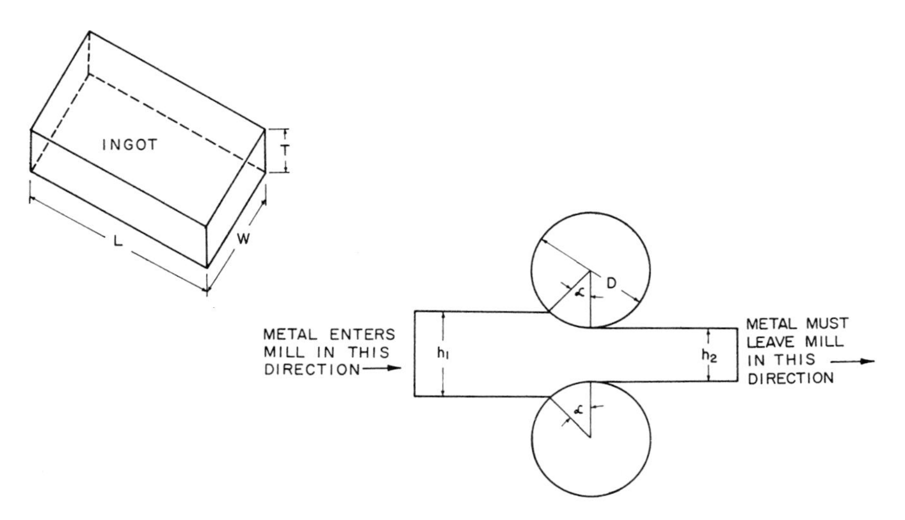

Fig. 5.5 Schematic illustration of the rolling process.

but before we can proceed further, we need the modeling equations for the system, which will provide the constraints that have to be observed.

Modeling Equations

The modeling equations to be used for describing the problem will be based on empirical data. For a fundamental treatment of rolling phenomena and further details of the rolling process the reader is referred to the literature [5, 6]. The empirical relationships to be quoted subsequently will help to define the basic concepts of rolling. We note, furthermore, that this example provides an interesting illustration of the fact that optimization can be readily carried out through the use of such empirical relationships.

Let us proceed now by defining some of the basic parameters that will appear in the constraints to be observed.

Limits on Inlet and Exit Velocity, Maximum Velocity, and Acceleration

The mill is so designed that the strip enters and leaves the system at a fixed velocity $v = v_e$.

In addition it is specified that there exists a maximum allowable velocity, v_a, dependent on the system parameters which cannot be exceeded and there is a limit on the acceleration and deceleration of the strip during its passage through the mill. For the purpose of calculation we shall consider that the

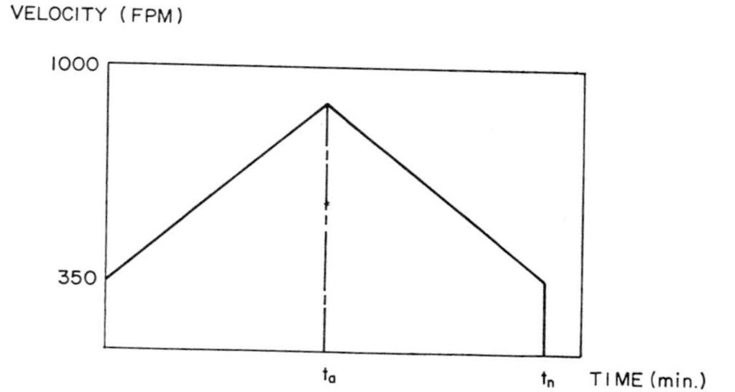

Fig. 5.6 Plot of velocity against time, for conditions where the maximum velocity has not been attained.

strip accelerates and decelerates at a uniform, fixed rate. Furthermore, let us assign the following numerical values to these parameters.

$$v_e = 350 \text{ ft/min}$$

$$\left|\frac{dv}{dt}\right| = 350 \text{ ft/min sec}$$

where v_a may be found from Figs. 5.8 or 5.9.

Figure 5.6 shows a velocity versus time plot for conditions where the retention time is such that the maximum allowable velocity is not attained. In contrast, Fig. 5.7 depicts a system where the maximum allowable velocity would have been exceeded; thus in this instance three distinct regions are observed: uniform acceleration, uniform velocity, and uniform deceleration.

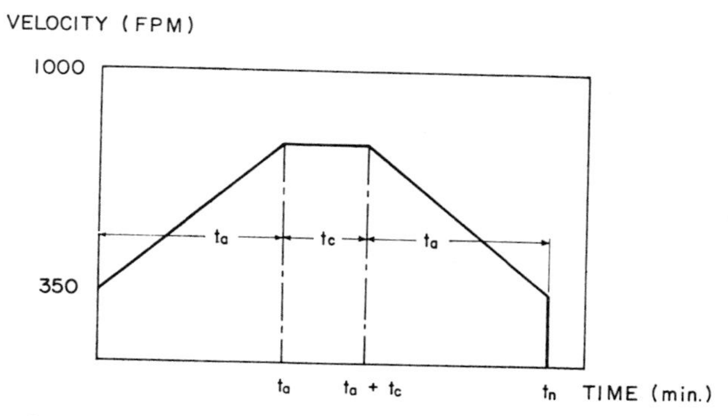

Fig. 5.7 Plot of velocity against time—the plateau corresponds to the maximum velocity.

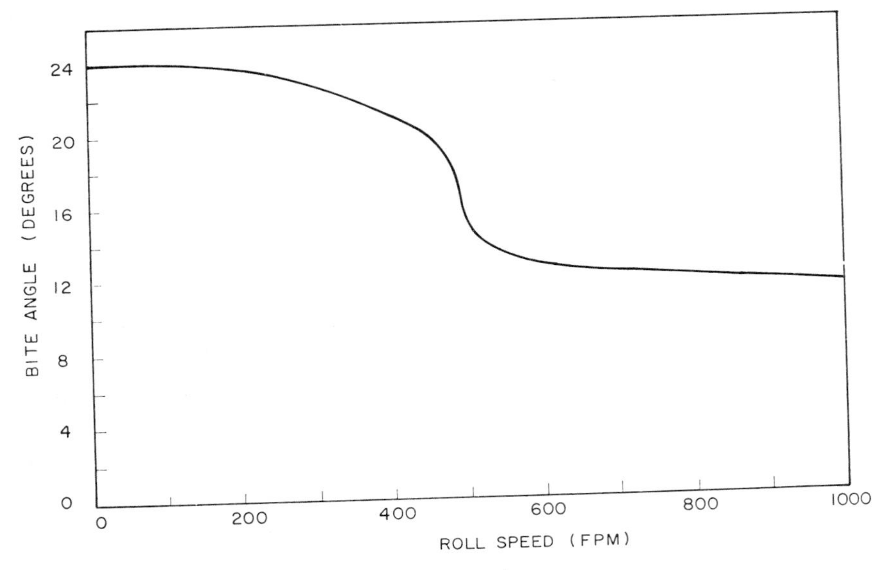

Fig. 5.8 Plot of the bite angle against the roll speed.

Bite Angle

The bite angle α has been shown in Fig. 5.5. There exists a relationship between the maximum allowable roll speed and the bite angle, which for the particular system considered here is given in Fig. 5.8. As will be shown subsequently, the bite angle is also a parameter in determining the total roll force.

Draft

The draft is defined as the reduction in width per pass, that is, $(h_1 - h_2)$ or $(h_{n-1} - h_n)$, the units of which will be inches in this particular example.

Because the draft is directly related to the bite angle (this relationship is shown in Fig. 5.10), the maximum allowable velocity and the draft are interrelated and the empirical relationship used in the present example is shown in Fig. 5.9.

Rolling Torque

The rolling torque is defined as

$$T_{\text{roll}} = 135\ DA\ \Delta H \text{ (lb-ft)} \tag{5.3.10}$$

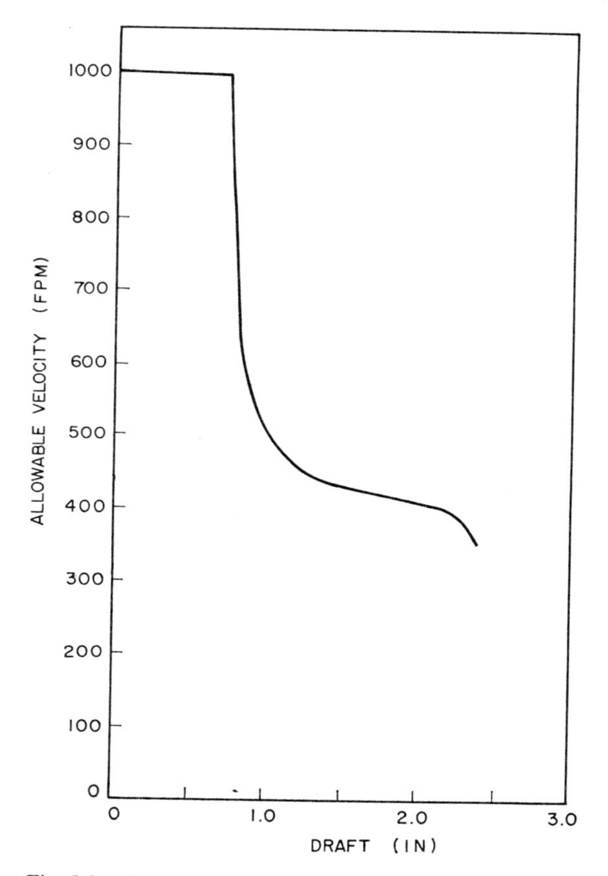

Fig. 5.9 Plot of the allowable velocity against the draft.

where

$$\Delta H = \text{incremental hp-hr/ton pass}$$
$$A = \text{exit area (in.}^2\text{)}$$
$$D = \text{roll diameter, in.}$$

The elongation caused by the rolling, E, and the power requirement, ΔH, are related, and for the system considered this relationship may be put in the following form.

$$\Delta H = 2.4(E - 1.0); \qquad \text{for} \quad E \leq 1.5$$

$$\Delta H = 1.2 + 2.4(E - 1.5); \qquad \text{for} \quad 1.5 \leq E \leq 2.0 \qquad (5.3.11)$$

and

$$\Delta H = -1.0 + 11.3 \log_{10}(E); \qquad \text{for} \quad E \geq 2.0$$

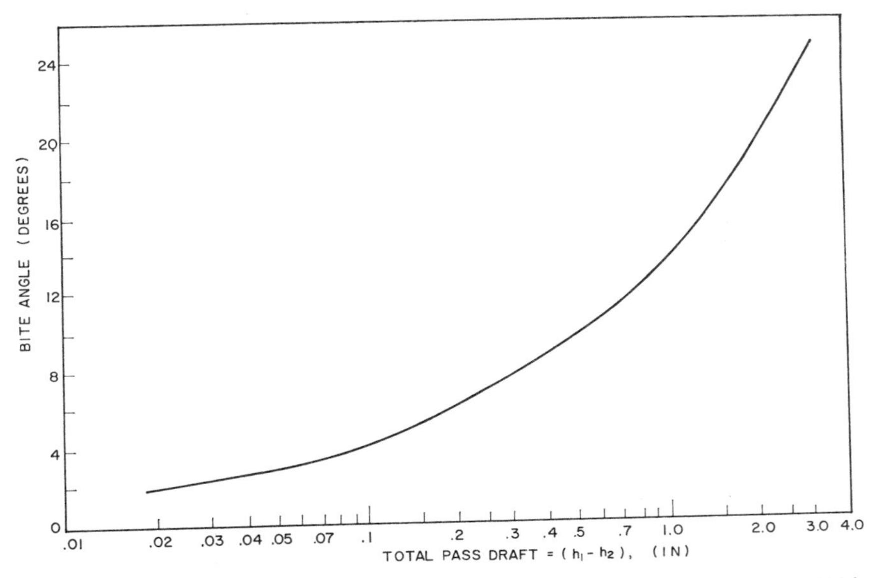

Fig. 5.10 Bite angle as a function of the total reduction in a pass for a roll diameter of 36 in.

We note that Eq. 5.3.11 represents an empirical expression valid for this particular case, where E, the elongation, is measured in inches.

For other systems the corresponding expression would have to be deduced from the "power curves," that is, from a plot of hp-hr/ton against the total elongation.

Available Torque

The available torque must be computed from an experimentally obtained relationship between the rpm of the rolls and the torque. For the system under consideration this relationship may be put in the following form.

$$T_{\text{avail}} = 2.56 \times 10^6 \text{ lb ft}, \quad \text{for rpm} < 55.0$$

$$T_{\text{avail}} = \left[2.25 - \frac{0.4(\text{rpm} - 55)}{50} \times \frac{0.9 \times 55}{\text{rpm}}\right] \times 10^6 \text{ lb ft}, \quad \text{for } 55 \leq \text{rpm} \leq 100 \tag{5.3.12}$$

and

$$T_{\text{avail}} = 0.9 \times 10^6 \text{ lb ft}, \quad \text{for rpm} > 100$$

Clearly, the torque stipulated through the setting of the draft, rpm, and the like cannot exceed T_{avail}.

Roll Force

Finally, the roll force F may be calculated from the following relationship

$$F = 0.0108R\left[\frac{\pi}{2}\sqrt{\frac{h_2}{R}}\tan^{-1}\sqrt{\frac{r}{1-r}} - \frac{\pi\alpha}{4} - \ln(1.05) + 0.5\ln\left(\frac{h_1}{h_2}\right)\right] \tag{5.3.13}$$

where

$$F = \text{roll force/in. width, } 10^6 \text{ lb}$$
$$R = \text{roll radius, in.}$$
$$r = \frac{h_1 - h_2}{h_1}$$

and α is the bite angle which may be obtained from Fig. 5.10. To prevent damage to the mill, one must specify the maximum allowable roll force, which is set in the present instance as

$$F < 3.6 \times 10^6 \text{ lb}$$

Let us now put the problem in our standard form. Let

$$x_n = \text{the thickness exiting stage } n = h_n$$
$$d_n = \text{the draft at stage } n = h_{n-1} - h_n$$

Then our state transformation is

$$x_n = x_{n-1} - d_n \tag{5.3.14}$$

with constraints on the thickness and the draft at each stage of the following form.

1. The velocity v takes the form shown in Figs. 5.6 and 5.7.

$$v = v_e + a\left|\frac{dv}{dt}\right| = 350(1 + a) \tag{5.3.15}$$

where

$$a = \begin{cases} 1; & 0 \leq t \leq t_a \\ 0; & t_a \leq t \leq t_a + t_c \\ -1; & t_a + t_c \leq t \leq t_n \end{cases}$$

The velocity is constrained by the relationships shown in Figs. 5.8 to 5.9.

$$v \leq v_a = f_1(d_n) \tag{5.3.16}$$

In addition, for practical reasons, v is not allowed to exceed 1000 ft/sec.

2. The rolling torque T_{roll} is given by Eq. 5.3.10, which becomes

$$T_{\text{roll}} = f_2(x_{n-1}, d_n) \tag{5.3.17}$$

and is constrained by the torque available (given by Eq. 5.3.12), which takes the form

$$T_{\text{roll}} \leq T_{\text{avail}} = f_3(v) \tag{5.3.18}$$

3. The roll force is given by Eq. 5.3.13, which takes the form

$$F_{\text{roll}} = f_4(x_{n-1}, d_n) \tag{5.3.19}$$

and is constrained (to prevent damage to the mill) so that

$$F_{\text{roll}} \leq 3.6 \times 10^6 \tag{5.3.20}$$

It is stressed that these relationships are empirical expressions, the precise form of which relates to the one particular application considered. The actual numerical values of the parameters chosen are likely to vary with different systems, although the general nature (if not the precise form) of these relationships could well remain the same.

The Objective Function

In words, our objective is to find a sequence of roll settings, $d_1, d_2, \ldots, d_n$, which will cause a reduction from a specified initial thickness, h_1, to a final thickness, h_n, in a minimum total rolling time, without violating the constraints given previously. The possible tradeoffs are between large drafts, d_n, which reduce the number of passes needed, but increase the rolling time per pass due to v_a constraints, and so forth, and smaller drafts which require more passes, but shorter rolling times per pass. The total rolling time can then be written

$$I = \sum_{n=1}^{N} t_n \tag{5.3.21}$$

where N, the number of passes, must be odd.

Before the objective can be expressed in a quantitative form, we have to calculate the total time per pass. This takes the form

$$t_n = 2t_a + t_c + t_f \tag{5.3.22}$$

where t_a, t_c are shown in Figs. 5.6–5.7 and t_f is the constant fixed time needed for the resetting of the screws in order to effect the reversal of the mill. For the present case we shall set $t_f = 0.033$ min. Thus our optimization problem is to minimize I given by Eq. 5.3.21 subject to the stage transformation (Eq. 5.3.14) and constraints (Eq. 5.3.15 to 5.3.20).

Solution. By beginning at the last stage and working backward, one can determine the optimum draft d_n satisfying the constraints at each stage, for a range of feed conditions. These are put into a table and interpolation used for each successive optimization. Because the constraints do not couple the

TABLE 5.3.5 OPTIMUM ROLLING SCHEDULE

Pass	x_{n-1}	Draft d_n	x_n	Roll force ($\times 10^6$)(lb)
1	24.0	2.2	21.8	2.91
2	21.8	2.2	19.6	2.95
3	19.6	2.2	17.4	3.00
4	17.4	2.2	15.2	3.06
5	15.2	2.2	13.0	3.14
6	13.0	1.9	11.1	2.96
7	11.1	1.9	9.2	3.08
8	9.2	1.9	7.3	3.25
9	7.3	1.9	5.4	3.52
10	5.4	1.6	3.8	3.52
11	3.8	1.3	2.5	3.60
12	2.5	0.7	1.8	2.74
13	1.8	0.7	1.125	3.38
Total time = 54.6 sec				

stages, no iterations need be done to satisfy them. This optimization continues in a back-to-front manner until the initial thickness, h_0, is a possible feed to the first stage. If the total number of stages is odd at this point, the optimal solution has been found. Otherwise, one more stage must be added and the optimum found for the feed thickness h_0.

This approach was used by Bayko and Smith for a raw stock with

$$\text{width} = 60.0 \text{ in.}$$

$$\text{entry thickness} = 24.0 \text{ in.}$$

$$\text{weight} = 10 \text{ tons}$$

The desired final thickness was 1.125 in.

Table 5.3.5 shows the optimum rolling schedule found by this dynamic programming algorithm. During the first few passes, the draft is constrained by the maximum allowable velocity. For the middle passes, the available torque constrains the draft. Toward the end of the rolling, the rolling force constrains the optimal draft. The optimal solution requires 54.6 sec and 13 passes to reach the final desired thickness.

These results can be compared with Table 5.3.6, which shows a typical rolling schedule used in an actual operating mill with the same parameters. We note that 13 passes are required to reach the final thickness and during

TABLE 5.3.6 REPRESENTATIVE NONOPTIMAL ROLLING SCHEDULE

Pass	x_{n-1}	Draft d_n	x_n	Roll force ($\times 10^6$)(lb)
1	24.0	1.5	22.5	2.325
2	22.5	2.3	20.2	3.084
3	20.2	2.2	18.0	3.046
4	18.0	2.2	15.8	3.103
5	15.8	2.1	13.7	3.083
6	13.7	1.9	11.8	2.972
7	11.8	1.8	10.0	2.967
8	10.0	1.8	8.2	3.097
9	8.2	1.7	6.5	3.164
10	6.5	1.6	4.9	3.300
11	4.9	1.75	3.15	4.049[a]
12	3.15	1.15	2.0	3.720[a]
13	2.0	0.875	1.125	4.073[a]
Total time = 59.3 sec				

[a] Roll force constraint violated.

the last three passes, the rolling force constraint was violated.† In spite of this, the rolling time was 59.3 sec, ~9% longer than the optimum.

This rather detailed example clearly illustrates the two-fold advantages of an optimization study: (1) to increase the productivity or profit of the operation, and (2) to devise means of insuring that all constraints can be met. Clearly, the calculated risk of exceeding the roll force constraint in order to prevent having to use two more passes in the rolling would not have been necessary if the optimal draft sequence were used. As can be seen, the economic penalties for *not* doing an optimization study were quite severe.

5.4 THE DISCRETE MAXIMUM PRINCIPLE

An alternative approach to dynamic programming, which is particularly well suited to the optimization of serial structures, makes use of the *discrete maximum principle*. The discrete maximum principle is loosely related to the *continuous maximum principle*, which will be discussed in Chapter 6, although as will be shown, it would be misleading to draw too close an analogy.

† Shortly after this study was completed, we understand that the actual rolling mill fractured due to the excessive rolling force applied.

Let us illustrate the use of the discrete maximum principle by considering the optimization of the serial structure depicted in Fig. 5.1 and represented by Eqs. 5.2.1 to 5.2.5. Thus we have

$$\mathbf{x}_n = \mathbf{f}_n(\mathbf{x}_{n-1}, \mathbf{u}_n) \qquad n = 1, 2, \ldots, N \tag{5.2.1}$$

which relates the output from stage n to the feed (i.e., output from the previous stage) and to the control variables $\mathbf{u}_n$ operating on the nth stage. We shall assume that the feed $\mathbf{x}_0$ is specified as $\mathbf{c}_0$.

Let us consider further that in addition to the equality constraints represented by Eq. 5.2.1, the operative inequality constraints may be expressed as

$$\mathbf{u}_{n*} \leq \mathbf{u}_n \leq \mathbf{u}_n^* \qquad n = 1, 2, 3, \ldots, N \tag{5.4.1}$$

which is simply a statement that the control variables are bounded between the limits $\mathbf{u}_{n*}$ and $\mathbf{u}_n^*$.

We note here that the types of constraints contained in Eqs. 5.2.3 to 5.2.4 could also be treated, but we shall not do so for the present.

As discussed in Section 5.2 the objective function for the whole cascade may be put in the following form.

$$I(\mathbf{x}_1, \mathbf{x}_2, \ldots, \mathbf{x}_n, \mathbf{u}_1, \mathbf{u}_2, \ldots, \mathbf{u}_n) = \sum_{n=1}^{N} F_n(\mathbf{x}_{n-1}, \mathbf{u}_n) + G(\mathbf{x}_N) \tag{5.2.5}$$

Let us form the *Lagrangian* for the system composed of Eqs. 5.2.1, 5.4.1, and 5.2.5. On recalling this approach from Section 2.3 and on noting that the objective function is given by Eq. 5.2.4 and that the equality constraints are defined by Eq. 5.2.1, we have the following Lagrangian,

$$L = \sum_{n=1}^{N} \{F_n(\mathbf{x}_{n-1}, \mathbf{u}_n) - \boldsymbol{\lambda}_n^T[\mathbf{x}_n - \mathbf{f}_n(\mathbf{x}_{n-1}, \mathbf{u}_n)]\} + G(\mathbf{x}_N) - \boldsymbol{\lambda}_0^T(\mathbf{x}_0 - \mathbf{c}_0) \tag{5.4.2}$$

where $\boldsymbol{\lambda}_n$ is the Lagrange multiplier associated with the equality constraints (Eq. 5.2.1), and $\boldsymbol{\lambda}_0$ is associated with the feed condition to stage 1.

On applying the Kuhn-Tucker stationary conditions of Section 2.4 for an optimum, we have

$$\frac{\partial L}{\partial \mathbf{x}_n} = \frac{\partial[F_{n+1}(\mathbf{x}_n, \mathbf{u}_{n+1}) + \boldsymbol{\lambda}_{n+1}^T \mathbf{f}_{n+1}(\mathbf{x}_n, \mathbf{u}_{n+1})]}{\partial \mathbf{x}_n} - \boldsymbol{\lambda}_n^T = \mathbf{0}$$

$$n = 0, 1, 2, \ldots, N-1 \tag{5.4.3a}$$

$$\frac{\partial L}{\partial \mathbf{x}_N} = \left(\frac{\partial G}{\partial \mathbf{x}_N} - \boldsymbol{\lambda}_N^T\right) = \mathbf{0} \tag{5.4.3b}$$

$$\frac{\partial L}{\partial \mathbf{u}_n} = \frac{\partial[F_n(\mathbf{x}_{n-1}, \mathbf{u}_n) + \boldsymbol{\lambda}_n^T \mathbf{f}_n(\mathbf{x}_{n-1}, \mathbf{u}_n)]}{\partial \mathbf{u}_n} = \mathbf{0} \qquad n = 1, 2, \ldots, N \tag{5.4.4}$$

for unconstrained $\mathbf{u}_n$, and the quantity H_n

$$H_n \equiv F_n(\mathbf{x}_{n-1}, \mathbf{u}_n) + \boldsymbol{\lambda}_n{}^T \mathbf{f}_n(\mathbf{x}_{n-1}, \mathbf{u}_n) \tag{5.4.5}$$

taking a maximum at the upper and lower bounds (Eq. 5.4.1). The quantity H_n can be loosely interpreted as the Lagrangian for the nth stage. Thus in a more compact form, our results can be restated in the following theorem.

THEOREM. (*Discrete Maximum Principle*) In order for $\mathbf{u}_n$, $n = 1, 2, \ldots, N$ to be the optimal control variable for the N stage problem defined by Eq. 5.2.1, 5.2.5, and 5.4.1, $\mathbf{u}_n$ must satisfy the conditions

$$\frac{\partial H_n}{\partial \mathbf{u}_n} = \mathbf{0} \qquad n = 1, 2, \ldots, N \tag{5.4.6}$$

for unconstrained $\mathbf{u}_n$; at constrained values, $\mathbf{u}_n$ has to maximize H_n (see *Example 2.4.1*). The Lagrange multipliers (sometimes termed adjoint variables), $\boldsymbol{\lambda}_n$, are defined by

$$\boldsymbol{\lambda}_n{}^T = \frac{\partial H_{n+1}}{\partial \mathbf{x}_n}; \qquad \boldsymbol{\lambda}_N{}^T = \frac{\partial G}{\partial \mathbf{x}_N} \qquad n = 0, 1, 2, \ldots, N-1 \tag{5.4.7}$$

Although there are a number of very interesting theoretical points to be explored (e.g., cf. [7–9]) here, we shall only discuss some computational algorithms and applications based on this theorem.

Equations 5.2.1, 5.4.6, and 5.4.7 represent a set of difference equations in $\mathbf{u}_n$, $\mathbf{x}_n$, $\boldsymbol{\lambda}_n$, which can be solved in a number of ways.

Control Vector Iteration

If $\mathbf{u}_n$ were known, Eqs. 5.2.1 and 5.4.7 are uncoupled and can be solved sequentially. This suggests the following computational algorithm.

1. Guess $\mathbf{u}_n$, $n = 1, 2, \ldots, N$.
2. Solve Eq. 5.2.1 forward from $\mathbf{x}_0$ to produce $\mathbf{x}_n$, $n = 1, 2, \ldots, N$.
3. Solve Eq. 5.4.7 backward from $\lambda_N{}^T = \partial G/\partial \mathbf{x}_N$ to produce $\boldsymbol{\lambda}_n$, $n = 1, 2, \ldots, N$.
4. Correct the guess of $\mathbf{u}_n$ by

$$\delta \mathbf{u}_n = \epsilon \left(\frac{\partial H_n}{\partial \mathbf{u}_n}\right)^{T\bullet} \qquad \epsilon > 0, \qquad n = 1, 2, \ldots, N \tag{5.4.8}$$

5. Go back to step 2 and iterate until convergence is obtained.

The validity of the correction (Eq. 5.4.8) can be seen by considering the first order variation in the objective (Eq. 5.2.5) under the constraint (Eq. 5.2.1).

$$\begin{aligned}\delta I &= I(\mathbf{u}_1, \mathbf{u}_2, \ldots, \mathbf{u}_N)_{\text{new}} - I(\mathbf{u}_1, \mathbf{u}_2, \ldots, \mathbf{u}_N)_{\text{old}} \\ &= [L(\mathbf{u}_1, \mathbf{u}_2, \ldots, \mathbf{u}_N)_{\text{new}} - L(\mathbf{u}_1, \mathbf{u}_2, \ldots, \mathbf{u}_N)_{\text{old}}]_{5.2.1\ \text{satisfied}}\end{aligned} \tag{5.4.9}$$

or

$$\delta I = \sum_n \left\{\frac{\partial L}{\partial \mathbf{u}_n}\,\delta\mathbf{u}_n + \frac{\partial L}{\partial \mathbf{x}_n}\,\delta\mathbf{x}_n + \frac{\partial L}{\partial \boldsymbol{\lambda}_n}\,\delta\boldsymbol{\lambda}_n\right\} + \text{higher order terms} \tag{5.4.10}$$

Now if Eq. 5.2.1 is satisfied, $\partial L/\partial \boldsymbol{\lambda}_n = \mathbf{0}$, and if Eq. 5.2.7 is satisfied, $\partial L/\partial \mathbf{x}_n = \mathbf{0}$; thus at each iteration in our algorithm

$$\delta I \simeq \sum_n \frac{\partial L}{\partial \mathbf{u}_n}\,\delta\mathbf{u}_n = \sum_n \frac{\partial H_n}{\partial \mathbf{u}_n}\,\delta\mathbf{u}_n \tag{5.4.11}$$

is valid to a first-order approximation. As we have seen in Chapter 3 (cf. the method of steepest ascent), the choice of $\delta\mathbf{u}_n$ given in Eq. 5.4.8 provides the largest value to δI.

Boundary Condition Iteration

If Eq. 5.4.6 could be solved to produce the explicit expression

$$\mathbf{u}_n = \mathbf{g}_n(\mathbf{x}_{n-1}, \boldsymbol{\lambda}_n) \tag{5.4.12}$$

then this could be substituted into Eqs. 5.2.1 and 5.4.7 to produce coupled equations in $\mathbf{x}_n$, $\boldsymbol{\lambda}_n$. Since the boundary conditions on these difference equations are split (i.e., $\mathbf{x}_0$ and $\boldsymbol{\lambda}_N$ are given), an iterative procedure is needed to solve them:

1. Guess $\mathbf{x}_N$ (or $\boldsymbol{\lambda}_0$).
2. Solve Eqs. 5.2.1 and 5.4.7 backward (forward) to yield $\mathbf{x}_n$, $\boldsymbol{\lambda}_n$, $n = 0, 1, \ldots, N$.
3. Compare $\mathbf{x}_0$ ($\boldsymbol{\lambda}_N$) computed with $\mathbf{x}_0$ ($\boldsymbol{\lambda}_n$) given and correct the guess of $\mathbf{x}_N$ ($\boldsymbol{\lambda}_0$).
4. Iterate until the boundary conditions are satisfied.

There are a large number of techniques (cf. [10]) for solving the two-point boundary value problem. Graphical methods work well for simple problems, while so called "shooting" techniques can be used in tackling complex systems.

Quasilinearization

Another approach for solving the two-point boundary value problem resulting from the use of Eq. 5.4.12 in Eqs. 5.2.1 and 5.4.7 is the method of quasilinearization [11]. The algorithm is as follows.

1. Guess $\mathbf{x}_n$, $\boldsymbol{\lambda}_n$, $n = 1, 2, \ldots, N$.
2. Linearize Eqs. 5.2.1 and 5.4.7 about this guess and solve the resulting linear two-point boundary value problem for an approximate solution satisfying the boundary conditions.
3. Relinearize about this new approximate solution and iterate until convergence is obtained.

Let us now review the characteristics of each of these approaches. The boundary condition iteration and quasilinearization approaches have several disadvantages. In case of complex problems it may be difficult or impossible to invert Eq. 5.4.6 to produce Eq. 5.4.12, thus making the approach useless.

A second problem arises from the fact that one must obtain perfect convergence in order to have a useful answer. This problem occurs because at each iteration an optimal control is produced—for the wrong problem. For boundary condition iteration, the boundary conditions are wrong and for quasilinearization the system equations are in error. If one were interested in the optimal controls for a wide variety of feed conditions, $\mathbf{x}_0$, then the boundary condition iteration approach could be used to advantage. The control vector iteration algorithm has the advantage that Eq. 5.4.12 is not necessary, and the right problem (but with a suboptimal result) is solved at each iteration. This latter property allows one to stop the computation at any iteration and have a suboptimal but useful solution.

In order to illustrate the discrete maximum principle, let us consider the following example.

EXAMPLE 5.4.1. We wish to effect the separation of a mixture composed of solid A and B, each with a given particle size distribution about the mean. Let the mean particle diameter and variance of material A be denoted by $\bar{d}_A$ and $\sigma_A{}^2$, respectively, and let the corresponding quantities for material B be denoted by $\bar{d}_B$ and $\sigma_B{}^2$.

Both these materials may be considered to have a log normal distribution, that is,

$$y(d) = \frac{1}{\log \sigma\sqrt{2\pi}} \exp\left\{-\left[\frac{(\log d - \log \bar{d})^2}{2 \log^2 \sigma}\right]\right\} \tag{5.4.13}$$

(where $y(d)$ is the fraction of the material with diameter between d and $d + \Delta d$), which holds throughout the processing sequence, although both

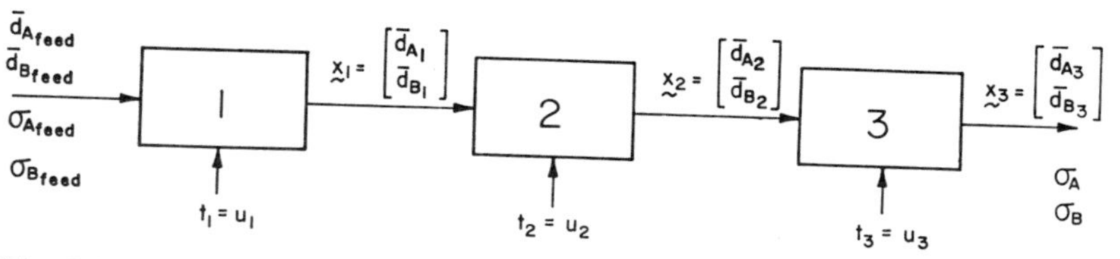

Fig. 5.11 Schematic representation of the grinding circuit in Example 5.4.1.

the mean particle size and the standard deviation will, of course, change during the processing.

Let us consider that the grindability of the two materials is different, and that our objective is to arrange the grinding of the solid mixture in a series of three continuously operating ball mills, so that we obtain the maximum differentiation in the size of the two solids. This would then facilitate easy separation by differential settling or elutriation.

The general scheme of operation is sketched in Fig. 5.11. In a continuous grinder the relationship between particle size, standard deviation, and holding time is assumed to be

$$\bar{d}_{A\,\text{out}} = \frac{1}{\dfrac{1}{\bar{d}_{A\,\text{in}}} + C_A t} + d_{A\infty} \tag{5.4.14}$$

for species A and

$$\bar{d}_{B\,\text{out}} = \frac{1}{\dfrac{1}{\bar{d}_{B\,\text{in}}} + C_B t} + d_{B\infty} \tag{5.4.15}$$

for species B.

Here t is the holding time in the ball mill, and the quantities C_A, C_B relate to the grindability of the material and may also depend on the type of equipment used. It is noted here that Eq. 5.4.14 and 5.4.15 are used for the purpose of illustration; in tackling real optimization problems relating to grinding, the relationships analogous to Eqs. 5.4.14 and 5.4.15 may well have to be deduced individually for each material and each type of grinding equipment [12, 13].

In the course of grinding, the standard deviation too will change, and let us express this as follows.

$$\sigma_{A\,\text{out}} = \sigma_{A\,\text{in}} \left(\frac{\bar{d}_{A\,\text{out}}}{\bar{d}_{A\,\text{in}}} \right)^{0.9} \tag{5.4.16}$$

$$\sigma_{B\,\text{out}} = \sigma_{B\,\text{in}} \left(\frac{\bar{d}_{B\,\text{out}}}{\bar{d}_{B\,\text{in}}} \right)^{0.85} \tag{5.4.17}$$

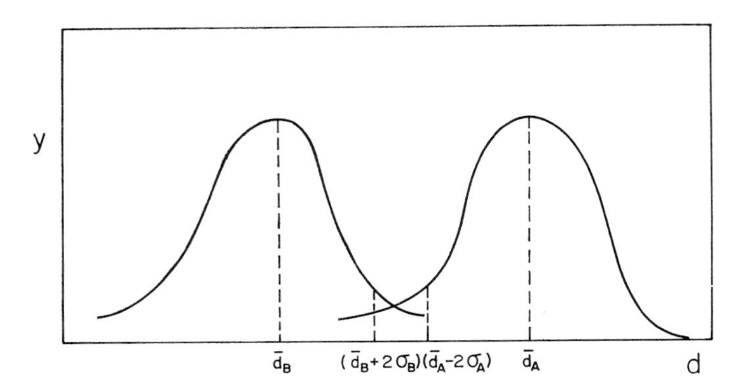

Fig. 5.12 Sketch of the expected particle size distribution of materials A and B leaving the third grinding stage, in Example 5.4.1.

Given this information, let us establish the optimum holding times in each of the three ball mills shown in Fig. 5.11, which gives us the maximum separation between species A and B.

We shall choose the feed material to have $\bar{d}_{A\,\text{feed}} = 2$ cm, $\bar{d}_{B\,\text{feed}} = 2.5$ cm, $\sigma_{A}{}^{2}{}_{\text{feed}} = 2$ cm^2, $\sigma_{B}{}^{2}{}_{\text{feed}} = 2.7$ cm^2. In addition, the grinding constants for each ball mill are

$$d_{A\infty} = 5 \times 10^{-4} \text{ cm}$$
$$d_{B\infty} = 3 \times 10^{-4} \text{ cm}$$
$$C_{A1} = 3 \text{ hr}^{-1} \qquad C_{B1} = 11 \text{ hr}^{-1}$$
$$C_{A2} = 4 \text{ hr}^{-1} \qquad C_{B2} = 12 \text{ hr}^{-1}$$
$$C_{A3} = 2 \text{ hr}^{-1} \qquad C_{B3} = 10 \text{ hr}^{-1}$$

Solution. The expected distribution of particle sizes leaving the third grinding stage is sketched in Fig. 5.12.

Since material B is easier to grind, as evidenced by the larger C values, one would expect $\bar{d}_B$ to be significantly smaller than $\bar{d}_A$ at the outlet of the third ball mill. In specifying a maximum separation, it is sufficient to optimize in terms of the mean particle size and the standard deviation, since for the given distribution function (Eq. 5.4.13), these quantities fully define the system.

In order to put this in the form of Eq. 5.2.1 let us define

$$\mathbf{x}_n = \begin{bmatrix} x_{1n} \\ x_{2n} \end{bmatrix} = \begin{bmatrix} \bar{d}_A \\ \bar{d}_B \end{bmatrix}_n \qquad n = 0, 1, 2, 3 \tag{5.4.18}$$

a vector which contains the average particle diameter exiting each stage. Normally, we would have to carry the standard deviations σ_A, σ_B along as state variables, but the particular form of Eqs. 5.4.16 and 5.4.17 allows us to represent them solely in terms of the particle diameters. The state equations then become

$$x_{1n} = \frac{1}{\dfrac{1}{x_{1n-1}} + C_{An} u_n} + d_{A\infty}; \qquad x_{10} = 2 \qquad n = 1, 2, 3 \quad (5.4.19)$$

$$x_{2n} = \frac{1}{\dfrac{1}{x_{2n-1}} + C_{Bn} u_n} + d_{B\infty}; \qquad x_{20} = 2.5 \qquad n = 1, 2, 3 \quad (5.4.20)$$

where u_n is the residence time in the nth stage and must be chosen optimally.

Since our objective is to maximize the separation between species A and B, we can express the standard deviation exiting the last stage by

$$\sigma_{A3} = \sqrt{2}\left(\frac{x_{1N}}{2}\right)^{0.9} \quad (5.4.21)$$

$$\sigma_{B3} = \sqrt{2.7}\left(\frac{x_{2N}}{2.5}\right)^{0.85} \quad (5.4.22)$$

and an objective which maximizes the separation while minimizing equipment size is

$$I = \frac{\left[x_{1N} - 2\sqrt{2}\left(\dfrac{x_{1N}}{2}\right)^{0.9}\right] - \left[x_{2N} + 2\sqrt{2.7}\left(\dfrac{x_{2N}}{2.5}\right)^{0.85}\right]}{x_{1N}} - \alpha \sum_{n=1}^{3} u_n \quad (5.4.23)$$

which must be maximized with respect to u_n.

The parameter α will be chosen at several levels to illustrate the type of separation possible at various values of equipment cost. This technique is often used to good advantage when it is impossible to define a very precise economic objective function. In order to put the objective in the form of Eq. 5.2.5, we let

$$G(\mathbf{x}_3) = \frac{\left[x_{13} - 2\sqrt{2}\left(\dfrac{x_{13}}{2}\right)^{0.9}\right] - \left[x_{23} + 2\sqrt{2.7}\left(\dfrac{x_{23}}{2.5}\right)^{0.85}\right]}{x_{13}} \quad (5.2.24)$$

and

$$F_n = -\alpha u_n \qquad n = 1, 2, 3 \quad (5.4.25)$$

so that

$$I = F_1 + F_2 + F_3 + G(\mathbf{x}_3) \quad (5.4.26)$$

Now we are in a position to apply the discrete maximum principle and a gradient procedure to the problem.

The relevant equations are

$$H_n = -\alpha u_n + \lambda_{1n}\left(\frac{1}{\dfrac{1}{x_{1n-1}} + C_{An}u_n} + d_{A\infty}\right) + \lambda_{2n}\left(\frac{1}{\dfrac{1}{x_{2n-1}} + C_{Bn}u_n} + d_{B\infty}\right)$$

$$n = 1, 2, 3 \quad (5.4.27)$$

where $\boldsymbol{\lambda}_n = \begin{bmatrix} \lambda_{1n} \\ \lambda_{2n} \end{bmatrix}$ is a vector of Lagrange multipliers for the nth stage, which are given by

$$\lambda_{1n} = \lambda_{1n+1}\left(\frac{1}{\dfrac{1}{x_{1n}} + C_{An+1}u_{n+1}}\right)^2\left(\frac{1}{x_{1n}}\right)^2 \qquad n = 0, 1, 2 \qquad (5.4.28)$$

$$\lambda_{2n} = \lambda_{2n+1}\left(\frac{1}{x_{2n} + C_{Bn+1}u_{n+1}}\right)^2\left(\frac{1}{x_{2n}}\right)^2 \qquad (5.4.29)$$

and

$$\lambda_{13} = \frac{1 - 0.9\sqrt{2}\left(\dfrac{x_{13}}{2}\right)^{-0.1}}{x_{13}} - \frac{G(\mathbf{x}_3)}{x_{13}}$$

$$\lambda_{23} = \frac{-\left[1 + \dfrac{1.7}{2.5}\sqrt{2}\left(\dfrac{x_{23}}{2.5}\right)^{-0.15}\right]}{x_{13}} \qquad (5.4.30)$$

Let us follow the control vector iteration procedure which was described earlier.

1. Guess $u_1 = 1$, $u_2 = 1$, $u_3 = 1$.
2. Solve Eqs. 5.4.19 and 5.4.20 to produce (for the first iteration)

$$x_{11} = 0.286 \qquad x_{12} = 0.134 \qquad x_{13} = 0.081$$
$$x_{21} = 0.0880 \qquad x_{22} = 0.0428 \qquad x_{23} = 0.030$$

and $I = -5.269$, where $\alpha = 1$ for this case.

3. Solve Eq. 5.4.28 to 5.4.30 backwards to produce (for the first iteration)

$$\lambda_{10} = 2.99 \times 10^{-2}, \qquad \lambda_{11} = 1.463, \qquad \lambda_{12} = 6.728$$
$$\lambda_{13} = 1.876, \qquad \lambda_{20} = -5.61 \times 10^{-3}, \qquad \lambda_{21} = -4.554$$
$$\lambda_{22} = -19.25, \qquad \lambda_{23} = -39.27$$

TABLE 5.4.1 OPTIMIZATION RESULTS FOR EXAMPLE 5.4.1

α	Iteration	Mill 1			Mill 2			Mill 3					
		x_{11}	x_{21}	u_1	x_{12}	x_{22}	u_2	x_{13}	x_{23}	u_3	σ_A	σ_B	I
1.0	1	1.1570	0.5759	0.1216	0.9561	0.4384	0.0455	0.9566	0.4387	0.0	0.7282	0.3743	−1.9307
	2	0.9417	0.4064	0.1875			0.0			0.0	0.7187	0.3512	−1.8891
	3	0.9417	0.4064	0.1875			0.0			0.0	0.7187	0.3512	−1.8891
0.1	1	0.6400	0.2329	0.3546			0.0			0.0	0.5079	0.2190	−1.6675
	2	0.6400	0.2329	0.3546			0.0			0.0	0.5079	0.2190	−1.6675
	3												
0.01	1	0.2526	0.0769	1.1553			0.0			0.0	0.2205	0.0855	−1.7284
	2	0.5766	0.2031	0.4119			0.0			0.0	0.4624	0.1950	−1.6332
	3	0.5766	0.2031	0.4119			0.0			0.0	0.4624	0.1950	−1.6332

4. Correct u_n by

$$\begin{aligned}\delta u_n &= \epsilon \frac{\partial H_n}{\partial u_n} \\ &= \epsilon\left[-\alpha - \lambda_{1n}\left(\frac{1}{\dfrac{1}{x_{1n-1}} + C_{An}u_n}\right)^2 C_{An} - \lambda_{2n}\left(\frac{1}{\dfrac{1}{x_{2n-1}} + C_{Bn}u_n}\right)^2 C_{Bn}\right]\end{aligned} \tag{5.4.31}$$

so that for the first iteration

$$u_1 = 1 - \epsilon[\alpha - 0.027]$$
$$u_2 = 1 - \epsilon[\alpha + 0.056]$$
$$u_3 = 1 - \epsilon[\alpha + 0.251]$$

where ϵ is chosen to maximize the improvement in I.

5. Return to step 2 and iterate.

The result of each iteration is shown in Table 5.4.1 where it is seen that the method stopped improving I in each case after two or three iterations. The problem was solved from several other starting points and the optimum found was very close to that shown in Table 5.4.1.

Examination of the optimal policy shown in the table shows that it is always optimal to use only a single grinding unit and the optimal residence time increases with decreasing cost of the residence time, α. For the case of $\alpha = 0.01$, the average particle size of material A is taken from 2 cm to 0.577 cm with the standard deviation being reduced from 1.41 cm to 0.462 cm. Similarly, the average particle size of material B is taken from 2.5 cm to 0.203 cm and with the standard deviation being reduced from 1.64 cm to 0.195 cm. Thus it is seen that although we have effected a degree of separation, there is still considerable overlap of the particle size distribution.

5.5 MULTILEVEL OPTIMIZATION

Previous discussion in this chapter was restricted to techniques for problems with serial structure. This section describes approaches which can treat problems with an arbitrary interacting structure. These techniques are related to the more general concept of multilevel or hierarchical systems optimization discussed so well by Mesarovic et al. [14]. It is perhaps useful to outline briefly some of the ideas of multilevel systems theory at this point.

Let us consider the system structure shown in Fig. 5.13. The process to be examined lies within the large black box and could be optimized by neglecting the structure within the box. However, suppose that the mathematical and possibly the topological structure of the system is such that it could be

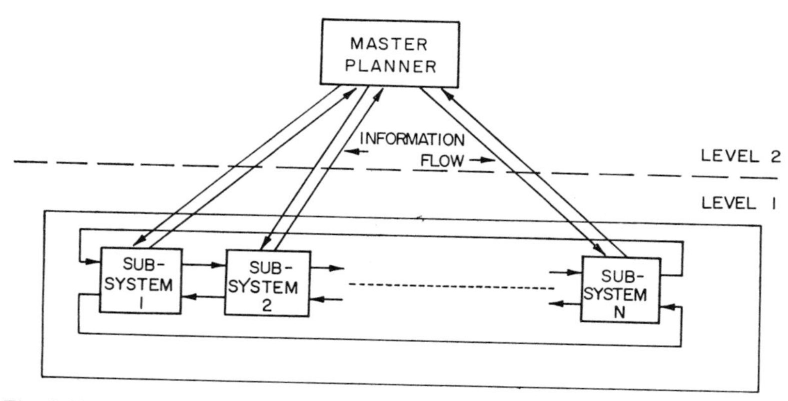

Fig. 5.13 A two-level decomposible structure.

decomposed into N subsystems, which may be linked so that information flows between them; in addition information is also exchanged between the subsystem and the master planner at level 2. As an example, the subsystems may be parallel manufacturing operations using a common pool of resources under control of the master planner at level 2.

A metallurgical example would be the operation of a large jobbing foundry, which uses a number of arc furnaces to produce a range of products. These arc furnaces are fed with both bought scrap and with the scrap which is generated internally. In considering the optimization of the system as a whole, we may proceed by considering the operation of the individual furnaces, that is, the first level. However, in the optimization of the whole operation, the "master planner" has to take into account the total scrap flow picture, the availability and size of given furnaces, and the additional complications arising from making special batches.

There are, of course, numerous examples of decomposable, hierarchical structures in the chemical process industry; these could include, for example, an oil refinery, a petrochemical complex, or a polymer processing operation. Here again, at the lower level we would be concerned with the operation of the individual processing units, such as distillation columns, heat exchangers, reactors, and mixer settlers. At the higher level we would have to reexamine the operation of these units within the perspective of their interrelations through connecting streams, common resources, and so on.

Further illustration of large, decomposable systems may be found in Examples 4.4.3 and 4.4.4. These were concerned with finding the optimum material flow to units (e.g., blast furnaces, basic oxygen furnaces, etc.) the operating characteristics of which were fixed. If we were to relax these constraints, that is, make the operation of the individual units subject to

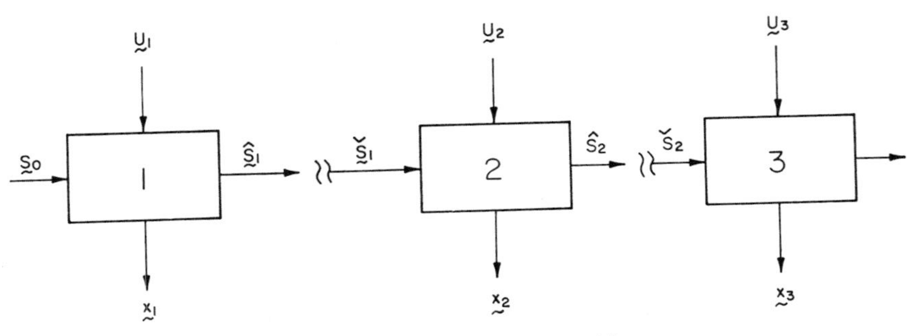

Fig. 5.14 Illustration of feasible on nonfeasible decomposition.

optimization, then the flow of the streams between these units would represent the interactions which have to be determined by the "master planner." A metallurgical problem of this type will be discussed in detail in Chapter 8, and a worked example, dealing with an oil refinery will be presented in Section 6 of this chapter.

By considering structured problems in this context, it is possible to devise strategies for the master planner (called coordination strategies) under which the subsystems can be optimized independently, and yet produce the optimal design for the entire system. In the following we shall place each decomposition optimization technique discussed into the framework of this multilevel approach. For purposes of classification, we shall identity the coordination strategies which result in a feasible solution of the system equations at each iteration as *feasible strategies* and all other strategies as *nonfeasible strategies*. While the concepts of feasible and nonfeasible strategies will be discussed in detail subsequently, at this stage it may be worthwhile to illustrate the difference between these two approaches.

Figure 5.14 shows a sketch of a system, which may be decomposed into three units. The input into the first unit is $\mathbf{s}_0$ and the characteristics of the material leaving unit 1 and entering unit 2 are defined by the vectors $\hat{\mathbf{s}}_1$ and $\check{\mathbf{s}}_1$, respectively. Analogous considerations apply to the flow of material between stages 2 and 3. The quantities $\mathbf{u}_1$, $\mathbf{u}_2$, . . . denote the control vectors for the various stages.

In a *feasible decomposition* procedure, we set $\hat{\mathbf{s}}_i, = \check{\mathbf{s}}_i$ in each of the iterations; thus the continuity of the vector is satisfied when going from one stage to the other.

In the *nonfeasible decomposition* procedure the constraint $\hat{\mathbf{s}}_i = \check{\mathbf{s}}_i$ is not met in the initial iterations, but toward the completion of the computation, $\hat{\mathbf{s}}_i$ and $\check{\mathbf{s}}_i$ are forced to coincide.

Both these techniques have their advantages and disadvantages, as will be discussed subsequently.

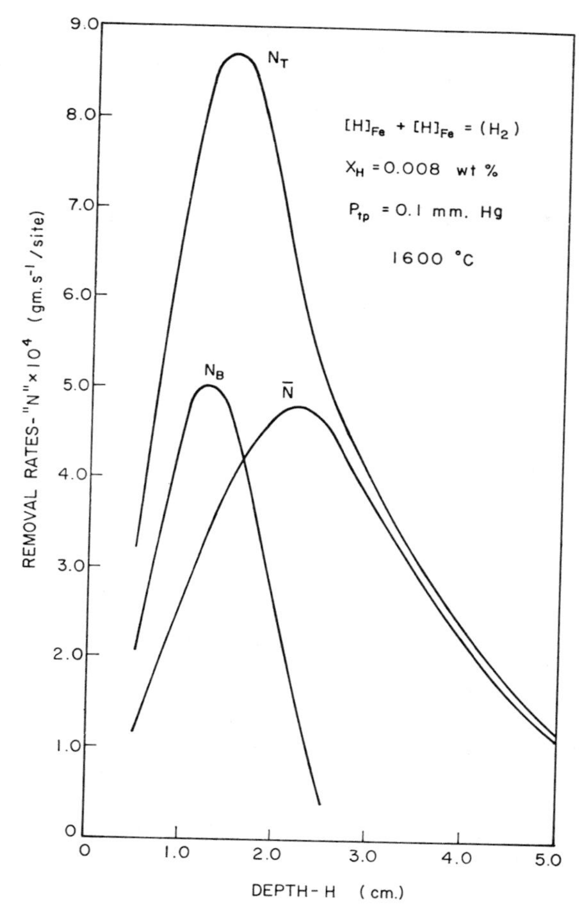

Fig. 5.15 The optimal depths for N_B, $\bar{N}$ and N_T in Example 3.1.2.

EXAMPLE 5.5.1. In order to illustrate the concepts of feasible and non-feasible decomposition, let us examine the ladle degassing problem of Example 3.1.2. This very simple problem has only one variable to adjust, the ladle depth. Suppose we define subsystem 1 as the mass transfer rate due to initial growth and rise of the bubble and subsystem 2 as the rate of mass transfer due to agitation of the free surface. Examination of Fig. 5.15 shows that if each subsystem were optimized separately with respect to bed depth, then subsystem 1 would choose $h_1 = 1.1$ cm and subsystem 2 would choose $h_2 = 2.2$ cm. Because physical reality requires these depths to be equal, then the master planner shown in Fig. 5.13 must impose additional constraints on the subsystems.

If the master planner chooses a *feasible strategy* by defining h himself, then there is no optimization to be performed by the subsystems in this simple example and only the unique value of N_B (for subsystem 1) and $\bar{N}$ (for subsystem 2) will be returned. The master planner then sums these and chooses new values of h until the optimum is reached. If the master planner uses a *nonfeasible strategy*, for example by imposing a cost C on the value of h_2 to system 2 (to give him incentive to reduce h_2) and a credit on the value of h_1 to subsystem 1 (to provide incentives for increasing h_1), then subsystem 1 would maximize

$$I_1 = N_B(h_1) + Ch_1$$

and subsystem 2 would maximize

$$I_2 = \bar{N}(h_2) - Ch_2$$

Clearly there is a particular price C which causes $h_1 = h_2$, so that the total objective

$$I = I_1 + I_2 = N_B(h_1) + \bar{N}(h_2) + C(\overset{\nearrow 0}{h_1 - h_2})$$

is the optimum total mass transfer rate, N_T, found at $h_1 = h_2 = 1.4$ cm. To summarize, this *nonfeasible* computational scheme is

1. Master planner chooses C.
2. Subsystems 1 and 2 maximize I_1 and I_2, respectively.
3. Master planner chooses new values of C to cause $h_1 \rightarrow h_2$.
4. Return to step 2 and iterate.

In effect, the master planner chooses at each iteration the Lagrange multiplier, C, for the constraint $h_1 = h_2$.

Even if all of the relevant concepts are not yet completely clear to the reader, this example should serve to contrast *feasible* and *nonfeasible* decomposition strategies. Let us now return to our general discussion of feasible and nonfeasible decomposition strategies.†

Having broadly defined these categories, let us consider details of approaches which exploit the structure of the problem.

The simplest type of constrained optimization problem is the linear programming (LP) problem:

$$\text{Max } \{I = \mathbf{c}^T\mathbf{x}\} \tag{5.5.1}$$

subject to

$$\mathbf{Ax} = \mathbf{b}; \qquad \mathbf{x} \geq \mathbf{0} \tag{5.5.2}$$

† Some of this material was published earlier in Ref. [42] and is used here with the permission of the Canadian Society of Chemical Engineering.

where slack variables have been used to convert the inequality constraints to equality constraints. If **A** has mainly block diagonal structure, then Eqs. 5.5.1 and 5.5.2 take the form

$$\text{Max}\left\{\sum_i \mathbf{c}_i{}^T\mathbf{x}_i\right\} \tag{5.5.3}$$

subject to

$$\begin{bmatrix} \mathbf{A}_1 & \mathbf{A}_2 & \cdot & \cdot & \cdot & \cdot & \cdot & \mathbf{A}_r \\ \mathbf{B}_1 & \mathbf{0} & \cdot & \cdot & \cdot & \cdot & \cdot & \mathbf{0} \\ \mathbf{0} & \mathbf{B}_2 & \cdot & \cdot & \cdot & \cdot & \cdot & \mathbf{0} \\ \mathbf{0} & \mathbf{0} & \mathbf{B}_3 & \cdot & \cdot & \cdot & \cdot & \mathbf{0} \\ \cdot & \cdot & & \cdot & & & & \cdot \\ \cdot & \cdot & & & \cdot & & & \cdot \\ \cdot & \cdot & & & & \cdot & & \cdot \\ \mathbf{0} & \mathbf{0} & & & & & & \mathbf{B}_r \end{bmatrix} \begin{bmatrix} \mathbf{x}_1 \\ \mathbf{x}_2 \\ \cdot \\ \cdot \\ \cdot \\ \cdot \\ \cdot \\ \mathbf{x}_r \end{bmatrix} = \begin{bmatrix} \mathbf{b}_0 \\ \mathbf{b}_1 \\ \cdot \\ \cdot \\ \cdot \\ \cdot \\ \cdot \\ \mathbf{b}_r \end{bmatrix} \tag{5.5.4}$$

where $\mathbf{x}_i$, $\mathbf{c}_i$, $\mathbf{b}_i$ represent the ith partition of $\mathbf{x}$, $\mathbf{c}$, $\mathbf{b}$, respectively. This formulation allows the problem to be broken up into r subproblems

$$\text{Max}\ \{\mathbf{c}_j{}^T\mathbf{x}_j\} \tag{5.5.5}$$

subject to

$$\mathbf{B}_j\mathbf{x}_j = \mathbf{b}_j \qquad j = 1, 2, \ldots, r \tag{5.5.6}$$

which can be solved independently subject to the coordinating problem that accounts for the subproblem interactions:

$$\underset{\mathbf{x}_j}{\text{Max}}\left(\sum_{j=1}^{r} \mathbf{c}_j{}^T\mathbf{x}_j\right) \tag{5.5.7}$$

subject to

$$\sum_{j=1}^{r} \mathbf{A}_j\mathbf{x}_j = \mathbf{b}_0 \tag{5.5.8}$$

If the matrices $\mathbf{A}_j$ are all zero, then the problem has no interactions and Eqs. 5.5.5 and 5.5.6 can be solved quite independently of one another. However, this case is rare because most problems have interactions. One approach to the solution of this problem, known as the Dantzig-Wolfe Decomposition Principle [15], makes use of the fact that the solution of Eqs. 5.5.7 and 5.5.8 can be obtained in terms of the vertices of the feasible regions for the subsystems (Eq. 5.5.6). The algorithm reduces to the iterative solution of the modified subproblem LP (at the first level).

$$\underset{\mathbf{x}_j}{\text{Max}}\ [\{\mathbf{c}_j - \boldsymbol{\sigma}_1\mathbf{A}_j\}\mathbf{x}_j] \tag{5.5.9}$$

subject to Eq. 5.5.6 where the fictitious price $\boldsymbol{\sigma}_1$ accounts for the interactions, and is found by taking the solutions of Eqs. 5.5.6 and 5.5.9 at the previous iteration and using them to solve an LP (at the second level) to generate a new set $\boldsymbol{\sigma}_1$. Several approaches for doing this are discussed in Ref. 16. An example of the application of the Dantzig-Wolfe Decomposition Procedure to an oil refinery operation will be discussed in the next section.

There are several decomposition approaches proposed for separable, nonlinear programs. Geoffrion [17], Bessiere and Sautter [18], and Yoshida [19] discuss the decomposition of nonlinear separable programming problems and techniques for their solution. Rosen and Ornea [20] and Ornea and Eldredge [21] discuss the feasible decomposition of nonlinear programming problems into linear subproblems by partition programming. This has the advantage that the very efficient LP algorithms can be used for subsystem optimization. The application of a decomposition approach to geometric programming problems has been recently discussed by Heymann and Avriel [22].

More general decomposition procedures, applicable to both nonlinear systems and linear systems without block diagonal structure have also been developed. These will be discussed in a general way and then the specific computational algorithms will be explored in detail.

Let us consider the general nonlinear programming problem given by Eqs. 1.2.1 to 1.2.3. Furthermore, let us assume that it is desired to divide this problem into r subsystems for optimization. This might be necessary because of storage limitations in the computer or owing to large geographical separations between the subsystem models. To accomplish this decomposition, auxiliary "interaction" variables, s_i, will be used to make the problem separable, and the formulation of the problem becomes

$$\operatorname*{Max}_{\mathbf{x}_j} \left\{ I(\mathbf{x}) = \sum_{i=1}^{r} I_i(\mathbf{x}_i, \mathbf{s}_i) \right\} \tag{5.5.10}$$

subject to

$$\mathbf{h}^i(\mathbf{x}_i, \mathbf{s}_i) = \mathbf{0} \qquad i = 1, 2 \cdots r \tag{5.5.11}$$

$$\mathbf{g}^i(\mathbf{x}_i, \mathbf{s}_i) \leq \mathbf{0} \qquad i = 1, 2 \cdots r \tag{5.5.12}$$

$$\mathbf{s}_i = \mathbf{f}_i(\mathbf{x}_j) \qquad j \neq i \qquad j, i = 1, 2 \cdots r \tag{5.5.13}$$

where the last equation defines the interaction variables, $\mathbf{s}_i$, and $\mathbf{x}_i$ is the ith partition of the vector $\mathbf{x}$.

Using the ideas of nonlinear programming, it is possible to write the Lagrangian of the problem

$$L = \sum_{i=1}^{r} \{ I_i(\mathbf{x}_i, \mathbf{s}_i) - \boldsymbol{\mu}_i^T \mathbf{g}^i(\mathbf{x}_i, \mathbf{s}_i) - \boldsymbol{\lambda}_i^T \mathbf{h}^i(\mathbf{x}_i, \mathbf{s}_i) + \boldsymbol{\pi}_i^T [\mathbf{s}_i - \mathbf{f}_i(\mathbf{x}_j)] \} \tag{5.5.14}$$

where $\boldsymbol{\mu}_i$, $\boldsymbol{\lambda}_i$, $\boldsymbol{\pi}_i$ are the Lagrange multipliers. From this formulation, it is obvious that all terms but the last can be separated so that the problem can be broken into r subsystems. However, it makes a great difference how the last term is treated. In order to illustrate this more clearly, let us apply the Kuhn-Tucker [23] necessary conditions to the problem

$$\frac{\partial L}{\partial \mathbf{x}_i} = \frac{\partial I_i}{\partial \mathbf{x}_i} - \left(\frac{\partial \mathbf{g}^i}{\partial \mathbf{x}_i}\right)^T \boldsymbol{\mu}_i - \left(\frac{\partial \mathbf{h}^i}{\partial \mathbf{x}_i}\right)^T \boldsymbol{\lambda}_i - \sum_{j \neq i}^{r} \left(\frac{\partial \mathbf{f}_j}{\partial \mathbf{x}_i}\right)^T \boldsymbol{\pi}_j = \mathbf{0} \tag{5.5.15}$$

$$\frac{\partial L}{\partial \mathbf{s}_i} = \frac{\partial I_i}{\partial \mathbf{s}_i} - \left(\frac{\partial \mathbf{g}^i}{\partial \mathbf{s}_i}\right)^T \boldsymbol{\mu}_i - \left(\frac{\partial \mathbf{h}^i}{\partial \mathbf{s}_i}\right)^T \boldsymbol{\lambda}_i^T + \boldsymbol{\pi}_i^T = \mathbf{0} \tag{5.5.16}$$

$$\left(\frac{\partial L}{\partial \boldsymbol{\mu}_i}\right)^T = \mathbf{h}^i(\mathbf{x}_i, \mathbf{s}_i) = \mathbf{0} \tag{5.5.17}$$

$$\left(\frac{\partial L}{\partial \boldsymbol{\lambda}_i}\right)^T = \mathbf{g}^i(\mathbf{x}_i, \mathbf{s}_i) = \mathbf{0} \qquad \forall \boldsymbol{\lambda}_i > \mathbf{0} \tag{5.5.18}$$

$$\left(\frac{\partial L}{\partial \boldsymbol{\pi}_i}\right)^T = \mathbf{s}_i - \mathbf{f}_i(\mathbf{x}_j) = \mathbf{0} \tag{5.5.19}$$

for $i = 1, 2, \ldots, r$ and unconstrained values of the variables. Now there are basically two ways in which the problem may be decomposed.

Feasible Decomposition

In this approach, the interaction variables $\mathbf{s}_i$ are specified at the second level and the r subsystem problems (at the first level) become

$$\underset{\mathbf{x}_j}{\text{Max}} \; \{I_i(\mathbf{x}_i, \hat{\mathbf{s}}_i)\} \tag{5.5.20}$$

subject to

$$\mathbf{h}^i(\mathbf{x}_i, \hat{\mathbf{s}}_i) = \mathbf{0} \tag{5.5.21}$$

$$\mathbf{g}^i(\mathbf{x}_i, \hat{\mathbf{s}}_i) \leq \mathbf{0} \tag{5.5.22}$$

$$\hat{\mathbf{s}}_j = \mathbf{f}_j(\mathbf{x}_i) \qquad j \neq i \tag{5.5.23}$$

where the ^ on $\hat{\mathbf{s}}_j$ reminds us that it is considered constant throughout the subsystem optimizations. The value of $\hat{\mathbf{s}}_i$ is adjusted at the second level until Eq. 5.5.16 is satisfied. One possible computational algorithm would be

(i) Choose $\hat{\mathbf{s}}_i$ at the second level.
(ii) Optimize each of the r subsystems (Eqs. 5.5.20 to 5.5.23) at the first level.

(iii) Correct $\hat{\mathbf{s}}_i$ at the second level so that

$$\hat{\mathbf{s}}_i^{\text{new}} = \hat{\mathbf{s}}_i^{\text{old}} + \epsilon\left[\frac{\partial L}{\partial \mathbf{s}_i}\right]^T \qquad \epsilon > 0$$

(iv) Go back to step (ii) and iterate.

Since the gradient $\dfrac{\partial L}{\partial \mathbf{s}_i}$ requires a knowledge of the Lagrange multipliers $\boldsymbol{\pi}_i$, $\boldsymbol{\lambda}_i$, $\boldsymbol{\mu}_i$ (which may be difficult to calculate), a nongradient direct search approach may be more desirable. One could adjust $\hat{\mathbf{s}}_i$ at each iteration so that I is increased by use of a direct search algorithm such as discussed in Chapter 3.

Nonfeasible Decomposition

In this approach we choose to fix the values of $\boldsymbol{\pi}_i$ for each subsystem and then allow each subsystem to select the optimal value of $\mathbf{s}_i$. This results in a nonfeasible solution at each iteration because Eq. 5.5.13 is violated. However, by adjusting the choice of $\boldsymbol{\pi}_i$ at the second level, feasibility is approached as the optimum is reached.

A possible algorithm for this case is

(i) Guess $\boldsymbol{\pi}_i$ at the second level.
(ii) Optimize at the first level of each r subsystems

$$\operatorname*{Max}_{\mathbf{x}_i, \mathbf{s}_i} \left\{ I_i(\mathbf{x}_i, \mathbf{s}_i) + \hat{\boldsymbol{\pi}}_i{}^T \mathbf{s}_i - \sum_{j \neq i}^{r} \hat{\boldsymbol{\pi}}_j{}^T \mathbf{f}_j(\mathbf{x}_i) \right\} \tag{5.5.24}$$

subject to

$$\mathbf{h}^i(\mathbf{x}_i, \mathbf{s}_i) = \mathbf{0} \tag{5.5.25}$$

$$\mathbf{g}^j(\mathbf{x}_i, \mathbf{s}_i) \leq \mathbf{0} \tag{5.5.26}$$

(iii) Adjust $\hat{\boldsymbol{\pi}}_i$ at the second level by

$$\hat{\boldsymbol{\pi}}_i^{\text{new}} = \hat{\boldsymbol{\pi}}_i^{\text{old}} + \epsilon \frac{\partial L^T}{\partial \boldsymbol{\pi}_i} \qquad \epsilon > 0 \tag{5.5.27}$$

(iv) Go back to step (ii) and iterate.

There are a number of other approaches possible for adjusting the values of $\mathbf{s}_i$ and $\boldsymbol{\pi}_i$ to produce the optimum. The gradient method suggested above was used by Brosilow et al. [24] and treated in great detail by Lasdon [25]. A Newton-Raphson second order corrective procedure was suggested by Bauman [26]. A third procedure used by Brosilow and Nunez [27] to optimize a catalytic cracker is the technique of adjusting $\boldsymbol{\pi}_i$ by making use of the saddle point property of the Lagrangian at the optimum. A large number of techniques for coordination have been discussed (e.g., [30–34]), and perhaps the best method has not yet been devised.

There is a very elegant and important theory concerning the methods of decomposition. However, due to limitations of space we shall only make note of some of the more important points here (cf. [25] for a more detailed discussion).

1. There is a dual problem to the one given in Eqs. 5.5.10 to 5.5.13 whose solution provides an upper bound on the optimal solution at each iteration.

2. Convergence of a convex optimization problem solved by these decomposition methods is assured only if each subproblem is also convex. Since convexity of the large problem does not imply convexity of the subproblems, it is important to check that this condition is fulfilled.

Perhaps it is useful to point out how the techniques of discrete dynamic programming and the discrete maximum principle fit into the present discussion. Aris *et al.* [33] and Wilde [34] have shown that dynamic programming can be applied to nonserial structures by using a feasible decomposition to specify the value of the cyclic loops and then applying serial dynamic programming algorithms to solve the subsystems.

The so-called "discrete maximum principle" leads to a nonfeasible technique if the state and adjoint variables are guessed at the second level and the resulting subsystem Lagrangians are made stationary with respect to the decision variables at the first level at every iteration. On the other hand, if the state and adjoint variables are determined exactly at the second level and only small corrections are made in the decision variables at the first level, then the procedure would be feasible.

One characteristic of this latter approach is that the subproblems are not optimized completely at each iteration because the internal prices (adjoint variables) are only valid for small improvements in the objective of the subsystem. Jackson [35] and Denn [36, 37] discussed the necessary framework for treating complex structures, while Lee [38, 39] has worked detailed examples illustrating a feasible algorithm resulting from the "discrete maximum principle" approach. As can be seen, discrete dynamic programming and the discrete maximum principle are special-purpose approaches for solving complex systems whose subproblems have mainly "serial" structure and thus should be considered for these applications.

5.6 CHEMICAL PROCESS EXAMPLES

At this stage it is desirable to illustrate the application of decomposition techniques using practical examples. In this section we shall select an

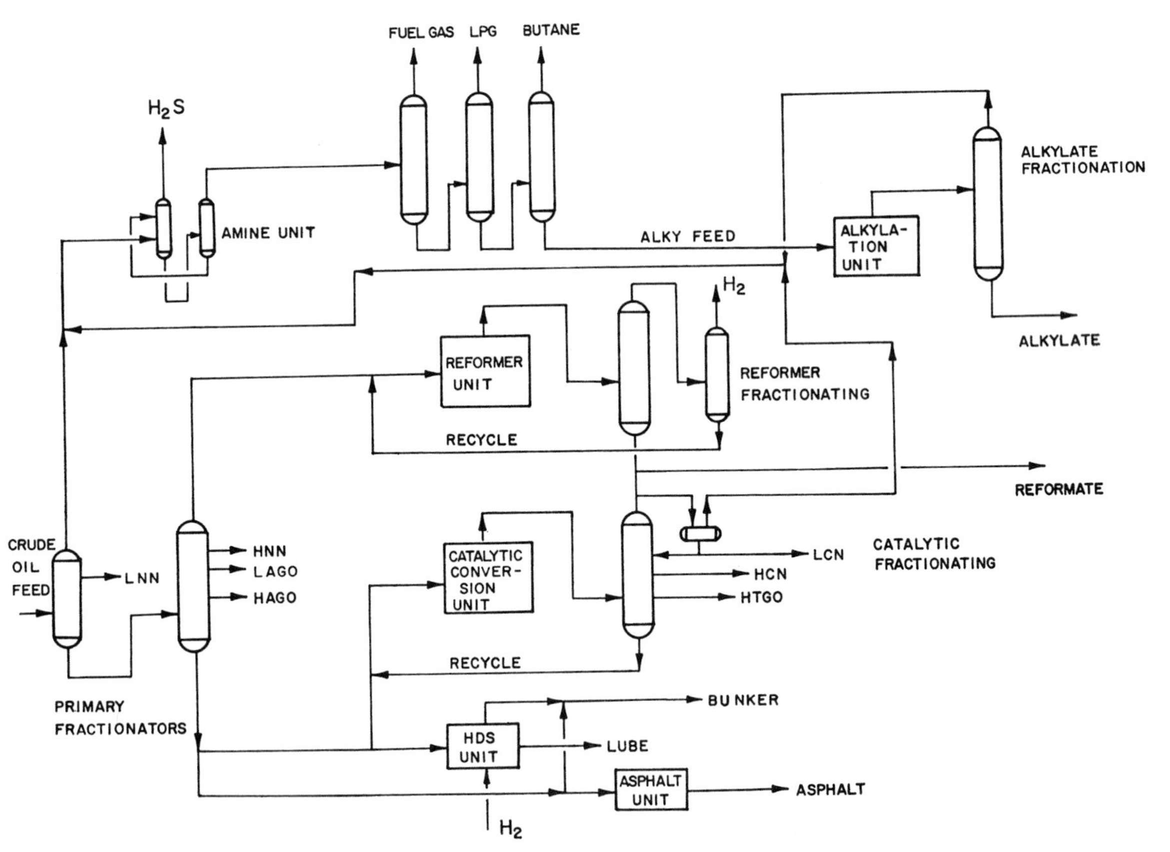

Fig. 5.16 Flow diagram for a simplified oil refinery.

example from the chemical process industry and note that a large-scale, nonlinear example dealing with a metallurgical system will be presented in Section 8.4.

Some General Observations

There have been only a limited number of applications of decomposition methods even to chemical process optimization. Brosilow et al. [24] treated three extractors in series with recycle by a nonfeasible decomposition, but found the optimization technique rather inefficient. Brosilow and Nunez [27] optimized an industrial catalytic-cracking, catalyst regenerator system by a feasible decomposition, and successfully found the known optimum. However, in neither of these studies was a computational comparison made between the decomposition approach and the black box approach. Plisken and Rzaev [40] reported on the optimization, by decomposition, of a very simple model of a styrene plant. They claimed (but gave no data to support this claim) computational efficiencies over the black box approach. Findeisen et al. [41] report the optimization, through decomposition, of a beet sugar plant. However, there is no discussion of computational efficiency, decomposition problems, and so on, in the paper.

In order to compare the programming effort and computational requirements for decomposition techniques versus the "black box" approach, two applications of these methods to chemical process optimization are discussed in detail [42].†

A Simplified Oil Refinery

A simplified linear model of an oil refinery was chosen as the first example to test the performance of the Dantzig-Wolfe Decomposition algorithm. The refinery flow sheet shown in Fig. 5.16 indicates the basic processing units available. The refinery model was obtained by breaking the two crude oils available into 17 pseudo components, $\mathbf{x}_l$, $\mathbf{x}_s$, and then expressing the 40 streams $\mathbf{w}_l$, $\mathbf{w}_s$, in the refinery in terms of these 17 pseudo components producing mass balances for each crude (light and standard)

$$\mathbf{M}_l \mathbf{w}_l = \mathbf{x}_l \tag{5.6.1}$$

$$\mathbf{M}_s \mathbf{w}_s = \mathbf{x}_s \tag{5.6.2}$$

The 48 product specifications, $\mathbf{p}$, can then be represented in terms of the stream variables $\mathbf{w}_l$, $\mathbf{w}_s$, plus 12 pseudo streams $\mathbf{w}_e$ independent of the

† Portions of these examples are used with the permission of the Canadian Society of Chemical Engineering.

crude oil type

$$\mathbf{p}(\lesseqgtr)\mathbf{A}_l\mathbf{w}_l + \mathbf{A}_s\mathbf{w}_s + \mathbf{A}_e\mathbf{w}_e \tag{5.6.3}$$

Finally, there is a constraint on the total crude oil capacity of the refinery

$$\mathbf{1}^T(\mathbf{x}_l + \mathbf{x}_s) \leq 30 \tag{5.6.4}$$

where $\mathbf{1}$ is a vector of "ones."
The optimization problem becomes

$$\underset{\mathbf{w}}{\text{Max}}\ \{I = \mathbf{c}^T\mathbf{w}\} \tag{5.6.5}$$

where the cost coefficients $\mathbf{c}$ have been chosen to reflect the product values as well as unit operating costs and $\mathbf{w}$ is the partitioned vector of dimension 92

$$\mathbf{w} = \begin{bmatrix} \mathbf{w}_l \\ \hline \mathbf{w}_s \\ \hline \mathbf{w}_e \end{bmatrix}$$

The constraints on the refinery are the total throughput constraint (Eq. 5.6.4), which can be rewritten

$$\mathbf{m}^T\mathbf{w} \leq 30 \tag{5.6.6}$$

where the 92 vector $\mathbf{m}$ is

$$\mathbf{m} = \begin{bmatrix} \mathbf{1}^T\mathbf{M}_l \\ \hline \mathbf{1}^T\mathbf{M}_s \\ \hline \mathbf{0} \end{bmatrix} \tag{5.6.7}$$

a product specification constraint

$$\mathbf{p}(\gtreqless)\mathbf{A}\mathbf{w}$$

where

$$\mathbf{A} = \left[\begin{array}{c|c|c} \mathbf{A}_l & \mathbf{0} & \mathbf{0} \\ \hline \mathbf{0} & \mathbf{A}_s & \mathbf{0} \\ \hline \mathbf{0} & \mathbf{0} & \mathbf{A}_e \end{array}\right] \tag{5.6.5}$$

and finally, the mass balances given by Eqs. 5.6.1 and 5.6.2. Thus we have 92 variables and 83 constraints for our problem.

This problem was first solved as a "black box" using the IBM MPS LP program. The solution required about 85 iterations of the simplex procedure (about 10 sec on the IBM 360/75) to produce the optimum profit of 120.45 M$/yr with 11.696 MB/yr of light crude and 18.304 MB/yr of the standard. The optimal values of the other streams can be found in Ref. 43.

Since the constraint Eq. 5.6.8 has block diagonal structure, it was natural to decompose the problem into two parts—one part for each crude type—and use the Dantzig-Wolfe decomposition procedure described previously to determine the optimum. However, it was necessary to have the auxiliary stream variables $\mathbf{w}_e$ in each subproblem. In addition, some of the constraints were particular to one type of crude so that subproblem 1 was a 52 variable, 62 constraint linear program, while subproblem 2 was a 52 variable 60 constraint linear program. The coordination at level 2 was a 4 variable, 3 constraint linear program. The decomposition procedure converged after 4 coordinations in about 30 sec computing time. At first glance this may seem excessive compared to the 10 sec required by the "black box" approach. However, this decomposition procedure was carried out with the general purpose MPS program as a subroutine solving the subproblems and with a graduate student solving the coordination LP by hand. This caused a great deal of unneeded initialization for each subproblem solution and thus caused the computing effort to be excessive. By determining the amount of superfluous initialization at each stage, it was possible to estimate that the net computing effort of the subproblem solutions was about 10 sec. Thus the computing times for "black box" optimization and a specially coded decomposition algorithm would be roughly equivalent for this problem.

The core memory requirements for this decomposition approach to oil refinery design are less than $1/n$ (where n is the number of crude oil types available) of that required for the "black box" optimization, so that relatively small computers can be used to solve problems of this type.

The Williams and Otto Chemical Plant

In order to test the performance of decomposition techniques on a nonlinear problem, the Williams and Otto Chemical Plant, discussed in Section 4.9, was decomposed into two subsystems as shown in Figure 5.17 by cutting flow streams between the decanter and distillation column and the recycle flow stream. This introduces two new variables, $\hat{F}_{RS}$ and $\hat{R}$. It should be pointed out that this decomposition is a physical one instead of simply arising from the mathematical structure discussed previously. As we shall see, this physical decomposition also determines the mathematical choice of subproblems.

When the subproblems were solved, it was discovered that a number of the variables in subsystem 2 had unbounded optimal values. This was an unfortunate example of the fact that subproblems may become unbounded (cf. Lasdon [16] p. 165 for ways of treating unbounded linear subproblems) or nonconvex when the entire problem is convex and bounded. Thus great care must be exercised in decomposing optimization problems. Some effort was expended to find another point of decomposition that would yield

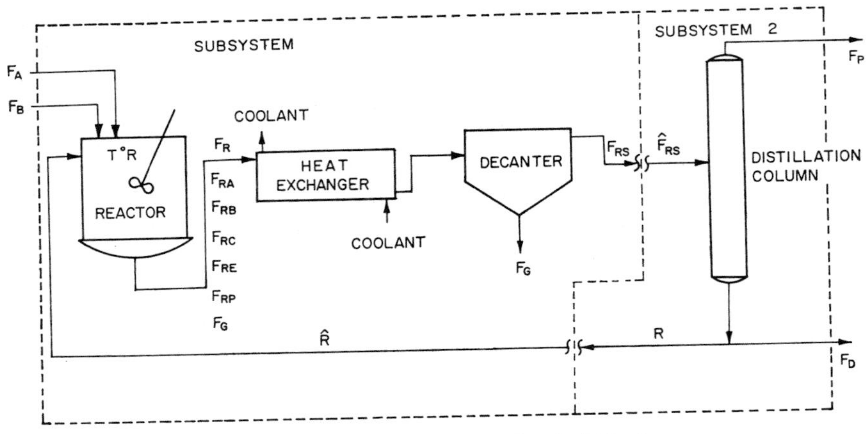

Fig. 5.17 Decomposition of the Williams and Otto chemical plant.

bounded subproblems, but without success. Thus it was decided to change the problem slightly to allow a comparison between the "black box" and decomposition approach. The following changes were made:

(i) Since the optimal value of F_P was found to be its upper bound, F_P was fixed at this value.

(ii) The utility cost was taken to be *utility cost* $= C_1R^{1.5} + C_2F_{RS}^{1.5}$.

These two changes were adequate to keep subsystem 2 bounded. The objective function then takes the nonlinear form

$$
\begin{aligned}
I = 100 \times [8400(0.3F_P + 0.0068F_D - 0.02F_A - 0.03F_B - 0.01F_G) \\
- 8400(C_1F_{RS}^{1.5} + C_2R^{1.5}) - (0.124)(8400)(0.3F_P + 0.0068F_D) \\
- 1{,}800{,}000]/1{,}800{,}000 \qquad (5.6.9)
\end{aligned}
$$

where

$$C_1 = (0.02)/(3600)^{1.5} \qquad \text{and} \qquad C_2 = (0.01)/(3600)^{1.5}$$

The modified Williams and Otto Plant optimization problem may be formulated as the desire to maximize I subject to the equality constraints given below.

Overall material balance:

$$h_1 = F_A + F_B + R - F_{RS} - F_G - \hat{F}_P = 0 \qquad (5.6.10)$$

Constraints on the separation efficiency of the distillation column:

$$h_2 = F_{RP} - 0.1F_{RE} - \hat{F}_P = 0 \qquad (5.6.11)$$

Material balance on component A:

$$h_3 = (-K_1F_{RA}F_{RB})V\rho/F_R^2 + R_A + F_A - F_{RA} = 0 \qquad (5.6.12)$$

Material balance on component B:

$$h_4 = (-K_1 F_{RA} F_{RB} - K_2 F_{RB} F_{RC}) V\rho / F_R^2 + R_B + F_B - F_{RB} = 0 \tag{5.6.13}$$

Material balance on component C:

$$h_5 = [(M_C/M_B) K_1 F_{RA} F_{RB} - (M_C/M_B) K_2 F_{RB} F_{RC} - K_3 F_{RC} F_{RP}] V\rho / F_R^2 + R_C - F_{RC} = 0 \tag{5.6.14}$$

Material balance on component E:

$$h_6 = (M_E/M_B K_2 F_{RB} F_{RC}) V\rho / F_R^2 + R_E - F_{RE} = 0 \tag{5.6.15}$$

Material balance on component P:

$$h_7 = [K_2 F_{RB} F_{RC} - (M_P/M_C) K_3 F_{RC} F_{RP}] V\rho / F_R^2 + R_P - F_{RP} - \hat{F}_P = 0 \tag{5.6.16}$$

Material balance on component G:

$$h_8 = [(M_G/M_C) K_3 F_{RC} F_{RP}] V\rho / F_R^2 - F_G = 0 \tag{5.6.17}$$

Definition of F_R:

$$h_9 = F_{RA} + F_{RB} + F_{RC} + F_{RE} + F_{RP} + F_G - F_R = 0 \tag{5.6.18}$$

Definition of F_{RS}:

$$h_{10} = F_R - F_G - F_{RS} = 0 \tag{5.6.19}$$

Definition of R:

$$h_{11} = R_A + R_B + R_C + R_E + R_P - R = 0 \tag{5.6.20}$$

The problem was first solved as a black box by the gradient projection algorithm OPTM from three starting points with the results shown in Table 5.6.1. As can be seen, nearly the same optimum is found from all three starting points, so that $I = 131.2\%$ might be considered the optimal return for this problem.

Next the problem was decomposed as shown in Fig. 5.17 and the subproblems coordinated in a *nonfeasible manner*. The particular *feasible* coordination technique discussed previously was not used in this problem because it required a knowledge of the Lagrange multipliers at the subsystem optima, and these were not available without solving a large set of linear equations at each iteration. However, other *feasible* coordination algorithms could have been used.

The subsystem objective for subsystem 1 becomes

$$I_1 = 100 \times [(8400)(\pi_1 F_{RS} - 0.02 F_A - 0.03 F_B - \pi_2 \hat{R} - 0.01 F_G) - (8400)(C_1 F_{RS}^{1.5} + C_2 \hat{R}^{1.5}) - 900{,}000]/1{,}800{,}000 \tag{5.6.21}$$

TABLE 5.6.1. RESULTS OF BLACK BOX OPTIMIZATION OF MODIFIED PLANT

Variable	Starting point 1	Starting point 2	Starting point 3
F_A	12,290	12,342	12,274
F_B	27,634	27,649	27,641
F_R	447,460	438,808	445,154
F_{RA}	52,807	52,228	52,189
F_{RB}	164,677	161,151	163,731
F_{RC}	10,088	9,983	9,943
F_{RE}	193,051	189,006	192,494
F_{RP}	24,068	23,664	24,012
F_G	2,768	2,777	2,793
F_{RS}	444,692	436,031	442,361
F_D	32,392	32,450	32,360
R	407,537	398,818	405,239
T	654	654	655
I	131.25%	131.23%	131.24%
$\Sigma_j h_j^2$	5×10^{-1}	2×10^{-1}	3×10^{-1}
Number of iterations	460	331	308
Execution time (sec)	70.2	55.8	50.4

and for subsystem 2,

$$\begin{aligned} I_2 = 100 \times [&(8400)(0.3F_P + 0.0068F_D + \pi_2 R - \pi_1 \hat{F}_{RS}) \\ &- (8400)(C_1 \hat{F}_{RS}^{1.5} + C_2 R^{1.5}) - (8400)(0.124) \\ &\times (0.3F_P + 0.0068F_D) - 900{,}000]/1{,}800{,}000 \end{aligned} \tag{5.6.22}$$

where

$$C_1 = (0.01)/(3600)^{1.5} \quad \text{and} \quad C_2 = (0.005)/(3600)^{1.5}$$

and π_1 and π_2 are Lagrange multipliers associated with the cut stream constraints

$$R - \hat{R} = 0 \tag{5.6.23}$$

$$F_{RS} - \hat{F}_{RS} = 0 \tag{5.6.24}$$

Subsystem 1 has 12 variables and 10 equality constraints generated in the obvious way from Fig. 5.4, while subsystem 2 has only an overall material balance and three variables.

The subsystem optima were found at each stage using the program

TABLE 5.6.2 TWO-LEVEL OPTIMIZATION OF THE MODIFIED W-O MODEL (INITIAL PRICES: $\pi_1 = 0.0010$, $\pi_2 = 0.0100$)

Number of coordination	$\pi_1 \times 10^2$	$\pi_2 \times 10^2$	$I_1(x^*)$	$I_2(x^*)$	Primal function	Dual function
0	0.1000	1.0000	−837.39	21,135.5	20,298.1	−388.1
1	0.5821	0.5168	757.4	438.5	1,195.9	−200.3
2	0.5821	0.5168	757.4	438.5	1,195.9	−200.3
3	0.5495	0.5495	−431.3	565.1	133.8	124.3
4	0.5469	0.5521	−443.5	575.2	131.7	122.2
5	0.5462	0.5528	−446.5	578.2	131.6	134.8
6	0.5462	0.5528	−446.6	578.2	131.6	134.8
7	0.5460	0.5530	−447.4	578.9	131.5	133.3

OPTM and the second level selection of π_1 and π_2 done by a gradient search. Briefly the algorithm was

1. Guess π_1, π_2.
2. Find the subsystem optima for these π_1, π_2.
3. Correct π_1, π_2 by the gradient step

$$\delta\pi_1 = \epsilon_1(\hat{F}_{RS} - F_{RS}) = -\epsilon_1 \frac{\partial L}{\partial \pi_1} \tag{5.6.25}$$

$$\delta\pi_2 = \epsilon_2(\hat{R} - R) = -\epsilon_2 \frac{\partial L}{\partial \pi_2} \tag{5.6.26}$$

4. Return to step 2 and iterate.

The starting point of the suboptimizations was equivalent to starting point 3 given previously. Two different sets of π_1, π_2 were tried as initial guesses. The first guess yielded the results shown in Table 5.6.2 where it was seen that reasonable convergence (illustrated by close values of the primal and dual problem solution) was obtained after about five iterations. The fact that the primal value slightly exceeded the dual value close to the optimum is thought to be due to imperfect convergence in the subsystem optimizations which results in errors in both the primal and dual function calculation. Brosilow and Nunez [27] also experienced this problem. The improvement per co-ordination was very slow after the third iteration, and at seven co-ordinations, Eqs. (5.6.24 and 5.6.25) were satisfied only within about 5%. Thus there are probably more efficient coordination procedures than the gradient method used here. The second set of initial values of π_1, π_2 yielded essentially the same results.

TABLE 5.6.3 OPTIMIZATION RESULTS. COMPARISON OF THE BLACK BOX AND DECOMPOSITION APPROACH

Variables	Black box[a]	Two-level[b]	Two-level[c]
F_A	12,274	12,215	12,258
F_B	27,641	27,664	27,530
F_R	445,154	450,404	472,445
F_{RA}	52,189	52,324	55,772
F_{RB}	163,731	166,543	173,498
F_{RC}	9,943	9,980	10,662
F_{RE}	192,494	194,571	204,545
F_{RP}	24,012	24,220	25,218
F_G	2,793	2,766	2,751
F_{RS}	442,361	458,547	451,619
F_D	32,360	33,502	32,000
R	405,239	421,257	414,729
T	655	655	654
I	131.2	133.3	131.4
Execution time			
(sec)	50.4	374.4	100.8
Σh_j^2	1.4×10^{-2}	25	9.3

[a] From starting point 3.
[b] From $\pi_1 = 0.0010$, $\pi_2 = 0.0100$.
[c] From $\pi_1 = \pi_2 = 0.0060$.

The comparison of the "black box" results of Chapter 4 with the decomposition approach is given in Table 5.6.3. It is seen that the decomposition approach yields a slightly higher profit, but this is due to the looser satisfaction of the equality constraints (Eqs. 5.6.24 and 5.6.25). The optimal operating conditions are all very similar. The computing times are clearly much longer for the decomposition approach, but this could be significantly improved by using a more efficient coordination algorithm.

The development of decomposition strategies, and in particular, coordination algorithms, is a very active area of research at present. It is expected that much more effective methods will emerge in the near future.

5.7 SUMMARY

In this chapter we have discussed various decomposition techniques for large structured systems. The techniques of dynamic programming and the discrete maximum principle seem admirably suited to serial structures with few state variables. The treatment of process optimization problems by multilevel decomposition methods was discussed for both linear and nonlinear problems. The Dantzig-Wolfe decomposition procedure for linear programs was applied to a simplified linear model of an oil refinery operating on different crude oil feedstocks. The optimal refinery operation was determined both by "black box" and decomposition procedures. While there were significant savings in core memory requirements using the technique of decomposition, there was no computational advantage over the "black box" approach. For a problem with a large number of crude oil choices, there may be a computational advantage. However, this advantage may be lost, owing to the time required to reload the subprograms into memory at each iteration. The nonfeasible decomposition approach to the nonlinear modified Williams-Otto chemical plant also gave similar conclusions. Though memory requirements may be reduced, the computer time required is certainly not competitive with the black box approach for this problem. It is felt, however, that further work is needed to develop more efficient coordination procedures to reduce the computational requirements of the decomposition approach.

In choosing between a black box or decomposition procedure, one must consider a number of points, for example:

1. *Size of the problem.* If the problem is sufficiently large, then decomposition may be required in order to handle the problem—although with the ever-increasing memory capabilities of modern computers, this is becoming less of a difficulty.

2. *Desire for modular structure.* If the process simulation routine relies on a modular structure, then decomposition may be needed to accomplish the optimization within this structure.

3. *Dominance of one processing unit.* If one of the processing units within the structure (for example, the chemical reactor, retort, autoclave, etc.) is either the most difficult to optimize (highly nonlinear, etc.) or most significant economically, then a decomposition procedure may be desirable to take advantage of this feature.

4. *Lack of quantitative process models.* If the process models are not exact mathematical descriptions, but are only available through semiquantitative predictions from experienced design engineers, then decomposition allows the iterative adjustment of subsystem goals to take maximum advantage of these imprecise process models. By breaking the complex structure into small enough pieces, the design engineers' semiquantitative understanding of each piece will allow an optimization where the "black box" approach would not be possible. This technique should have a particular appeal in the optimization of complex metallurgical operations because of the lack of availability of good process models for the individual units.

These considerations, combined with computational experience, should allow the optimal selection of an approach to large-scale process optimization problems.

PROBLEMS

1. A reversible first-order exothermic reaction is to be carried out in a series of three well-mixed, stirred vessels. The reactant concentration in the nth stage is denoted by c_n and can be calculated by the mass balance

$$c_n = c_{n-1} - \theta_n\{k_1(T_n)c_n - k_2(T_n)[1 - c_n]\} \qquad n = 1, 2, 3$$

where θ_n, T_n are the holding time and temperature in the nth stage. The problem is to choose θ_n, T_n at each stage so as to minimize a combination of exit reactant concentration and total reaction residence time.

$$I = c_3 + \alpha \sum_{n=1}^{3} \theta_n$$

The rate constants are given by

$$k_i = A_i \exp\{-E_i/RT_n\}$$

where $\ln A_1 = 22.0\ (\text{sec}^{-1})$ $\quad E_1 = 25{,}000$
$\ln A_2 = 38.0\ (\text{sec}^{-1})$ $\quad E_2 = 50{,}000$

In addition, the holding times are constrained to lie between $0 \leq \theta_n \leq 1$ hr.

a. Use discrete dynamic programming to determine the optimal holding times and temperatures at each stage for $\alpha = 1.0$, and feed to the first stage having concentration $c_o = 1.0$.

b. Repeat the analysis for $\alpha = 10$, $\alpha = 0.1$ to show the influence of reactor capital costs.

2. Solve problem 1 using the discrete maximum principle. If possible, compare the advantages and disadvantages of the discrete maximum principle versus discrete dynamic programming for this problem.

3. Suppose a fraction of the product stream, γ, is recycled in the serially structured system of Section 5.2 so that the feed of reactant to the first stage is

$$\mathbf{x}_0 = (1 - \gamma)\mathbf{c}_0 + \gamma \mathbf{x}_N$$

a. Suggest a discrete dynamic programming algorithm for solving this problem (*Hint:* see Ref. 33).

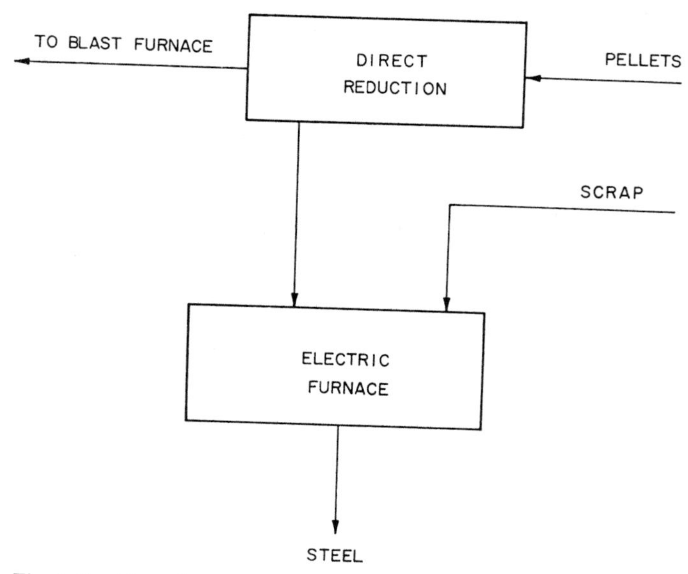

Fig. 5.18 Flow diagram for the system in Problem 4.

b. Revise the discrete maximum principle for this problem. Show that with recycle the last term in Eq. 5.4.2 is

$$-\boldsymbol{\lambda}_0^T(\mathbf{x}_0 - (1 - \gamma)\mathbf{c}_0 - \gamma\mathbf{x}_N)$$

and Eq. 5.4.3b becomes

$$\frac{\partial G}{\partial \mathbf{x}_N} - \boldsymbol{\lambda}_N^T + \gamma\boldsymbol{\lambda}_0^T = \mathbf{0}$$

Suggest a computational algorithm for solving this recycle optimization problem.

c. If 20% of the product stream (with reactant concentration x_3) is recycled in problem 1, determine the optimal θ_n, T_n policy, using one of the algorithms suggested in *a* and *b* above.

4. Apply a decomposition technique to optimize a section of an integrated steel plant, sketched in Fig. 5.18. Raw (unreduced) pellets are introduced into a direct reduction unit, which produces totally or partially reduced pellets. These pellets may be fed into a blast furnace (for which the operation receives credit) or may be fed into an electric furnace to produce steel. An alternative feed to the electric furnace is scrap. The problem, then, is to maximize the profit from the operation sketched here. The following information is available.

Direct Reduction Unit

Nominal Capacity: 2000 tons/day
The following relationship holds between the throughput M (tons/day) and the fraction of iron in the reduced pellets exiting the unit, f

$$f = \exp\left\{\frac{-M}{20{,}000}\right\} \qquad M \leq 2200 \tag{1}$$

Cost of unreduced pellets: $18/ton
Prorated production costs: $5/ton (input)
Fixed costs: $7000/day
Credit allowable for pellets introduced into the blast furnace:

$$\$/\text{ton} = 55Bf[1 - (1 - f)^{1/3}] \tag{2}$$

where B represents the tons/day of pellets sent to the blast furnace

Electric Furnace

Capacity:

$$\left(\text{scrap} + \frac{EF}{f^2}\right) \leq 3000 \text{ tons/day}$$

Cost of scrap: $30/ton
Fixed cost: $11,000/day
Prorated operational cost in

$$\$/\text{ton steel produced:} \quad \frac{10}{1 - (1 - f)^4} \tag{3}$$

where EF represents the tons/day of pellets fed to the Electric Furnace and Eq. 3 reflects the fact that it is very expensive to process partially reduced pellets in an electric furnace. Credit for steel produced in the electric furnace: $80/ton. The availability of scrap is limited to 1500 tons/day.

REFERENCES

1. F. Habashi, *Principles of Extractive Metallurgy*, Vol. 2, Gordon and Breach, New York 1970.
2. Richard Bellman, *Dynamic Programming*, Princeton University Press, 1957.
3. R. Aris, *Discrete Dynamic Programming*, Blaisdell, 1964.
4. R. E. Larson, *State Increment Dynamic Programming*, American Elsevier, 1968.

5. H. Ford, *Met. Rev.* **2,** 1 (1957).
6. R. Hill, *The Mathematical Theory of Plasticity*, Clarendon Press, Oxford, 1950.
7. H. Halkin, *Adv. Control. Syst.* **4,** 173 (1964).
8. F. Horn and R. Jackson, *I and EC Fund.* **4,** 110, 487 (1965).
9. M. M. Denn, and R. Aris, *Ind. Eng. Chem. Fund.* **4,** 240 (1965).
10. L. Fox, *The Numerical Solution of Two Point Boundary Value Problems*, Oxford, 1957.
11. E. Stanley Lee, *Quasilinearization and Invariant Imbedding*, Academic Press, 1968.
12. See Rex Bull, Lecture Notes, Colorado School of Mines.
13. G. A. Grandy et al., in *Decade of Digital Computing in the Mineral Industry*, A. Weiss, Ed., 765, A. I. M. E. (1969).
14. M. D. Mesarovic, D. Macko, and Y. Takahara, *Theory of Multi-Level Hierarchical Systems*, Academic Press, 1970.
15. G. B. Dantzig and P. Wolfe, *Econometrica* **29** (October 1961).
16. Leon S. Lasdon, *Optimization Theory for Large Systems*, Macmillan, 1970.
17. A. M. Geoffrion, *Oper. Res.*, 375 (1970).
18. F. Bessiere and E. A. Sautter, *Manage. Sci.* **15,** 1 (1968).
19. O. Yoshida, *Electr. Eng. Jap.* **87,** 29 (1968).
20. J. B. Rosen and J. C. Ornea, *Manage. Sci.* **10,** 160 (1963).
21. J. C. Ornea and G. G. Eldredge, Paper 4.15, AIChE-IChE Joint Meeting, London (1965).
22. M. Heymann and M. Avriel, *JOTA* **3,** 392 (1969).
23. H. W. Kuhn and A. W. Tucker, *Proc. Second Berkeley Symp. Math. Stat.* 1950, p. 481.
24. C. B. Brosilow, L. S. Lasdon, and J. D. Pearson, *Proc* 1965 *JACC*, Rochester, N.Y.
25. L. S. Lasdon, *IEEE Trans. Syst. Sci. Cybern.* **SSC4,** 86 (1968).
26. E. J. Bauman, *Adv. Control. Syst.* **6,** 159 (1968).
27. C. B. Brosilow and E. Nunez, *Can. J. Ch. E.* **46,** 205 (1968).
28. W. Findeisen, *IEEE Trans. System. Sci. Cybern.* **SSC4** 155 (1968).
29. C. B. Brosilow and L. S. Lasdon, *AIChE-IChE Symp. Ser.* **4,** 45 (1965).
30. R. Kulikowski, *Automatica* **6,** 315 (1970).
31. S. Kodoma and E. Bamba, *Elect. Eng. Jap.* **88,** 69 (1968).
32. M. D. Mesarovic, D. Macko, and Y. Takahara, *Automatica* **6,** 261 (1970).
33. R. L. Aris, L. Nemhauser, and D. J. Wilde, *AIChE J.*, **10,** 913 (1964).
34. D. J. Wilde, *Chem. Eng. Prog.* **61,** 3, 86 (1965).
35. R. Jackson, *Chem. Eng. Sci.* **19,** 19 (1964).
36. M. M. Denn and R. Aris, *I and EC Fund.* **4,** 248 (1965).
37. M. M. Denn, *Chem. Eng. Sci.* **21,** 703 (1966).
38. E. S. Lee, *AIChE J.* **15,** 393 (1969).
39. E. S. Lee, *Can. J. Ch. E.* **47** 431 (1969).
40. L. G. Pliskin and T. G. Rzaev, *Autom. Remote. Control.*, 1864 (1968).
41. W. Findeisen, J. Pulaczowski, and A. Manitius, *Automatica* **6,** 581 (1970).
42. B. S. Jung, W. Mirosh, and W. H. Ray, *Can. J. Ch. E.* **49,** 844 (1971).
43. W. Mirosh, *MASc Eng. Rep.*, University of Waterloo, 1970.

6 Trajectory Optimization of Lumped Parameter Systems

6.1 INTRODUCTION

In all the problems that we have discussed in the preceding chapters the variables to be selected by optimization were constant parameters, such as vessel size, residence time, flow rate of a stream, (constant) operating temperature, and the like. However, in many chemical or metallurgical operations the control variables may also be a function of some independent variables such as time, or distance.

As an example, let us consider the operation of a basic oxygen furnace, or perhaps the oxygen-argon process for stainless steel production. In both processes, it may be desirable to operate the system at an oxygen blowing rate which is a function of time. It follows, therefore, that our objective may be to find the *optimal oxygen flow rate* for each instance of time.

A conceptually similar problem would be to consider the steady-state operation of a moving bed reactor (e.g. direct reduction unit or chemical reactor with decaying catalyst), and set as our objective to find the optimal temperature along the reactor—to be achieved by judicious heating or cooling.

These two problems mentioned have the common feature that there exists only one independent variable, i.e., time or one space dimension. These problems, which we term *lumped parameter problems*, form the subject matter of this chapter. Problems where we have more than one independent variable, for example, both time and space, as in the optimization of a slab reheating furnace, will be treated in the next chapter which is devoted to *distributed parameter systems*.

In this chapter we shall develop necessary conditions for optimality, and computational techniques for control variable synthesis. Then several examples will be worked to illustrate the ideas developed.

6.2 LUMPED PARAMETER OPTIMIZATION PROBLEMS

The general class of problems we wish to optimize can be represented by the modeling equation

$$\frac{d\mathbf{x}(t)}{dt} = \mathbf{f}(\mathbf{x}(t), \mathbf{u}(t)) \qquad 0 \le t \le t_f \tag{6.2.1}$$

where $\mathbf{x}(t)$ is an n dimensional vector of the state variables and $\mathbf{u}(t)$ is an m dimensional vector of control variables which we wish to choose optimally.

The boundary conditions will depend on the physical nature of the problem. If we specify only the initial state, then

$$\mathbf{x}(0) = \mathbf{x}_0 \tag{6.2.2}$$

An example of such an initial condition would be to specify the composition of the initial charge in a batch chemical reactor. Thus if we were to fix the initial conditions only, the result would be a straightforward *initial value problem*.

In other practical systems we may also desire to specify the final state

$$\mathbf{x}(t_f) = \mathbf{x}_f \tag{6.2.3}$$

(an example of which would be the requirement that the final product be of a given composition in the batch reactor); then we have a *two-point boundary value problem*. Thus, we must find an optimal control $\mathbf{u}(t)$ which also causes $\mathbf{x}(t_f) = \mathbf{x}_f$.

Other possible conditions may require that some components of $\mathbf{x}$ are specified at the initial time and others at the final time. Alternatively, one may wish that some *transversality conditions*

$$\boldsymbol{\psi}(\mathbf{x}(t_f)) = \mathbf{0} \tag{6.2.4}$$

be satisfied at the final time. In physical terms such a transversality condition may mean that rather than requiring a given final composition of the charge, we may wish to specify some relationship between the concentration of the components. This might correspond to the situation where there are trade-offs possible in the final product specifications.

There are other boundary conditions which arise from practical problems, but we shall not discuss them here.

In order to specify what is meant by optimal, we must select an objective functional† $I[\mathbf{u}(t)]$,

$$I[\mathbf{u}(t)] = G(\mathbf{x}(t_f)) + \int_0^{t_f} F(\mathbf{x}, \mathbf{u})\, dt \tag{6.2.5}$$

which we wish to maximize or minimize. We shall see that Eq. 6.2.5 is sufficiently general that it allows the treatment of a wide class of practical problems.

Although the definitions of $G(\mathbf{x}(t_f))$ and $\int_0^{t_f} F(\mathbf{x}, \mathbf{u})\, dt$ have been given implicitly above, it may be helpful to illustrate the form that these functions can take for a given application.

If we were to consider the behavior of a two phase gas-liquid system, such as the operation of the oxygen-argon blown stainless steel converter, then the components of the objective functional may take the following form:

$$G(\mathbf{x}(t_f)) = [\mathbf{x}(t_f) - \mathbf{x}_s]_{t_f}^T [\mathbf{x}(t_f) - \mathbf{x}_s]_{t_f} \tag{6.2.6}$$

where $\mathbf{x}_s$ is a vector describing the desired end composition; thus in this instance the function $G(\mathbf{x}(t_f))$ is just the square deviation from the desired end composition. The function $\int_0^{t_f} F(\mathbf{x}, \mathbf{u})\, dt$ may be used to describe the sum of the loss of chromium into the slag, the loss of metal due to splashing, and the total heat lost during the "heat"; thus we may write

$$\int_0^{t_f} F(\mathbf{x}, \mathbf{u})\, dt = C_1 \int_0^{t_f} (\text{rate of Cr loss})\, dt + C_2 \int_0^{t_f} \begin{bmatrix} \text{rate of metal} \\ \text{loss due to} \\ \text{splashing} \end{bmatrix} dt + C_3 \int_0^{t_f} (\text{rate of heat loss})\, dt \tag{6.2.7}$$

† It is perhaps helpful to make clear that our objective now is a functional (the transformation of a function into a value for I) rather than a function (the transformation of a parameter into a value for I) as in previous chapters.

Here C_1, C_2, and C_3 are the appropriate cost factors. It is noted that the influence of the state Eq. 6.2.1. appears implicitly in the three integrals appearing on the right-hand side of Eq. 6.2.7.

In this instance the slag-metal compositions, together with the temperature and the variables denoting the relationship between blowing rate and rate of ejection would constitute the *state variables*, whereas the blowing rate, the composition of the oxygen-argon mixture, and slag additives would constitute *the control variables*. The appropriate thermodynamic relationships, Fourier's equation, and the Stefan-Boltzmann radiation law would enter the problem as state variable constraints, both of these latter being concerned with the heat loss calculations.

As an added example of the form that

$$\int_0^{t_f} F(\mathbf{x}, \mathbf{u})\, dt$$

may take, in practice it is often convenient to leave $\mathbf{x}(t_f)$ in Eq. 6.2.3 unspecified and to use the objective functional to force $\mathbf{x}$ to a given, desired final value. For example, the minimization of

$$\int_0^{t_f} (x - x_s)^2\, dt$$

will cause x to approach x_s in a very short time.

As in parameter optimization problems, there arise constraints of the form

$$\mathbf{g}(\mathbf{x}, \mathbf{u}) \leq \mathbf{0} \tag{6.2.8}$$

$$\mathbf{h}(\mathbf{x}, \mathbf{u}) = \mathbf{0} \tag{6.2.9}$$

and there are techniques for handling these.† However, because of the tremendous complexity that constraints of this form add, and because a great many practical problems only involve constraints of the form

$$\mathbf{u}_* \leq \mathbf{u} \leq \mathbf{u}^* \tag{6.2.10}$$

we shall only be concerned with upper and lower bounds on our control for the present.

In many practical problems, one may wish to choose t_f (e.g., the batch time) optimally as well. As seen in the next section, this presents no theoretical difficulties.

† For example, see the text by Bryson and Ho [16].

6.3 NECESSARY CONDITIONS FOR OPTIMALITY

As we derived the necessary conditions for optimality of parameter optimization problems in Chapter 2, so we now derive necessary conditions for optimality of lumped parameter trajectory optimization problems. Readers who are primarily interested in the final results—the necessary conditions for optimality—may turn to page 220 without serious loss of continuity.

Let us consider the system with n state variables $x_i(t)$, m control variables $u_j(t), j = 1, 2, \ldots, m$ and with dynamic behavior described by the ordinary differential equations

$$\frac{d\mathbf{x}}{dt} = \mathbf{f}(\mathbf{x}, \mathbf{u}); \qquad \mathbf{x}(0) = \mathbf{x}_0 \tag{6.3.1}$$

We wish to find the control vector $\mathbf{u}(t)$, $0 \leq t \leq t_f$, such that the objective functional given by Eq. 6.2.5 is maximized. Suppose that we have a set of nominal values for the control variables

$$\bar{\mathbf{u}}(t) = \begin{bmatrix} \bar{u}_1(t) \\ \bar{u}_2(t) \\ \cdot \\ \cdot \\ \cdot \\ \bar{u}_m(t) \end{bmatrix}$$

which we think may be optimal. Let us express any other control as a perturbation about $\bar{\mathbf{u}}(t)$

$$\mathbf{u}(t) = \bar{\mathbf{u}}(t) + \delta\mathbf{u}(t) \tag{6.3.2}$$

and represent the state $\mathbf{x}(t)$ resulting from $\mathbf{u}(t)$ as a perturbation about the state $\bar{\mathbf{x}}(t)$ caused by the control $\bar{\mathbf{u}}(t)$; that is,

$$\mathbf{x}(t) = \bar{\mathbf{x}}(t) + \delta\mathbf{x}(t) \tag{6.3.3}$$

By checking the value of I in Eq. 6.2.5 for all perturbations $\delta\mathbf{u}(t)$ we could determine whether $\bar{\mathbf{u}}(t)$ is optimal. However, there are variations $\delta\mathbf{x}(t)$ which are produced by the perturbations $\delta\mathbf{u}(t)$ so that one must consider whether Eq. 6.3.1 is satisfied. If the perturbations $\delta\mathbf{u}(t)$ are chosen small enough, that is,

$$|\delta\mathbf{u}(t)| \leq \epsilon_1$$

then a first-order expansion about $\bar{\mathbf{u}}(t)$ would be adequate to represent the system. Thus we linearize Eqs. 6.3.1 and 6.2.5 about the nominal controls

$\bar{\mathbf{u}}$ to obtain Eqs. 6.3.4 and 6.3.5, respectively

$$\frac{d(\delta\mathbf{x})}{dt} = \frac{\partial\bar{\mathbf{f}}}{\partial\mathbf{x}}\,\delta\mathbf{x} + \frac{\partial\bar{\mathbf{f}}}{\partial\mathbf{u}}\,\delta\mathbf{u}; \qquad \delta\mathbf{x}(0) = \delta\mathbf{x}_0 \tag{6.3.4}$$

$$\begin{aligned}\delta I &= I[\bar{\mathbf{u}}(t) + \delta\mathbf{u}(t)] - I[\bar{\mathbf{u}}(t)] \\ &= \frac{\partial\bar{G}}{\partial\mathbf{x}}\,\delta\mathbf{x}(t_f) + \int_0^{t_f}\left(\frac{\partial\bar{F}}{\partial\mathbf{x}}\,\delta\mathbf{x} + \frac{\partial\bar{F}}{\partial\mathbf{u}}\,\delta\mathbf{u}\right)dt + \left[\bar{F}(t_f) + \frac{\partial\bar{G}}{\partial\mathbf{x}}\,\bar{\mathbf{f}}(t_f)\right]\delta t_f\end{aligned} \tag{6.3.5}$$

where the overbar reminds us that the partial derivatives are evaluated along the *nominal trajectory* $\bar{\mathbf{u}}(t)$, $\bar{\mathbf{x}}(t)$.

The last term in Eq. 6.3.5 arises because we may wish to choose t_f optimally; thus variations δt_f are allowed as well as variations $\delta\mathbf{u}(t)$.

Let us now adjoin to the objective functional the linearized constraint Eqs. 6.3.4 and 6.3.5 by using the adjoint variables (i.e., time dependent Lagrange multipliers) $\boldsymbol{\lambda}(t)$. If we require that Eq. 6.3.4 be satisfied everywhere, then the subtraction of

$$\int_0^{t_f}\left\{\boldsymbol{\lambda}^T(t)\left[\frac{d(\delta\mathbf{x}(t))}{dt} - \frac{\partial\bar{\mathbf{f}}}{\partial\mathbf{x}}\,\delta\mathbf{x} - \frac{\partial\bar{\mathbf{f}}}{\partial\mathbf{u}}\,\delta\mathbf{u}\right]dt = 0\right. \tag{6.3.6}$$

from Eq. 6.3.5 yields

$$\begin{aligned}\delta I = {} & \frac{\partial\bar{G}}{\partial\mathbf{x}}\,\delta\mathbf{x}(t_f) + \left[\bar{F}(t_f) + \frac{\partial\bar{G}}{\partial\mathbf{x}}\,\bar{\mathbf{f}}(t_f)\right]\delta t_f \\ & + \int_0^{t_f}\left(\frac{\partial\bar{F}}{\partial\mathbf{x}} + \boldsymbol{\lambda}^T\frac{\partial\bar{\mathbf{f}}}{\partial\mathbf{x}}\right)\delta\mathbf{u} + \left.\left(\frac{\partial\bar{F}}{\partial\mathbf{u}} + \boldsymbol{\lambda}^T\frac{\partial\bar{\mathbf{f}}}{\partial\mathbf{u}}\right)\delta\mathbf{u}\right\}dt \\ & - \int_0^{t_f}\boldsymbol{\lambda}^T(t)\,\frac{d(\delta\mathbf{x})}{dt}\,dt\end{aligned} \tag{6.3.7}$$

By integrating the last term by parts, we obtain

$$\begin{aligned}\delta I = {} & \left[\bar{F}(t_f) + \frac{\partial\bar{G}}{\partial\mathbf{x}}\,\bar{\mathbf{f}}(t_f)\right]\delta t_f + \boldsymbol{\lambda}^T(0)\,\delta\mathbf{x}_0 + \left[\frac{\partial\bar{G}}{\partial\mathbf{x}} - \boldsymbol{\lambda}^T(t_f)\right]\delta\mathbf{x}(t_f) \\ & + \int_0^{t_f}\left\{\left[\frac{\partial\bar{H}}{\partial\mathbf{x}} + \frac{d\boldsymbol{\lambda}^T}{dt}\right]\delta\mathbf{x} + \frac{\partial\bar{H}}{\partial\mathbf{u}}\,\delta\mathbf{u}\right\}dt\end{aligned} \tag{6.3.8}$$

where H (sometimes called the *Hamiltonian*) is defined by

$$H \equiv F(\mathbf{x}, \mathbf{u}) + \boldsymbol{\lambda}^T\mathbf{f}(\mathbf{x}, \mathbf{u}) \tag{6.3.9}$$

Equation 6.3.8 represents the influence of variations $\delta\mathbf{u}(t)$ on δI, both directly and through $\delta\mathbf{x}(t)$. In order to express the direct influence of

$\delta\mathbf{u}(t)$ alone, let us define the heretofore arbitrary functions $\boldsymbol{\lambda}(t)$ such that they satisfy

$$\frac{d\boldsymbol{\lambda}^T}{dt} = -\frac{\partial \bar{H}}{\partial \mathbf{x}} \tag{6.3.10}$$

In effect, this allows the influence of the system equations (Eq. 6.3.1) to be transmitted by $\boldsymbol{\lambda}(t)$ and is felt in $\partial\bar{H}/\partial\mathbf{u}$ which carries $\boldsymbol{\lambda}(t)$.

The remaining terms outside the integral in Eq. 6.3.8 will depend on the boundary conditions of the physical system. Let us first consider the case where t_f is specified (so that $\delta t_f = 0$), $\mathbf{x}_0$ is fixed, and $\mathbf{x}(t_f)$ is unspecified. In this case the variations $\delta\mathbf{x}_0 = \mathbf{0}$, and $\delta\mathbf{x}(t_f)$ are completely arbitrary. However, the condition

$$\lambda_i(t_f) = \frac{\partial \bar{G}}{\partial x_i} \tag{6.3.11}$$

will shift the influence of $\delta\mathbf{x}(t)$ to $\boldsymbol{\lambda}(t)$ and cause it to arise in $\partial\bar{H}/\partial\mathbf{u}$. Notice that Eq. 6.3.11 completes the definition of $\boldsymbol{\lambda}(t)$ when combined with Eq. 6.3.10. If only some of the components of $\mathbf{x}(t_f)$ are unspecified, then $\delta x_i(t_f) = 0$ for those specified and Eq. 6.3.11 holds for those unspecified. Similarly, if some components of $\mathbf{x}_0$ were to be unspecified, then $\lambda_i(0) = 0$ would hold for those components.

If in addition we wish to choose t_f optimally, then the first term in Eq. 6.3.8 remains. Now if all the $x_i(t_f)$ are unspecified, then Eq. 6.3.11 must hold and in addition

$$H(t_f) = \bar{F}(t_f) + \frac{\partial \bar{G}}{\partial \mathbf{x}} \bar{\mathbf{f}}(t_f) \tag{6.3.12}$$

must vanish when t_f is chosen optimally.

In the case where some of the $x_i(t_f)$ are fixed at x_{if}, then by a Taylor series expansion

$$x_i(t + \delta t_f) = x_i(t_f) + \bar{f}_i(t_f)\,\delta t_f = \bar{x}_i(t_f) = x_{if} \tag{6.3.13}$$

as shown in Fig. 6.1.

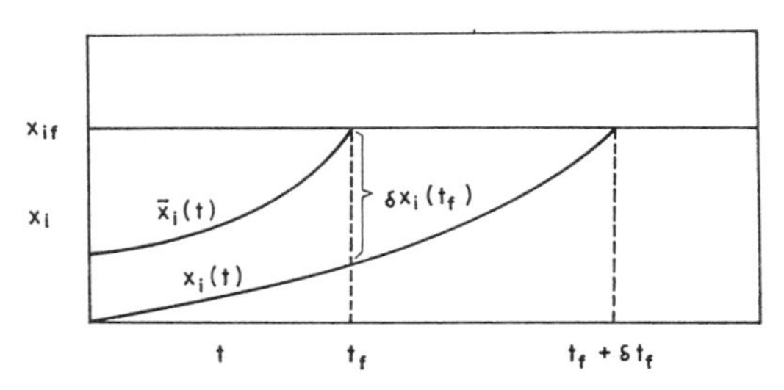

Fig. 6.1 The expansion of $x_i(t)$ about t_f.

Thus

$$\delta x_i(t_f) = x_i(t_f) - \bar{x}_i(t_f) = -\bar{f}_i(t_f)\, \delta t_f \qquad (6.3.14)$$

and again we see that

$$H(t_f) = \bar{F}(t_f) + \boldsymbol{\lambda}^T(t_f)\bar{\mathbf{f}}(t_f) \qquad (6.3.15)$$

must vanish for t_f to be optimal. Notice that in this case the $\lambda_i(t_f)$ associated with fixed $x_i(t_f)$ are unspecified while those associated with unspecified $x_i(t_f)$ are given by Eq. 6.3.11.

Having removed the boundary condition terms from Eq. 6.3.8, we obtain

$$\delta I = \int_0^{t_f} \left[\frac{\partial \bar{H}}{\partial \mathbf{u}}\, \delta\mathbf{u}(t) \right] dt \qquad (6.3.16)$$

and we now see the direct influence of variations $\delta\mathbf{u}(t)$ on δI. A necessary condition for optimality of $\bar{\mathbf{u}}(t)$ is that $\delta I \leq 0$ for all possible small variations $\delta\mathbf{u}(t)$. It is clear from Eq. 6.3.16 that the only way this can be true is that

$$\frac{\partial \bar{H}}{\partial \mathbf{u}} = \mathbf{0} \qquad (6.3.17)$$

at every t.

Suppose that some components of $\bar{\mathbf{u}}(t)$ include segments along the constraints u_i^*, u_{i*}. Obviously, variations $\delta u_i(t)$ can only be negative at the upper bound u_i^* and only positive along lower bounds u_{i*}. An examination of Eq. 6.3.16 shows that a necessary condition for optimality at upper and lower bounds is

$$\text{for} \quad \bar{u}_i(t) = u_i^*, \quad \frac{\partial \bar{H}}{\partial u_i} \geq 0$$

and

$$\text{for} \quad u_i(t) = u_{i*}, \quad \frac{\partial \bar{H}}{\partial u_i} \leq 0 \qquad (6.3.18)$$

Equations 6.3.18 can be reduced to the requirement that H *have a local maximum at the constraints.*

The results derived here may now be summarized as follows:

THEOREM 1 (WEAK MAXIMUM PRINCIPLE). In order for a control $\bar{\mathbf{u}}(t)$, $\mathbf{u}_* \leq \bar{\mathbf{u}}(t) \leq \mathbf{u}^*$ to be optimal in the sense that it maximizes the objective I in Eq. 6.2.5 while satisfying the system Eqs. 6.3.1, it is necessary that Eq. 6.3.17 be satisfied for unconstrained portions of the path and H defined by Eq. 6.3.9 be maximized along constrained portions of the control trajectory.

Thus given

$$\frac{d\mathbf{x}}{dt} = \mathbf{f}(\mathbf{x}, \mathbf{u}), \quad \mathbf{x}(0) = \mathbf{x}_0 \tag{6.3.1}$$

and

$$I[\mathbf{u}(t)] = G(\mathbf{x}(t_f)) + \int_0^{t_f} F(\mathbf{x}, \mathbf{u})\, dt \tag{6.2.5}$$

the necessary condition for $\bar{\mathbf{u}}(t)$ to maximize

$$I[\mathbf{u}(t)]$$

is that

$$\frac{\partial \bar{H}}{\partial \mathbf{u}} = \mathbf{0} \tag{6.3.17}$$

on the unconstrained portion of the path and

$$H \equiv F(\mathbf{x}, \mathbf{u}) + \boldsymbol{\lambda}^T \mathbf{f}(\mathbf{x}, \mathbf{u}) \tag{6.3.9}$$

be at the maximum on the constrained portion of the path. Here H is the Hamiltonian defined by Eq. 6.3.9, and $\boldsymbol{\lambda}$ is the time dependent Lagrange multiplier, which is defined by

$$\frac{d\boldsymbol{\lambda}^T}{dt} = -\frac{\partial \bar{H}}{\partial \mathbf{x}} \tag{6.3.10}$$

and

$$\lambda_i(t_f) = \frac{\partial \bar{G}}{\partial x_i} \tag{6.3.11}$$

for those state variables unspecified at $t = t_f$.

A much stronger version of these necessary conditions, whose derivation is available elsewhere [1–3], is summarized in the following theorem.

THEOREM 2 (STRONG MAXIMUM PRINCIPLE). In order for a control $\mathbf{u}(t)$ (constrained to lie in some constraint set Ω) to be optimal for the problem given by Eqs. 6.2.5 and 6.3.1, it is necessary that H be maximized by $\mathbf{u}(t)$ almost everywhere.

In addition it is necessary that the Hamiltonian, $H(t)$, remain constant along the optimal trajectory and $H(t)$ take the constant value of zero when the terminal time, t_f, is unspecified (cf. Eq. 6.3.15).

This much stronger result can also be shown sufficient for optimality under certain convexity assumptions. For further details see the work of Lee and Markus [2].

The results developed in this section are very similar to those arising from dynamic programming or the classical calculus of variations. While the relationship can be made quite explicit, we shall not pursue the discussion further here. The reader is referred to Dreyfus [4] and Leitman [5] for a treatment of these relationships.

EXAMPLE 6.3.1. Consider the radiant heating of a small billet or slab having a uniform temperature distribution so that the modeling equations are

$$\frac{dT}{dt'} = C_1(T_s^4 - T^4) \qquad T(0) = T_0 \tag{6.3.19}$$

where T_s is the radiant source temperature bounded by $T_* \leq T_s \leq T^*$ and T_0 is the initial temperature. Let us determine the optimal source temperature $T_s(t')$ so as to bring the billet to temperature T_1 in minimum time while minimizing the heat losses. This objective can be expressed as

$$\underset{T_s(t)}{\text{Min}} \left\{ I[T_s(t)] = t_f' + C_2 \int_0^{t_f'} T_s(t)^4 \, dt' \right\} \tag{6.3.20}$$

where t_f' is left free and C_2 denotes the relative value of heat losses to operating time.

Solution. Let us define the variables

$$x_0 = T_0, \quad x = T, \quad u = T_s^4, \quad t = C_1 t',$$
$$C = \frac{C_2}{C_1}, \quad x_s = T_1, \quad u_* = T_*^4, \quad u^* = T^{*4}$$

so that our problem becomes

$$\underset{u(t)}{\text{Min}} \left\{ I[u] = \int_0^{t_f} [1 + Cu(t)] \, dt \right\} \tag{6.3.21}$$

subject to

$$\frac{dx}{dt} = u - x^4; \quad x(0) = x_0; \quad x(t_f) = x_s \tag{6.3.22}$$

and

$$u_* \leq u \leq u^* \tag{6.3.23}$$

We can now define the Hamiltonian

$$H = 1 + Cu(t) + \lambda(u - x^4) = 1 - \lambda x^4 + (\lambda + C)u \tag{6.3.24}$$

and adjoint variables $\lambda(t)$ by

$$\frac{d\lambda}{dt} = 4\lambda x^3 \tag{6.3.25}$$

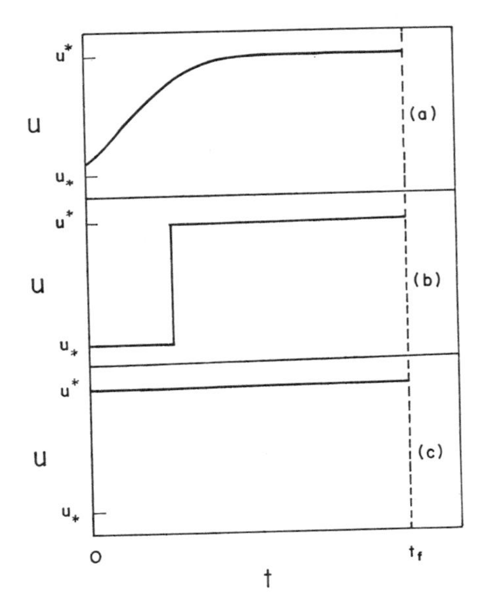

Fig. 6.2 Possible optimal policies for the billet reheating problem: (*a*) gradual increase of u from u_* to u^*, (*b*) stepwise increase from the lower bound to the upper bound, and (*c*) optimum at the upper bound.

From the fact that H is linear in u, it is clear that

$$u(t) = \begin{cases} u^* & \text{if } (\lambda + C) < 0 \\ u_* \leq u \leq u^* & \text{if } (\lambda + C) = 0 \\ u_* & \text{if } (\lambda + C) > 0 \end{cases} \tag{6.3.26}$$

because this is the only policy which minimizes H and thus satisfies the maximum principle. We note at this juncture, that had our objective been to *maximize* rather than *minimize* H, then Eq. 6.3.26 would have taken the following form:

$$u(t) = \begin{cases} u_* & \text{if } (\lambda + C) < 0 \\ u_* \leq u \leq u^* & \text{if } (\lambda + C) = 0 \\ u^* & \text{if } (\lambda + C) > 0 \end{cases}$$

Let us now deduce the exact optimal policy. If $u_* < u < u^*$ somewhere on the optimal policy, say, the region $0 \leq t \leq t_1$ in Fig. 6.2*a* (which indicates a slow increase in source temperature with time until the final value is reached), then from Eq. 6.3.26 $\lambda + C = 0$ and $d\lambda/dt = 0$ on $0 \leq t \leq t_1$. However, Eq. 6.3.25 does not allow this, because $d\lambda/dt = 0$ implies $\lambda = 0$, a contradiction. Thus the policy shown in Fig. 6.2*a* (as well as any other policy where $u_* < u < u^*$) is not optimal.

Examination of Eq. 6.3.25 shows that $(\lambda + C)$ can change sign only once and that is when $\lambda(0) + C > 0$ and $\lambda(0) < 0$. This would produce the "optimal" policy shown in Fig. 6.2*b*. However, the fact that H must be identically zero along the optimal trajectory when t_f is unspecified, leads to

$$u_{\text{opt}} = \frac{-1 + \lambda x^4}{C + \lambda} \tag{6.3.27}$$

Clearly if $\lambda < 0$ and $\lambda + C > 0$, then $u_{\text{opt}} < 0$ which is physically impossible. Therefore, the policy given in Fig. 6.2*b* cannot be optimal.

The only remaining possibility for the optimal policy is shown in Fig. 6.2*c*, in which the radiant heat source is kept at its maximum value until the billet reaches the desired temperature.

All that remains is to evaluate Eq. 6.3.19 with $T_s = T^*$ in order to determine the actual minimum time.

Although this problem may be a simple one, it illustrates the application of the maximum principle to a metallurgical problem. In addition, it shows that in some simple cases the optimal trajectory may be deduced without performing any calculations.

6.4 COMPUTATIONAL TECHNIQUES

Just as there were computational approaches for parameter optimization developed in Chapters 3 and 4 based on the necessary conditions described in Chapter 2, so we shall introduce computational procedures for trajectory optimization based on the necessary conditions of the previous section. We shall divide the methods into several classes and describe each of them in turn.

6.5 CONTROL VECTOR ITERATION PROCEDURES

The control vector iteration procedures are very similar in philosophy to the gradient type methods discussed in Chapter 3. Basically one makes use of Eq. 6.3.16.

$$\delta I = \int_0^{t_f} \left[\sum_{i=1}^{m} \frac{\partial \bar{H}}{\partial u_i} \delta u_i(t) \right] dt \tag{6.5.1}$$

Suppose that $\bar{u}_i(t)$ is not optimal, so that $\partial \bar{H}/\partial u_i \neq 0$; how can we determine a correction $\delta u_i(t)$ so as to improve I (*i.e.*, cause $\delta I > 0$)? It can be shown [2] that by choosing $\delta u_i(t)$ to be corrected in the gradient direction at each time t, this produces the greatest local improvement in I. Thus on selecting

$$\delta u_i(t) = \epsilon \frac{\partial \bar{H}}{\partial u_i}, \qquad \epsilon > 0 \tag{6.5.2}$$

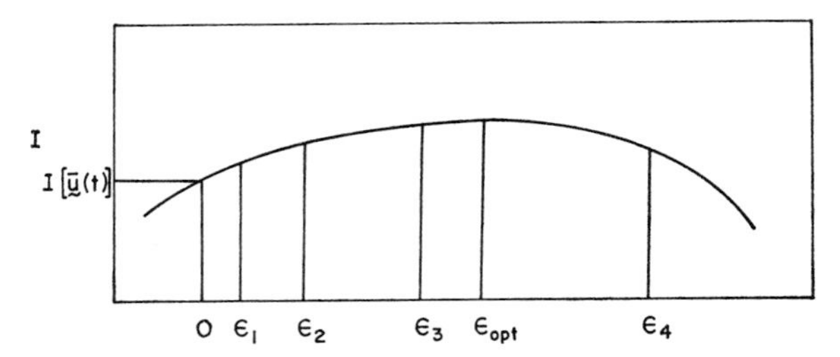

Fig. 6.3 Determination of the optimal ϵ by a quadratic approximation.

one obtains

$$\delta I = \epsilon \int_0^{t_f} \left\{ \sum_{i=1}^{m} \left[\frac{\partial \bar{H}}{\partial u_i} \right]^2 \right\} dt > 0 \tag{6.5.3}$$

which guarantees $\delta I > 0$ for ϵ small enough that the linear approximation is not violated.

These results can be incorporated in the following modified gradient method [6]:

(i) Guess $\bar{\mathbf{u}}(t)$, $0 \leq t \leq t_f$.

(ii) With this value of $\bar{\mathbf{u}}(t)$, integrate the state Eqs. 6.2.1 forward in time to produce $\bar{\mathbf{x}}(t)$, $0 \leq t \leq t_f$.

(iii) With these values of $\bar{\mathbf{x}}(t)$, $\bar{\mathbf{u}}(t)$, integrate the adjoint Eqs. 6.3.10 backward in time, $0 \leq t \leq t_f$.

(iv) Correct $\bar{\mathbf{u}}(t)$ by Eqs. 6.5.2 where ϵ is chosen arbitrarily. Evaluate I for this new control $\mathbf{u}(t)$.

(v) If $I[\mathbf{u}(t)] > I[\bar{\mathbf{u}}(t)]$ double ϵ and repeat step (iv); if $I[\mathbf{u}(t)] < I[\bar{\mathbf{u}}(t)]$, halve ϵ and repeat step (iv). Do this until a concave function $I(\epsilon)$ is formed (cf. Fig. 6.3).

(vi) Fit a quadratic $I(\epsilon)$ to these results and predict the optimal value of ϵ, i.e., ϵ_{opt}.

(vii) Let

$$\bar{\mathbf{u}}^{\text{new}} = \bar{\mathbf{u}}^{\text{old}} + \epsilon_{\text{opt}} \frac{\partial \bar{H}}{\partial \mathbf{u}} \tag{6.5.4}$$

and return to step (ii).

(viii) Iterate until convergence is attained.

Experience has shown that these methods will lead to rapid progress in the first few iterations, but tend to become very slow as the optimum is approached. Even though convergence to the optimum can be proved theoretically, the rate of convergence can be so slow that the exact optimum is never found in a finite number of iterations. For this reason, several second-order methods have been proposed (e.g., [7–8]) which are similar to those for parameter optimization. In addition, the conjugate gradient procedures discussed earlier have been extended to trajectory optimization problems [9, 10] and have often shown improved convergence properties over the standard gradient methods. However, we shall not discuss these here.

As an illustration of control vector iteration techniques let us consider the establishment of the optimal temperature profile in a packed bed reactor in which a solid oxide is reduced by hydrogen.

EXAMPLE 6.5.1. Solid nickel oxide particles are reduced by hydrogen in a continuous (moving) packed bed reactor, as sketched in Fig. 6.4. The bed

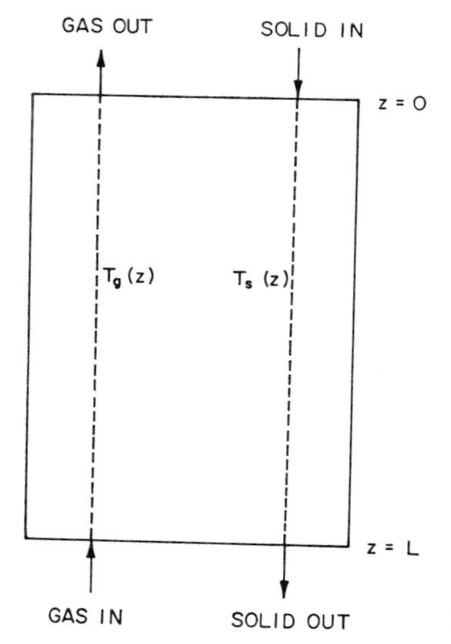

Fig. 6.4 The moving bed reactor in Example 6.5.1.

height is fixed at L, the inlet temperatures of the solid and gas streams are also fixed, and information is available on the kinetics of the reaction, which is considered to be a unique function of the temperature and the fraction of solids reacted.

We shall further assume that gas phase mass transfer is negligible and the reactant gas is present in an excess, so that no hydrogen "starvation" can occur.

Accordingly, the problem is to find the optimal axial temperature distribution in the solid phase of the bed, which will maximize the yield of metallic nickel.

Kinetic Data

The kinetic data to be used in the optimization are shown in Fig. 6.5 on a plot of conversion/time against the conversion, for various fixed temperatures. These data were, in part, deduced from the thesis by J. W. Evans [11].

Such plots could be readily obtained from a "cross plot" of experimental data, which would be typically represented on plots of fractional weight loss against time.

It is noted that a characteristic feature of the curves shown in Fig. 6.5 is that at low conversions, i.e., at low values of fractional reaction, the higher the temperature, the faster the reaction.

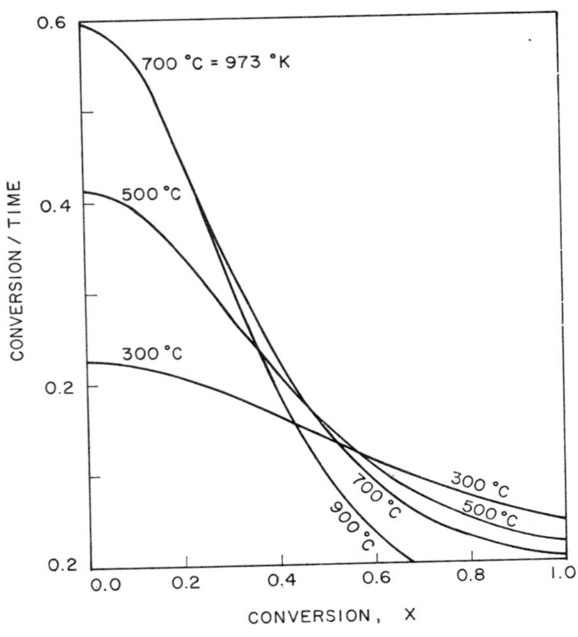

Fig. 6.5 Plot of the rate of conversion (time^{-1}) against the conversion for Example 6.5.1.

In contrast, at higher values of conversion, the rate of reaction is slower at high temperatures.

In physical terms such behavior is caused by the sintering of the reacted zone at higher temperatures and suggests that a definite optimal policy exists.

We note at this stage that the mathematical manipulation that follows provides a good illustration of the concept of the optimal temperature progression and of the technique of control vector iteration. However, one may take exception on physical grounds to the use of the kinetic information for systems where sintering occurs. Comments on this fact will be given at the end of the example.

Before proceeding further, let us define the following parameters:

C_p^g	heat capacity of gas (Btu/lb °F)
C_p^s	heat capacity of solid (Btu/lb °F)
$G_g(z)$	mass flow rate of gas (lb/min ft^2)
$G_s(z)$	mass flow rate of solid (lb/min ft^2)
L	height of bed (ft)
ρ_B	porosity of bed
A	interfacial area/unit bed volume
h	convective heat transfer coefficient
W_{Ni}	molecular weight of Ni
W_{NiO}	molecular weight of NiO
ρ_m	density of the bed
X	fractional conversion of solid
t	time (min)
r	rate of reaction (min)$^{-1}$

Formulation

For the purpose of computation it is convenient to express the kinetic data given in Fig. 6.5 in a numerical form.

By standard curve fitting techniques we obtain the following relationship:

$$t = 1.67 \times T_1 e^{2.5X^2T_2} \tag{6.5.5}$$

where T_1 and T_2 are coefficients, defined as

$$T_1 = 1.0 - 1.4\left(\frac{973}{T_s}\right) + 1.4\left(\frac{973}{T_s}\right)^2$$

and

$$T_2 = 1.0 - \ln\left[5.36\left(\frac{973}{T_s}\right)^{1/3} - 4.35\right]$$

Equation 6.5.5 which is valid for $573 \leq T_s \leq 973$°K, provides the desired empirical relationship between t, T_s, and X. The corresponding reaction rate $r(T_s, X)$ may now be obtained by differentiation, and we have

$$r(T_s, X) = \frac{dX}{dt} = \frac{1}{1.67T_1 e^{2.5X^2T_2}(1 + 5X^2T_2)} \tag{6.5.6}$$

Because the kinetic expressions are written in terms of conversion, solid temperature, and time, the reader may find this representation inconsistent with the fact that we have considered a steady-state system. The quantity *time* appearing in these equations may be regarded as a *residence time* of the solid in the reactor, and is thus equivalent to

$$t \equiv \frac{z\rho_m}{G_s}$$

In addition to the kinetic expression we also need a statement of the heat balance on the gas, the heat balance on the solids, and an expression of the conservation of the solids.

These may be written as

$$-\frac{d}{dz}(G_s C_p^s T_s) = hA(T_s - T_g) - r\,\Delta H G_s \tag{6.5.7}$$

(heat balance on the solids)

$$-\frac{d}{dz}(G_g C_p^g T_g) = hA(T_s - T_g) \tag{6.5.8}$$

(heat balance on the gas)

and

$$-\frac{d}{dz}(G_s) = \rho_m(1 - \rho_B)r\left(1 - \frac{W_{\mathrm{Ni}}}{W_{\mathrm{NiO}}}\right) \tag{6.5.9}$$

(conservation of the solids)

The conservation of all reactant species yields

$$G_s + G_g = G_{sL} + G_{so} \tag{6.5.10}$$

Our actual objective of the calculation is to find *the optimal solid temperature profile* within the system. This calculation may then be performed by using Eqs. 6.5.6 and 6.5.9, with the appropriate initial conditions.

The heat balance equations, Eqs. 6.5.7 and 6.5.8 would, of course, be needed if or when we had to devise the means for establishing the desired temperature profile in the solid stream; e.g., by outside heating or cooling, or by the introduction of side streams, and the like.

Thus the governing equations are given as

$$\frac{dG_s}{dz} = -\rho_m(1 - \rho_B)r\left(1 - \frac{W_{\mathrm{Ni}}}{W_{\mathrm{NiO}}}\right) \tag{6.5.9}$$

and

$$r(T_s, X) = \frac{1}{1.67T_1 e^{2.5X^2T_2}(1 + 5X^2T_2)} \tag{6.5.6}$$

Let us define

$$\xi = \frac{z}{L}$$

and note that

$$X \equiv \frac{G_{so} - G_s}{G_{so}\left(1 - \dfrac{W_{\text{Ni}}}{W_{\text{NiO}}}\right)}$$

so that we have

$$\frac{dX}{d\xi} = \left[\frac{\rho_m(1 - \rho_B)L}{G_{so}}\right] r(T_s, X) \tag{6.5.11}$$

Since the residence time of the solid may be defined as

$$\theta = \left[\frac{\rho_m(1 - {}_B\rho)L}{G_{so}}\right]$$

Eq. 6.5.11 becomes

$$\frac{dX}{d\xi} = \theta r(T_s, X) \tag{6.5.12}$$

where $r(T_s, X)$ is defined by Eq. 6.5.6.

The optimization problem then is to choose $T_s(\xi)$ so as to maximize the exit conversion $X(1)$.

Let us choose the following parameters, $L = 20$ ft, $\rho_m = 464$ lb/ft^3, $W_{\text{Ni}} = 58.7$, $W_{\text{NiO}} = 74.7$, $\rho_B = 0.3$, $G_{so} = 1301$ lb/min ft^2 which produces a residence time θ of 5 min.

Solution. Let us now redefine variables to put the problem in the form of Eq. 6.3.1; let $u = T_s$, $x = X$, $c = \theta/1.67$, $t = \xi$ so that Eq. 6.5.6 is written as

$$\frac{dx}{dt} = \frac{c \exp\left[-2.5x^2T_2(u)\right]}{T_1(u)[1 + 5x^2T_2(u)]}; \qquad 0 \leq t \leq 1; \quad x(0) = 0 \tag{6.5.13}$$

where $T_1(u)$, $T_2(u)$ are defined by Eq. 6.5.5 and

$$573 \leq u \leq 973$$

The objective functional to be maximized is

$$I[u] = x(1)$$

Thus our Hamiltonian for this problem becomes

$$H = \frac{\lambda(t)c \exp\left[-2.5x^2T_2(u)\right]}{T_1(u)[1 + 5x^2T_2(u)]} \tag{6.5.14}$$

and the adjoint variable $\lambda(t)$ is defined by

$$\frac{d\lambda}{dt} = -\frac{\partial H}{\partial x} = \lambda c H\left\{5xT_2(u) + \frac{10xT_2(u)}{[1 + 5x^2T_2(u)]}\right\} \tag{6.5.15}$$

where $\lambda(1) = 1$.

The control vector iteration algorithm can now be applied to this problem as follows:

(i) Guess $\bar{u}(t) = 573°K$ $0 \leq t \leq 1$.

(ii) Integrate Eq. 6.5.13 $0 \leq t \leq 1$ to produce the conversion profile $n = 1$ shown in Fig. 6.6.

(iii) Integrate the adjoint Eq. 6.5.15 backwards in time to produce $\lambda(t)$ $0 \leq t \leq 1$.

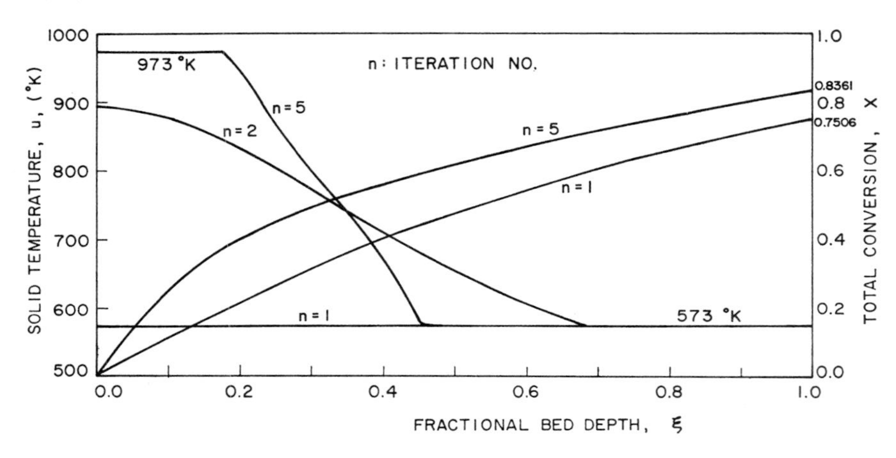

Fig. 6.6 Trajectory optimization of the NiO reduction process in Example 6.5.1, with a starting point $T_s = 573°K$.

(iv) Correct $\bar{u}(t)$ by

$$u(t) = \bar{u}(t) + \epsilon \frac{\partial H}{\partial u}(t) \tag{6.5.16}$$

where

$$\frac{\partial H}{\partial u} = -H\left\{\frac{d \ln T_1}{du} + \frac{dT_2}{du}\left[2.5x^2 + \frac{5x^2}{(1 + 5x^2T_2)}\right]\right\} \tag{6.5.17}$$

and stored values of $x(t)$, $\lambda(t)$ are used in Eq. 6.5.17. The step-size ϵ is chosen to produce the maximum increase in I at each iteration. This was done by the procedure described in Fig. 6.3.

(v) Iterate until the maximum improvement in I is less than 0.0001.

The optimization was attempted also from the starting point $T_s = u = 973°K$. These results are shown in Fig. 6.7. As can be seen the same optimal temperature policy and concentration trajectory was found from both initial guesses.

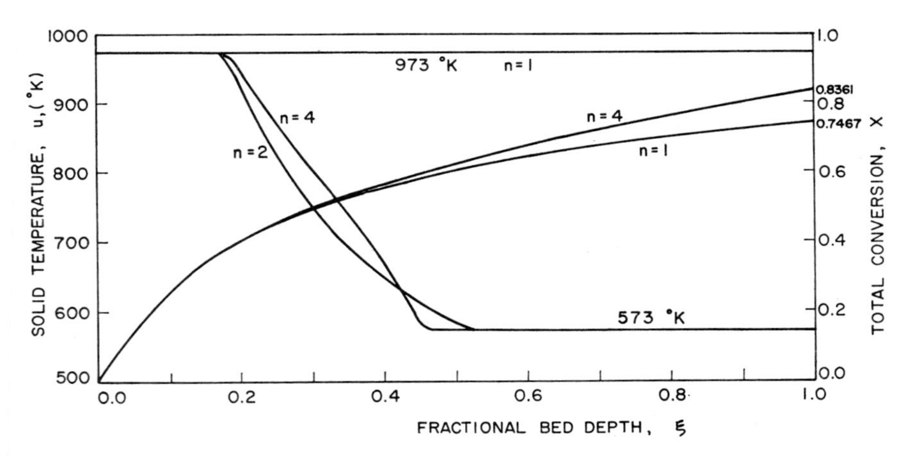

Fig. 6.7 Trajectory optimization of the NiO reduction process in Example 6.5.1, with a starting point $T_s = 973°K$.

The progress of the iterations are detailed in Figs. 6.6 and 6.7, and Table 6.5.1. These show quite clearly the advantages of the optimal profile over the initial guesses—the improvement in final conversion being approximately 10%.

At this stage it may be worthwhile to pause for a moment, and reexamine what we have done in terms of the physics of the system. We took a set of kinetic data, viz. the curves shown in Fig. 6.5, then fitted an empirical equation to these data, relating the conversion to the fraction unreacted and the temperature, and thus obtained Eq. 6.5.5 or 6.5.13. This latter equation was then appropriately manipulated to obtain the optimal temperature progression.

In using the kinetic expression, that is, Eq. 6.5.5, we implicitly assumed that the rate of reaction could be uniquely expressed as a function of the temperature and of the fraction reacted. This would be a reasonable assumption for systems where temperature affects the equilibrium conditions (equilibrium constant) and the kinetic coefficients.

However, the situation is basically different in our case. Here the effect of raising the temperature is to promote sintering of the reacted shell, and once sintering has taken place, this effect cannot be "undone" by simply lowering the temperature. In other words, the physical nature of the system precludes our ability to move from one curve to another in Fig. 6.5, especially from a curve at a higher temperature to a lower one.

It follows that our kinetic model was incorrect for a system where the irreversible sintering causes the drop in the overall rate at higher temperatures and at high levels of conversion.

TABLE 6.5.1 OPTIMIZATION RESULTS FOR EXAMPLE 6.5.1

Initial trial temperature (°K)	Iteration†															Computer time required (sec)
	$n = 1$			$n = 2$			$n = 3$			$n = 4$			$n = 5$			
	m	ϵ	$x(1)$	m	ϵ	$x(1)$	m	ϵ	$x(1)$	m	ϵ	$x(1)$	m	ϵ	$x(1)$	
573	1	0	0.7506	4	6.95×10^5	0.8278	11	2.39×10^6	0.8358	14	2.94×10^6	0.8361	19	3.97×10^6	0.8361	25.141
973	1	0	0.7467	12	5.05×10^6	0.8358	5	2.51×10^6	0.8361	10	4.42×10^6	0.8361				15.301

† m = number of searches required to obtain optimum ϵ (including quadratic fitting).

Had we considered a system where the conflict is brought about between a fast reaction and unfavorable equilibrium conditions at high temperatures, then it would have been permissible to represent the curves in Fig. 6.5 by Eqs. 6.5.5; indeed, this would have been one of the classical optimization problems [12].

As matters stand, however, the example given is an illustration of optimization with an improper model. Even though the optimization itself went well, the inadequacy of the model makes the results meaningless. This should serve to re-emphasize the need for sufficient model development before optimization proper is undertaken. This particular problem will be re-examined in Chapter 8, where it will be shown that the use of an optimal temperature progression would offer only small improvements over operation at a fixed optimum temperature.

We shall now treat an example which has an adequate model for the optimization to be performed.

EXAMPLE 6.5.2. Let us consider a batch chemical reactor in which we can control the reaction temperature exactly† and in which we wish to carry out the following reaction:

$$A \xrightarrow{k_1} B \xrightarrow{k_2} C \tag{6.5.18}$$

We consider the kinetics and the temperature dependence of the rate constants known, and our objective is to find the *optimal temperature progression*, for a fixed batch time, which will maximize the production of the intermediate "B." We note that this is one of the classical optimization problems relating to reactor design and that the general scheme given by Eq. 6.5.18 is of considerable practical importance in a number of chemical processing operations, e.g., the oxidation of hydrocarbons, or the chlorination of aromatics. In all these cases we may wish to maximize the production of an intermediate and thus wish to prevent the reaction from going to completion.

In order to define the problem, let us assume that the reaction is of second order with respect to the first step and of first order with respect to the second step. Thus the material balance on the reacting species may be written as:

$$\frac{dc_1}{dt} = -k_1(T)c_1^{\,2} \qquad c_1(0) = 0 \tag{6.5.19}$$

$$\frac{dc_2}{dt} = k_1(T)c_1^{\,2} - k_2(T)c_2 \qquad c_2(0) = 0 \tag{6.5.20}$$

where

$$c_1 = [A], \quad c_2 = [B], \quad k_i(T) = A_{io} \exp\left(-\frac{E_i}{RT}\right) \qquad i = 1, 2$$

† In many practical situations such close temperature control is, in fact, quite feasible.

Let us consider that the temperature is bounded by

$$T_* \leq T(t) \leq T^* \tag{6.5.21}$$

The object of the optimization is to find the temperature program $T(t)$ which maximizes the amount of species B present after one hour of reaction. Thus our objective becomes

$$\underset{T(t)}{\text{Max}} \, [I = c_2(1)] \tag{6.5.22}$$

The additional parameters of the system required to define the problem are given as follows:

$A_{10} = 4000.0 \text{ sec}^{-1}$ liter/mole $\qquad T_* = 298°\text{K}$

$A_{20} = 6.2 \times 10^5 \text{ sec}^{-1}$ $\qquad T^* = 398°\text{K}$

$E_1 = 5000$ cal/g mole $\qquad$ batch time: 1 hr

$E_2 = 10{,}000$ cal/g mole

Solution. For this problem, the Hamiltonian is

$$H = (\lambda_2 - \lambda_1)k_1(T)c_1^2 - \lambda_2 k_2(T)c_2 \tag{6.5.23}$$

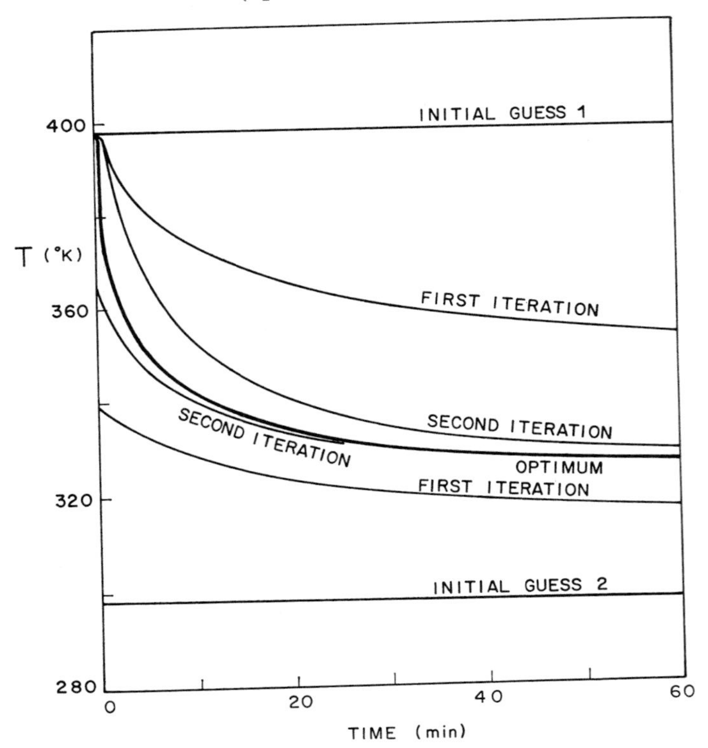

Fig. 6.8 The optimal temperature program in Example 6.5.2.

where the adjoint variables λ_1, λ_2 are given by

$$\frac{d\lambda_1}{dt} = -\frac{\partial H}{\partial c_1} = 2(\lambda_1 - \lambda_2)k_1(T)c_1 \qquad \lambda_1(1) = 0 \tag{6.5.24}$$

$$\frac{d\lambda_2}{dt} = -\frac{\partial H}{\partial c_2} = \lambda_2 k_2(T) \qquad \lambda_2(1) = 1 \tag{6.5.25}$$

and the gradient $\partial H/\partial T$ is

$$\frac{\partial H}{\partial T} = \frac{1}{RT^2}[(\lambda_2 - \lambda_1)E_1 k_1 c_1^{\,2} - E_2 \lambda_2 k_2(T) c_2] \tag{6.5.26}$$

The modified gradient algorithm was applied to the problem from two different initial guesses of $T(t)$. The resulting optimal temperature program is shown in Fig. 6.8 together with some of the intermediate iterations. Figure 6.9 shows the optimal yield, $c_2(1)$, as a function of the number of

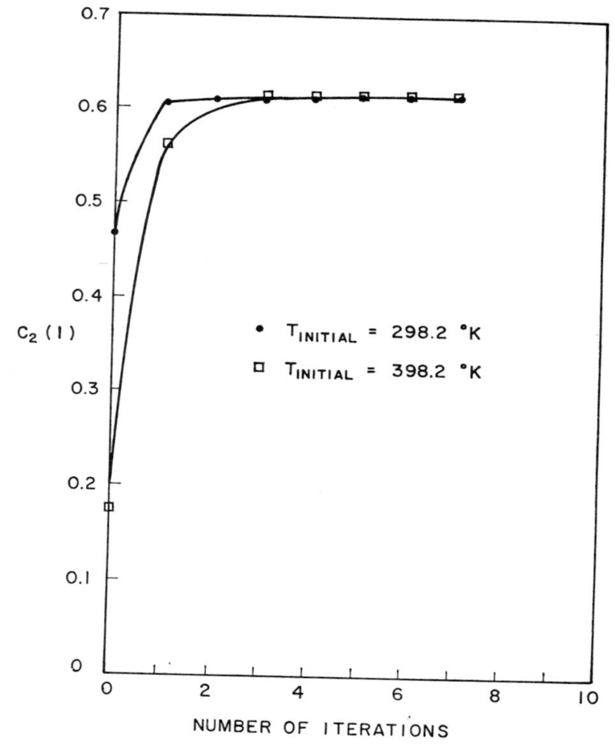

Fig. 6.9 Plot of $c_2(1)$ against the number of iterations.

iterations—for two initial guesses of the temperature program. As can be seen, the same optimal temperature program is found in both cases within 3 to 4 iterations. An inspection of the graph shows quite clearly that the optimal policy would produce very marked improvements in the yield of the desired intermediate species B. This improvement is found to be ~30% and 300% compared to operation of constant temperatures corresponding to the initial guesses of 398.2°K and 298°K, respectively. This example is thus an illustration of situations where optimization may produce significant improvements in performance.

We would like to re-emphasize here that successful trajectory optimization requires accurate modeling equations for the system, as was clearly illustrated by this example and the preceding problem discussed under Example 6.5.1. Although accurate modeling equations are likely to be available for many chemical processing operations, the development of appropriate quantitative kinetic information must necessarily precede the optimization proper in the majority of metallurgical systems. The trajectory optimization of metallurgical operations is an area where much useful (and very profitable) work is yet to be done.

6.6 DIRECT SUBSTITUTION APPROACHES

In Chapter 2 we have seen that by the solution of the necessary conditions for optimality, one could generate candidates for the optimal parameters. In a similar way, we can produce candidates for the optimal control trajectory by attempting to solve the necessary conditions for optimality given in Section 6.3 by substitution.

The first step in such a procedure is to eliminate the control vector, $\mathbf{u}(t)$, by solving Eq. 6.3.17 for $\mathbf{u}(t)$ explicitly:

$$u_i(t) = g_i(\mathbf{x}, \boldsymbol{\lambda}) \qquad i = 1, 2, \ldots, m \tag{6.6.1}$$

It is clear that this may not always be possible; however, it can be done for simple problems. If Eq. 6.6.1 is then substituted into Eqs. 6.3.1 and 6.3.10, the result is a set of $2n$ equations

$$\frac{d\mathbf{x}}{dt} = \mathbf{f}_1(\mathbf{x}, \boldsymbol{\lambda}); \quad \mathbf{x}(0) = \mathbf{x}_0 \tag{6.6.2}$$

$$\frac{d\boldsymbol{\lambda}}{dt} = \mathbf{f}_2(\mathbf{x}, \boldsymbol{\lambda}); \quad \boldsymbol{\lambda}(t_f) = \left(\frac{\partial G}{\partial \mathbf{x}}\right)_{tf} \tag{6.6.3}$$

with split boundary conditions.† This two-point boundary value problem (TPBVP) has a solution which produces the optimal values of $\mathbf{x}(t)$, $\boldsymbol{\lambda}(t)$, and when these are substituted into Eq. 6.6.1 one obtains the optimal control $\mathbf{u}(t)$.

What has been effected by the elimination of $\mathbf{u}(t)$ is the trading of a trajectory optimization problem for a TPBVP. TPBVP are notoriously difficult to solve, even numerically, and thus most of the techniques associated with this approach are techniques for solving TPBVP. Let us discuss several types of these techniques.

Boundary Condition Iteration

This approach tries to find, by some iterative procedure, the missing boundary conditions $\mathbf{x}(t_f)$ or $\boldsymbol{\lambda}(0)$ so that Eqs. 6.6.2 and 6.6.3 can be integrated together in the same direction of time. For simple scalar cases a mapping can be done of guessed values of $\lambda(0)$ *vs* the resulting values of $\lambda(t_f)_{\text{calc}} - (\partial G/\partial x)_{t_f}$ as sketched in Fig. 6.10. Obviously this graphical technique will

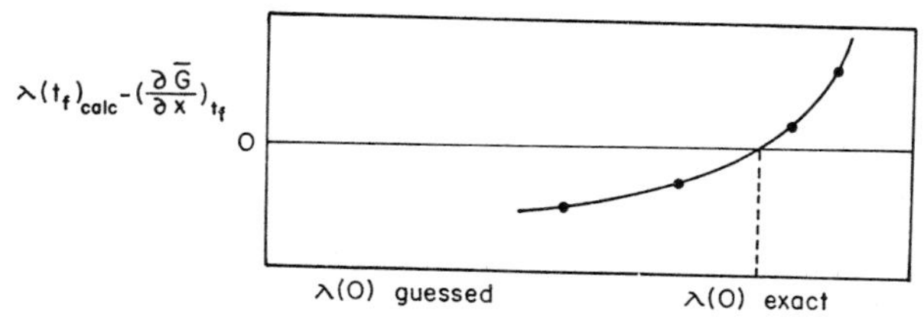

Fig. 6.10 Solution of a two-point boundary value problem by a mapping procedure.

not work well for multivariable problems. However, there are a number of techniques proposed for solving these problems [13] by perturbation methods or by minimizing the error in the boundary conditions by a direct search.

There is a basic difficulty with the boundary condition iteration approach which often arises in practical problems. The numerical integration of Eqs. 6.6.2 and 6.6.3 in the same direction is very often unstable. The reason for this behavior is that the state equations are usually stable when integrated forward, but are unstable in the reverse direction. Similarly, the adjoint equations are usually unstable when integrated forward, but are stable in the reverse direction. This can cause great numerical difficulties which are quite independent of the choice of the proper boundary conditions. Nevertheless, the method has been used successfully for some problems.

† The boundary conditions for $\boldsymbol{\lambda}(t_f)$ given here assume that all the $\mathbf{x}(t_f)$ are unspecified. In some instances we may end up with two sets of boundary conditions for Eqs. 6.6.2 and none for Eq. 6.6.3, as will be illustrated by the example given at the end of the section. However, these problems, too, are two-point boundary value problems.

Let us illustrate the direct substitution approach by considering a slight variation on the problem posed in Example 6.3.1.

EXAMPLE 6.6.1. As before, let us consider the radiant heating of a small billet, as described in Eq. 6.3.19:

$$\frac{dT}{dt} = C_1(T_s^4 - T^4), \qquad T(0) = T_0 \tag{6.3.19}$$

where T_s is the source temperature, T_0 is the initial temperature of the billet, and T_1 is the final desired temperature.

Our objective is to find the optimal source temperature $T_s(t)$ so as to bring the billet to the desired temperature in a minimum time, while minimizing the rate of wear of the refractory roof.

The rate of wear may be expressed as

$$C_2 e^{T_s^4} \tag{6.6.4}$$

where C_2 is a constant.

Thus the objective may be written as

$$\underset{T_s(t)}{\text{Min}} \left\{ I[T_s(t)] = t'_f + C_2 \int_0^{t'_f} e^{T_s^4(t)}\, dt' \right\} \tag{6.6.5}$$

where, as before, t'_f is left free and C_2 denotes the relative value of the roof erosion to operating time.

Let us define the variables,

$$x_0 = T_0, \quad x = T, \quad u(t) \equiv u = T_s^4, \quad t = C_1 t', \quad C = \frac{C_2}{C_1},$$

$$x_1 = T_1 \quad \text{and} \quad u_* = T_*^4, \quad u^* = T^{*4}$$

where, as before, u^* and u_* denote the upper and lower limits of the operating temperature, respectively.

The problem may now be written as:

$$\underset{u(t)}{\text{Min}} \left[I(u) = \int_0^{t_f} 1 + Ce^u\, dt \right] \tag{6.6.6}$$

subject to

$$\frac{dx}{dt} = u - x^4; \quad x(0) = x_0; \quad x(t_f) = x_1 \tag{6.6.7}$$

Let us follow the procedure set out in Eqs. 6.6.1 to 6.6.3.

The Hamiltonian is given as

$$H = \lambda(u - x^4) + Ce^u + 1 \tag{6.6.8}$$

Thus from Eq. 6.3.17 we have that

$$\frac{\partial \bar{H}}{\partial u} = 0, \tag{6.3.17}$$

that is,

$$\lambda + Ce^u = 0,$$

$$u = \ln\left(-\frac{\lambda}{C}\right) \tag{6.6.9}$$

On recalling Eq. 6.3.10, i.e.,

$$\frac{d\boldsymbol{\lambda}^T}{dt} = -\frac{\partial \bar{H}}{\partial \mathbf{x}} \tag{6.3.10}$$

by differentiation we obtain

$$\frac{d\lambda}{dt} = 4\lambda x^3 \tag{6.6.10}$$

Thus the optimal $u(t)$ is defined by

$$-\frac{dx}{dt} = x^4 + \ln\left(-\frac{\lambda}{C}\right) \tag{6.6.11}$$

$$x = x_0, \qquad t = 0$$

$$x = x_1, \qquad t = t_f$$

and

$$\frac{d\lambda}{dt} = 4\lambda x^3 \tag{6.6.10}$$

Equations 6.6.10 and 6.6.11 may be solved numerically, e.g., by the boundary condition iteration technique described above. The result would then be

$$x = x(t)$$

$$\lambda = \lambda(t)$$

from which $u = u(t)$ is readily obtained from Eq. 6.6.9.

We note that the applicability of the technique depends critically on the types of functional relationships involved.

Had the rate of wear, i.e., Eq. 6.6.4 been given by an alternative expression, e.g.,

$$\text{Rate of wear} = C_2' e^{T_s} \tag{6.6.12}$$

then H would have taken the following form:

$$H = \lambda(u - x^4) + C \exp\,[(u)^{1/4}] \tag{6.6.13}$$

thus

$$\frac{\partial H}{\partial u} = 0 = \lambda + \tfrac{1}{4} C u^{-3/4} \exp\,[(u)^{1/4}] \tag{6.6.14}$$

Clearly, Eq. 6.6.14 cannot be solved explicitly for u. While one could proceed with solving the problem for certain parametric relationships between u and λ, this is likely to be cumbersome in the majority of cases.

Quasilinearization

This technique (cf. [14]) linearizes Eqs. 6.6.2 and 6.6.3 about some reference trajectory $\hat{\mathbf{x}}$, $\hat{\boldsymbol{\lambda}}$, applies the principle of superposition to convert the TPBVP to an initial value problem, and solves this initial value problem numerically to produce the solution to this linearized TPBVP. Since the linearized solution will, in general, be in error, this procedure must be continued with a better estimate of $\hat{\mathbf{x}}$, $\hat{\boldsymbol{\lambda}}$ until the solution of the linearized equations converges to the solution of the nonlinear equations. This approach may have the numerical stability problems of the previous technique, and it may fail to converge unless a reasonably good guess of $\hat{\mathbf{x}}(t)$, $\hat{\boldsymbol{\lambda}}(t)$ is available; however, it has been used successfully on a number of practical problems [14].

6.7 CONTROL VECTOR PARAMETERIZATION

An alternative approach is termed control vector parameterization. For example one could represent $u_i(t)$ by a set of trial functions $\phi_{ij}(t)$, that is,

$$u_i(t) = \sum_{j=1}^{s} a_{ij}\phi_{ij}(t) \tag{6.7.1}$$

and use parameter optimization techniques to determine the optimal set of coefficients a_{ij}. A second approach, which seems to have a large number of advantages, is to generate u_i in a feedback form, i.e., expand in a set of trial functions of the state variables

$$u_i(t) = \sum_{j=1}^{p} b_{ij}\Phi_{ij}(x_1, x_2, \ldots, x_n) \tag{6.7.2}$$

and determine the optimal constants b_{ij}. Limited computational experience [15] has shown that this second parameterization scheme has much better convergence properties than the previous one.

Both of these approaches have the advantages that no adjoint equations need be solved, and standard parameter optimization techniques such as those presented in Chapter 3 can be used to determine the coefficients. An even greater advantage, as demonstrated in Fig. 6.11, is the ability of the technique to optimize complex process models by allowing a parameter

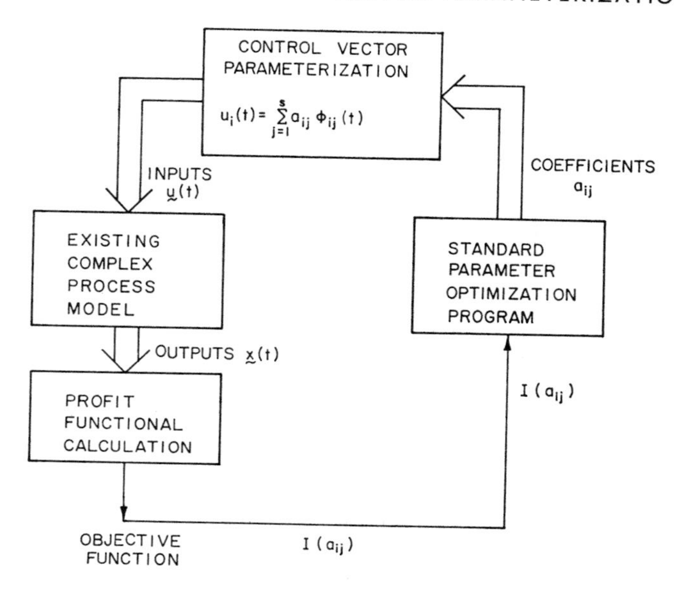

Fig. 6.11 The use of control vector parameterization with an existing complex process model.

estimation scheme to select the experiments to be performed on this model. This avoids having to modify existing process models in order to perform optimization—a significant practical advantage.

The principal disadvantage of parameterization methods is that the functional form of the optimal control must be specified in advance. This requires much more physical insight than is needed by the previous methods discussed. In the absence of a physical feeling for the general shape of the optimal control, a very general functional form (Eq. 6.7.1) must be used and the optimization performed with respect to a large number of coefficients. On the other hand, if one has good reasons to suspect a particular form of the optimal policy (e.g., a falling temperature profile in Example 6.5.2), then a simple functional form with only a few coefficients should be adequate. A *word of caution* is in order however: surprises do arise from optimization (that is why we do it), and in practical problems one should use several types of functional forms to insure that the functional form chosen is, in fact, general enough.

Examples given below and in Chapter 8 will illustrate the application of these ideas to particular problems.

Several other parameterization approaches have been reported and seem to have some merit. See Ref. 15 for a discussion and comparison of these methods.

EXAMPLE 6.7.1. Let us solve the moving packed bed optimization problem posed in Example 6.5.1 by control vector parameterization in the state variable, x.

Solution. Let us suppose that the solid temperature $u(t)$ can be represented by the feedback control law.

$$u = b_0 + b_1 x + b_2 x^2 \tag{6.7.3}$$

where it is understood that if u predicted by Eq. 6.7.3 violates the upper- and lower-bound constraints $u_* \le u \le u^*$, then the constraint value is used. Thus our procedure is

(i) Guess b_0, b_1, b_2.

(ii) Substitute Eq. 6.7.3 into Eq. 6.5.13 and calculate $I = x(1)$ for the particular values b_0, b_1, b_2 chosen.

(iii) Use a multivariable search technique from Section 3.3 to choose a new set of b_0, b_1, b_2.

(iv) Go back to step (ii) and iterate until I is maximized.

This technique was applied from the initial guesses, $b_0 = 973$, $b_1 = 200$, $b_2 = 200$ and determined the optimal parameters $\hat{b}_0 = 1415$; $\hat{b}_1 = -622$; $\hat{b}_2 = -1278.0$ using the pattern search program, PATERN. The results, plotted in Fig. 6.12 give an optimal objective value, $I = 0.8352$ compared

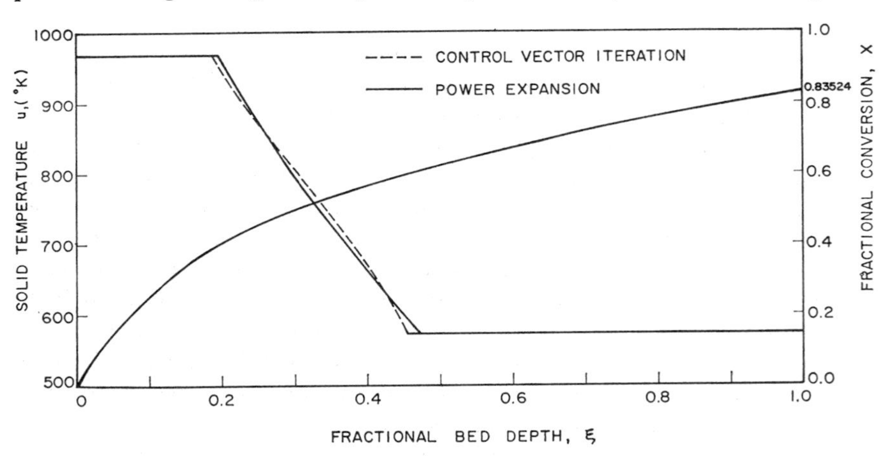

Fig. 6.12 The use of control vector parameterization in Example 6.7.1, nickel oxide reduction.

to the value $I = 0.8361$ found by rigorous means in Section 5.5. This close correspondence to the exact optimum, and the fact that only about 20 sec of computing time was required, makes this a very attractive approach.

One additional advantage of control vector parameterization illustrated by this example is that the programming effort is minimal and optimization requires less sophistication than the other trajectory optimization procedures.

6.8 DISCUSSION

Let us now discuss briefly the advantages and disadvantages of these various computational approaches. The control vector iteration procedure has the advantage that it can be applied with little algebraic manipulation even to the most complex problems. In addition, because the state equations are solved exactly at each stage, each iteration produces a feasible solution. This has the attraction that one may stop at any iteration with a suboptimal, but reasonably good useable solution.

The direct substitution procedures have the disadvantage that a fair amount of algebraic manipulation is required to produce the TPBVP, and then sophisticated procedures are needed for solution. The complexity of these methods makes them difficult for the novice to apply. One advantage of the boundary condition iteration approach is that every iteration produces an optimum solution—to the wrong problem. If $\mathbf{x}(t_f)$ is the boundary condition to be adjusted so that Eqs. 6.6.2 and 6.6.3 are integrated backward together, then a calculated value of $\mathbf{x}(0)$ is produced at each iteration. Thus each iteration produces the optimal solution for that calculated initial condition $\mathbf{x}(0)$. This property would be useful if one wished to obtain the optimal policies for a variety of initial conditions.

The control vector parameterization procedure seems to be the most attractive for the novice. Very little sophistication is required and standard techniques for parameter optimization may be applied. The one major disadvantage seems to be that there is no guarantee that the parameterized optimal control will converge to the exact optimum unless the trial functions are chosen in a sufficiently general way. The number of trial functions needs to be as small as possible to minimize the number of coefficients to be optimized, and yet the functional form must be capable of representing the exact optimum. Thus care must be exercised in the choice of trial functions.

A final word on the practical problems of *convergence* is in order. As in parameter optimization problems, trajectory optimization algorithms always stop progressing before the exact optimum is reached. However, efficient algorithms will usually stop very close to the true optimum. Thus, to insure that the optimum has indeed been found, one must be able to produce the same "optimal policy" from several initial guesses. This would seem to insure that, at least, a local optimum has been found. One should be aware that multiple optima are possible and in rare cases these have been found in real problems. Thus even though several starting points *must always be used* to insure that the algorithm has converged, this is not an absolute guarantee that the global optimum has been found.

6.9 SOME SPECIAL CONSIDERATIONS

A discussion of trajectory optimization would be somewhat misleading without mention of several important special cases that arise frequently in practical problems and require special methods of solution.

Linear Systems

If one has a problem in which the model has the linear form

$$\frac{d\mathbf{x}}{dt} = \mathbf{A}\mathbf{x} + \mathbf{B}\mathbf{u}; \quad \mathbf{x}(0) = \mathbf{x}_0 \tag{6.9.1}$$

then there are special techniques often requiring no iteration which can be applied to these problems. Since this is a rather broad area, we shall not develop these special approaches here; the interested reader is advised to consult Refs. 2 and 16.

Systems Linear in the Control

If one has a problem in which the control appears linearly in the Hamiltonian

$$H = h_0(\mathbf{x}, \boldsymbol{\lambda}) + \sum_{i=1}^{m} h_i(\mathbf{x}, \boldsymbol{\lambda})\, u_i(t) \tag{6.9.2}$$

then the structure of the optimal control policy is clear without further computation. For example, if $\mathbf{u}(t)$ is constrained by $\mathbf{u}_* \leq \mathbf{u} \leq \mathbf{u}^*$, then from the strong maximum principle (Theorem 2, Section 6.3) we see that the

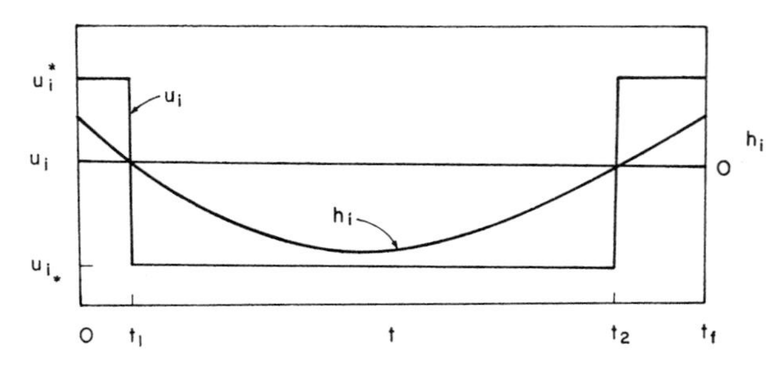

Fig. 6.13 Bang-Bang control policy.

optimal control has the form:

$$u_i(t) = \begin{cases} u_i^* & \text{if} \quad h_i > 0 \\ u_{i*} & \text{if} \quad h_i < 0 \end{cases} \tag{6.9.3}$$

This behavior, plotted in Fig. 6.13, is called a *bang-bang* control policy. The points t_1, t_2 where $h_i(t)$ changes sign are called switching times.

There is another special situation which occurs when $h_i = 0$ over some interval of time. An examination of Fig. 6.14 shows us that $h_i = 0$ over

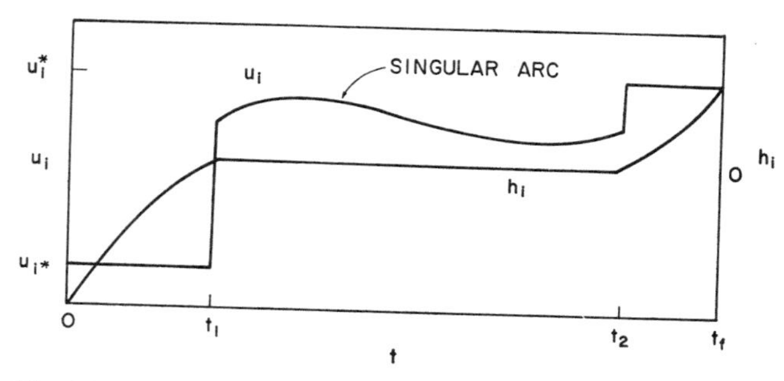

Fig. 6.14 Control policy with a singular arc.

$t_1 \leq t \leq t_2$. Since this causes the control to vanish from H in that interval, it is not clear how we can determine the optimal value of $u_i(t)$ over $t_1 \leq t \leq t_2$. The control over this interval is called a *singular arc*, and these especially difficult problems are called *singular control problems*. The reader is referred to [16, 17] for a deeper discussion of these problems and the techniques available for their solution.

PROBLEMS

1. A certain set of parallel reactions

$$A \xrightarrow{k_1} B$$

$$A \xrightarrow{k_2} C$$

is carried out in a tubular reactor of length L with a plug flow velocity v. Both reactions are irreversible, first order, with velocity constants given by

$$k_i = A \exp\left[-E_i/RT\right] \qquad i = 1, 2$$

The material balances along the tube $0 \leq z \leq L$ are

$$v\frac{dA}{dz} = -(k_1 + k_2)A \qquad A(0) = A_f$$

$$v\frac{dB}{dz} = k_1 A \qquad B(0) = 0$$

The optimization problem is to find the temperature profile $T(z)$ which maximizes the production of B. The problem can be simplified by defining

$$x_1 = \frac{A}{A_f}, \quad x_2 = \frac{B}{A_f}, \quad u = \frac{k_1 L}{v}, \quad y = \frac{z}{L}, \quad p = \frac{E_2}{E_1}$$

$$\beta = \frac{L A_Z}{v\left(\dfrac{L}{v} A_1\right)^p}$$

so that our objective is to choose $u(y)$, $0 \leq y \leq 1$ so as to maximize

$$I = x_2(1)$$

while satisfying the material balances

$$\frac{dx_1}{dy} = -(u + \beta u^p)x_1 \qquad x_1(0) = 1$$

$$\frac{dx_2}{dy} = ux_1 \qquad x_2(0) = 0$$

The data needed for optimization are the practical bounds on temperature, $0 \leq u \leq 6$; the ratio of activation energies, $p = 2.0$; and the value of β, $\beta = 0.5$. It should be noted that the use of this dimensionless form reduces the number of parameters required considerably, so that the optimization results can apply to all systems having these parameters.

(i) Apply the maximum principle to this problem and calculate the optimal temperature profile $u(y)$ using a modified gradient procedure. Check your results from two initial guesses of $u(y)$.

(ii) Apply the control vector parameterization procedure to the problem by finding the coefficients a_0, a_1, a_2 in the expression

$$u(y) = a_0 + a_1 y + a_2 y^2$$

which maximize I. Use a multivariable search routine (cf. Appendix C) to determine a_0, a_1, and a_2. Compare your results with that found by the gradient method.

(iii) Assume, as is likely, that the optimal profile cannot be imposed exactly, but one must be satisfied with three isothermal sections. Determine, through a control vector parameterization procedure, the optimal temperature (b_1, b_2, b_3) of each of the isothermal sections

$$u = \begin{cases} b_1 & 0 \le y \le \frac{1}{3} \\ b_2 & \frac{1}{3} \le y \le \frac{2}{3} \\ b_3 & \frac{2}{3} \le y \le 1 \end{cases}$$

2. We wish to develop the design for the optimal temperature progression for a kiln, in which small ceramic specimens are being fired. In a laboratory scale experimental study it was found that the maximum temperature program which avoids damage to the material is represented by the following data:

Time (hr)	T_{msp}, Maximum specimen temperature (°C)
0	20
1	50
2	100
3	300
4	450
5	800
6	900
7	980
8	1050
9	1100

We may assume that the specimens are small enough so that we can neglect the temperature gradients within them, and that heat is transferred exclusively by thermal radiation.

The rate at which heat is transferred to the specimens is given by the following expression

$$\dot{q} = 0.3\sigma(T_f^4 - T_{sp}^4)$$

where $\dot{q}$ = the heat flux absorbed by the specimen

0.3 = a numerical factor that combines the emissivities and the view factors

σ = Boltzmann's constant = 1.37×10^{-12} cal sec^{-1} cm^{-2} °K^{-4}

$T_f(t)$ = the time dependent (or space dependent) furnace wall temperature and

T_{sp} = the temperature of the specimen

$\dot{q}$ and T_{sp} are further related by the following heat balance equation:

$$\frac{dT_{sp}}{dt}\rho C_p = R\dot{q}$$

where ρ = density of the specimens, taken as 3.1 g/cm^3
C_p = specific heat of the specimens, taken as 0.4 cal/g°C
R = is the ratio exposed surface area/volume which is taken as 2 cm^{-1}

The additional constraint that has to be observed is that

$$T_f < 1300°\text{C}$$

The problem is then:

(i) to find the furnace temperature progression, $T_f(t)$ such that the material is brought to the final temperature, 1100°C, rapidly, while not violating the maximum temperature schedule given in the table. Also it is desired that the source temperature not be excessive in order to conserve heating costs. An objective functional, which if minimized, produces the desired profile is

$$I = \int_0^9 [(T_{sp} - T_{msp})^2 + 0.0001(T_f - 800°\text{K})^2]\, dt$$

Use one of the computational techniques of Section 6.4 to determine the optimal source temperature program, $T_f(t)$.

(ii) If it were possible only to use three discrete temperature levels in the furnace, for three time periods, rather than having the furnace temperature as a continuous function of time, determine the optimal stepwise furnace temperature program.

Hint: As a first step, fit a polynomial or an exponential function to the data points in the table.

3. Assume that you wish to determine the optimal temperature profile described in Problem 1 except that in this case a fraction γ of the product stream is recycled and mixed with fresh feed. This causes the boundary conditions on x_1 and x_2 to become

$$x_1(0) = (1 - \gamma) + \gamma x_1(1)$$
$$x_2(0) = \gamma x_2(1)$$

(i) Apply the maximum principle to this problem. In particular, examine Eq. 6.3.8 and show that only the boundary conditions on the adjoint variables change from Problem 1. Show that the new boundary conditions on $\boldsymbol{\lambda}$ become

$$\lambda_1(1) = \gamma\lambda_1(0)$$

$$\lambda_2(1) = 1 + \gamma\lambda_2(0)$$

(ii) Develop a computational procedure for determining the optimal temperature profile in this case. Other computational algorithms have been reported by J. D. Paynter and S. G. Bankoff, *Can. J. Ch. E.* **44,** 340 (1966); F. J. M. Horn and R. C. Lin, *I. and E.C. Process Design* **6,** 21 (1967).

(iii) Test your computational algorithm by calculating the optimal temperature profile for $\gamma = 0.1, 0.2, 0.5$. Compare your results with those of Problem 1 to see the effect of product recycle for this problem.

REFERENCES

1. L. S. Pontryagin et al., *Mathematical Theory of Optimal Processes,* Wiley, 1962.
2. E. B. Lee and L. Markus, *Foundations of Optimal Control Theory,* Wiley, 1967.
3. M. M. Denn, *Optimization by Variational Methods,* McGraw-Hill, 1969.
4. S. E. Dreyfus, *Dynamic Programming and the Calculus of Variations,* Academic Press, 1965.
5. G. Leitmann, *An Introduction to Optimal Control,* McGraw-Hill, 1966.
6. R. E. Kopp and H. G. Moyer, *Adv. Control Syst.* **4** (1966).
7. D. H. Jacobson, *Int. J. Control* **7,** 175 (1968).
8. L. Padmanabhan and S. G. Bankoff, *Proc. 1969 Joint Autom. Control Conf.* **1969,** p. 34.
9. L. S. Lasdon, S. K. Mitter, and A. D. Waren, *IEEE Trans. Auto. Cont.* **AC12,** 132 (1967).
10. B. Pagurek and C. M. Woodside, *Automatica* **4,** 337 (1968).
11. J. W. Evans, Ph.D. Thesis, State University of New York at Buffalo, 1970.
12. O. Levenspiel, *Chemical Reaction Engineering,* Wiley, 1962, p. 217.
13. L. Fox, *The Numerical Solution of Two Point Boundary Value Problems,* Oxford, 1957.
14. E. S. Lee, *Quasi-Linearization and Invariant Imbedding,* Academic Press, 1968.
15. G. A. Hicks and W. H. Ray, *Can. J. Ch. E.* **49,** 522 (1971).
16. A. W. Bryson and Y-C Ho, *Applied Optimal Control,* Blaisdell, 1969.
17. C. D. Johnson, *Adv. Control Syst.* **2,** (1964).

7 Trajectory Optimization of Distributed Parameter Systems

7.1 INTRODUCTION

In a number of practical problems, either the variables to be optimized or the state variables are distributed in both space and time. In this chapter we shall consider the optimization of these *distributed trajectories*. Although distributed parameter systems can have models which are integro-differential equations or differential-difference equations as well as partial differential equations, in this chapter we shall be concerned with partial differential equation representations, resulting naturally from transport models of metallurgical or chemical processes.

In the next section we shall discuss a simple heat transfer example that illustrates the principles. In Section 3 we shall broaden our analysis to a more general class of distributed parameter models, for which the necessary conditions of optimality are developed in Section 4. Section 5 deals with some computational approaches which can be used to synthesize the optimal trajectories. Finally, more complex examples are given to demonstrate these computational procedures.

7.2 A HEAT TRANSFER PROBLEM

Consider the problem of reheating a steel slab by thermal radiation (for rolling) in a batch furnace as sketched in Fig. 7.1. For proper rolling characteristics, it is desirable that the slab have a specified temperature distribution $T_s(z)$. Thus our problem is to control the heat flux to the

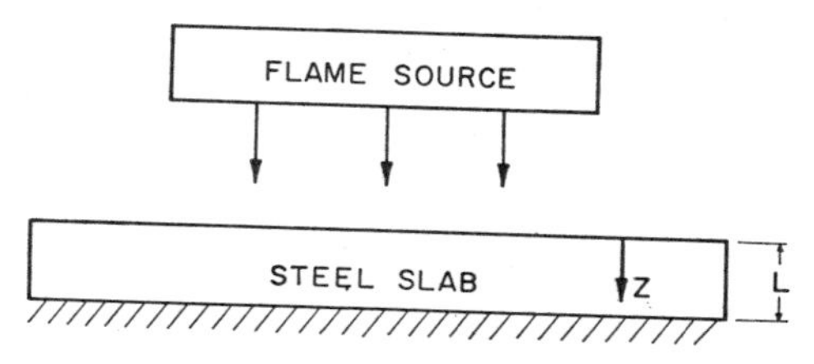

Fig. 7.1 Radiant heating of a slab.

surface of the slab in such a way as to approach this desired temperature distribution in some optimal fashion.

To be more precise, let us consider the modeling equations for the slab

$$\frac{\partial T(z,t)}{\partial t} = \frac{1}{\beta(T)} \frac{\partial\left[\dfrac{\alpha(T)\,\partial T(z,t)}{\partial z}\right]}{\partial z} \qquad \begin{matrix} 0 \le z \le L \\ 0 \le t \le \theta \end{matrix} \tag{7.2.1}$$

$$\frac{\partial T(0,t)}{\partial z} = v(t) \tag{7.2.2}$$

$$\frac{\partial T(L,t)}{\partial z} = 0 \tag{7.2.3}$$

$$T(z,0) = T_0(z) \tag{7.2.4}$$

which reflect the fact that negligible heat is lost at the sides and bottom of the slab, and by adjusting the flame, one can control the heat flux at the upper surface between bounds

$$v_* \le v(t) \le v^* \tag{7.2.5}$$

In order to quantify our wish to approach the desired temperature distribution $T_s(z)$ in a short period of time, let us choose the objective

$$I[v(t)] = \int_0^\theta \int_0^L [T(z,t) - T_s(z)]^2 \, dt \, dz \tag{7.2.6}$$

which we shall minimize. This objective has the advantage that it will drive $T(z, t)$ toward the desired final temperature distribution $T_s(z)$ with the minimum integral square error, in a short time, irrespective of the choice of θ.

Thus, if we could find the control $v(t)$ which minimizes I in Eq. 7.2.6, we would have the optimal trajectory of this distributed parameter system. In the remainder of this chapter we shall develop the machinery for determining such optimal trajectories for these kinds of problems.

7.3 A GENERAL CLASS OF PROBLEMS

Now that we can appreciate the type of physical problems which are to be tackled in this chapter, we shall broaden the scope of our interest to include the more general class of problems described by the following partial differential equations:

$$\mathbf{A}\frac{\partial \mathbf{x}}{\partial t} = \mathbf{f}\left(\mathbf{x}, \frac{\partial \mathbf{x}}{\partial z}, \frac{\partial^2 \mathbf{x}}{\partial z^2}, \mathbf{u}\right) \tag{7.3.1}$$

$$0 \leq t \leq \theta$$

$$0 \leq z \leq L$$

where $\mathbf{x}(t, z)$ is an n vector of state variables, $\mathbf{u}(t, z)$ is an m vector of control variables, and $\mathbf{A}$ is an $n \times n$ matrix. To prevent matters from becoming too complex, we shall restrict ourselves to two independent variables $0 \leq t \leq \theta$, $0 \leq z \leq L$, although the analysis could be extended to more independent variables in a straightforward way [1, 2].

Equation 7.3.1 is the general representation of a very large number of practical problems. The drying of porous materials, the behavior of chemical or hydrometallurgical reactor systems, and heat transfer problems such as described in the last section are only a few examples of problems having this form. We shall describe a number of specific examples in Chapter 8.

The boundary conditions associated with Eq. 7.3.1 depend on the particular problem being considered; however, normally there is an initial state

$$\mathbf{x}(z, 0) = \mathbf{w}(z) \tag{7.3.2}$$

which may be available as a control variable. For example, the initial temperature distribution in the slab of the previous section might be subject to control by preheat. The system boundary conditions are usually split (for obvious physical reasons) and can take a variety of forms. We shall consider three separate cases of boundary conditions here:

CASE 1. Some state variables, x_s, may have boundary conditions of the form

$$\frac{\partial x_s}{\partial z} = g_s(\mathbf{x}, \mathbf{v}(t)) \qquad \text{at} \quad z = 0 \tag{7.3.3a}$$

$$\frac{\partial x_s}{\partial z} = h_s(\mathbf{x}, \mathbf{y}(t)) \qquad \text{at} \quad z = L \tag{7.3.4a}$$

as, for example, when there is convective or radiant heat transfer at the surface.

CASE 2. Other state variables, x_r, may have boundary conditions of the form

$$x_r(0, t) = \text{const.} \tag{7.3.3b}$$

$$x_r(L, t) = \text{const.} \tag{7.3.4b}$$

CASE 3. Still others, x_p, may take the form

$$x_p(0, t) = v_p(t) \tag{7.3.3c}$$

$$x_p(L, t) = y_p(t) \tag{7.3.4c}$$

which allows the surface conditions to be controlled in an optimal fashion. Here $\mathbf{v}(t)$ is a control operating at $z = 0$, and $\mathbf{y}(t)$ is a control operating at $z = L$. It can be seen that the boundary conditions given by Eqs. 7.2.2 and 7.2.3 are just special cases of this general form.

The trajectory optimization problem for this system can be stated in the most general way, as the desire to maximize the functional

$$\begin{aligned} I[\mathbf{u}(z,t), \mathbf{v}(t), \mathbf{y}(t), \mathbf{w}(z)] &= \int_0^L G_1(\mathbf{x}(\theta, z), \mathbf{w}(z))\, dz \\ &\quad + \int_0^\theta G_2(\mathbf{x}(L, t), \mathbf{x}(0, t), \mathbf{y}, \mathbf{v})\, dt \\ &\quad + \int_0^L \int_0^\theta G\left(\mathbf{x}, \mathbf{u}, \frac{\partial \mathbf{x}}{\partial z}, \frac{\partial^2 \mathbf{x}}{\partial z^2}\right) dz\, dt \end{aligned} \tag{7.3.5}$$

by choosing the trajectories of $\mathbf{u}(z, t)$, $\mathbf{v}(t)$, $\mathbf{y}(t)$, $\mathbf{w}(z)$.

In the next section we derive necessary conditions for optimality for this system.

7.4 NECESSARY CONDITIONS FOR OPTIMALITY

Just as was done in Chapter 6 for lumped parameter systems, we shall use first variations to derive a weak maximum principle for the system described by Eqs. 7.3.1 to 7.3.5; however, the details of the derivation are not necessary for an understanding of the remainder of the chapter, and the reader may skip to the results on p. 259 if he wishes.

As in the case of the maximum principle for ordinary differential equations, let us assume that we have a nominal set of optimal control trajectories $\bar{\mathbf{u}}(z, t)$, $\bar{\mathbf{v}}(t)$, $\bar{\mathbf{y}}(t)$, $\bar{\mathbf{w}}(z)$, and let us consider the effect of variations $\delta\mathbf{u}$, $\delta\mathbf{v}$, $\delta\mathbf{y}$, $\delta\mathbf{w}$ about these nominal trajectories. We shall begin by expanding Eq. 7.3.1 about the nominal trajectories to yield the perturbation equations

$$A_{ij}\frac{\partial(\delta x_j)}{\partial t} = \frac{\partial \bar{f}_i}{\partial x_j}\,\delta x_j + \frac{\partial \bar{f}_i}{\partial u_k}\,\delta uk + \frac{\partial \bar{f}_i}{\partial(\dot{x}_j)}\,\delta(\dot{x}_j) + \frac{\partial \bar{f}_i}{\partial(\ddot{x}_j)}\,\delta(\ddot{x}_j) \quad (7.4.1)$$

where the overbar signifies that the quantity is evaluated along the nominal trajectory. In addition

$$\dot{x}_j \equiv \frac{\partial x_j}{\partial z}\,; \quad \ddot{x}_j \equiv \frac{\partial^2 x_j}{\partial z^2}$$

and we use the convention that a repeated subscript denotes a sum over that index; for example,

$$\frac{\partial f_i}{\partial x_k}\,\delta x_k \equiv \sum_{k=1}^{n}\frac{\partial f_i}{\partial x_k}\,\delta x_k = \frac{\partial f_i}{\partial x_1}\,\delta x_1 + \frac{\partial f_i}{\partial x_2}\,\delta x_2 + \cdots + \frac{\partial f_i}{\partial x_n}\,\delta x_n$$

Equation 7.4.1 can be rewritten as

$$A_{ij}\frac{\partial(\delta x_j)}{\partial t} = \left(\frac{\partial \bar{f}_i}{\partial x_k}\right)\delta x_k + \left(\frac{\partial \bar{f}_i}{\partial u_k}\right)\delta u_k + \frac{\partial \bar{f}_i}{\partial(\dot{x}_j)}\frac{\partial(\delta x_j)}{\partial z} + \frac{\partial \bar{f}_i}{\partial(\ddot{x}_j)}\frac{\partial^2(\delta x_j)}{\partial z^2} \quad (7.4.2)$$

Expanding the objective (Eq. 7.3.5) in the same way yields

$$\begin{aligned}\delta I = &\int_0^L \left[\frac{\partial \bar{G}_1}{\partial x_k(z,\theta)}\,\delta x_k(z,\theta) + \frac{\partial \bar{G}_1}{\partial w_j(z)}\,\delta w_j(z)\right] dz \\ &+ \int_0^\theta \left[\frac{\partial \bar{G}_2}{\partial x_k(L,t)}\,\delta x_k(L,t) + \frac{\partial \bar{G}_2}{\partial x_k(0,t)}\,\delta x_k(0,t)\right. \\ &\qquad\left. + \frac{\partial \bar{G}_2}{\partial y_j(t)}\,\delta y_j(t) + \frac{\partial \bar{G}_2}{\partial v_j(t)}\,\delta v_j(t)\right] dt \\ &+ \int_0^\theta\int_0^L \left[\frac{\partial \bar{G}}{\partial x_k}\,\delta x_k + \frac{\partial \bar{G}}{\partial u_i}\,\delta u_i + \frac{\partial \bar{G}}{\partial(\dot{x}_j)}\frac{\partial(\delta x_j)}{\partial z} + \frac{\partial \bar{G}}{\partial(\ddot{x}_j)}\frac{\partial^2(\delta x_j)}{\partial z^2}\right] dt\,dz\end{aligned} \quad (7.4.3)$$

Now let us use a distributed Lagrange multiplier (called an adjoint variable) $\lambda_k(z, t)$ to form the quantity

$$\int_0^\theta \int_0^L \left\{ \lambda_i(z,t) \left[A_{ij} \frac{\partial(\delta k_j)}{\partial t} - \left(\frac{\partial \bar{f}_i}{\partial x_k}\right) \delta x_k - \left(\frac{\partial \bar{f}_i}{\partial u_k}\right) \delta u_k - \frac{\partial \bar{f}_i}{\partial(\dot{x}_j)} \frac{\partial(\delta x_j)}{\partial z} - \frac{\partial \bar{f}_i}{\partial(\ddot{x}_j)} \frac{\partial^2(\delta x_j)}{\partial z^2} \right] \right\} dt\, dz = 0 \quad (7.4.4)$$

which can be subtracted from Eq. 7.4.3 to yield

$$\begin{aligned} \delta I = & \int_0^\theta \left[\frac{\partial \bar{G}_2}{\partial x_k(L,t)} \delta x_k(L,t) + \frac{\partial \bar{G}_2}{\partial x_k(0,t)} \delta x_k(0,t) + \frac{\partial \bar{G}_2}{\partial y_j(t)} \delta y_j(t) + \frac{\partial \bar{G}_2}{\partial v_j(t)} \delta v_j(t) \right] dt \\ & + \int_0^L \left[\frac{\partial \bar{G}_1}{\partial x_k(z,\theta)} \delta x_k(z,\theta) + \frac{\partial \bar{G}_1}{\partial w_j(z)} \delta w_j(z) \right] dz \\ & + \int_0^\theta \int_0^L \left\{ \left(\frac{\partial H}{\partial x_k}\right) \delta x_k + \left(\frac{\partial H}{\partial u_i}\right) \delta u_i + \frac{\partial H}{\partial(\dot{x}_j)} \frac{\partial(\delta x_j)}{\partial z} + \frac{\partial H}{\partial(\ddot{x}_j)} \frac{\partial^2(\delta x_j)}{\partial z^2} - \lambda_i \left[A_{ij} \frac{\partial(\delta x_j)}{\partial t} \right] \right\} dt\, dz \quad (7.4.5) \end{aligned}$$

where the quantity H (known as the *Hamiltonian*) is defined as

$$H = \bar{G} + \lambda_i \bar{f}_i \quad (7.4.6)$$

If we integrate the last three terms by parts so that

$$\int_0^\theta \int_0^L \left(\frac{\partial H}{\partial \dot{x}_j} \frac{\partial(\delta x_j)}{\partial z} \right) dz\, dt = \int_0^\theta \left\{ \left[\frac{\partial H}{\partial \dot{x}_j} \delta x_j \right]_0^L - \int_0^L \frac{\partial}{\partial z} \left(\frac{\partial H}{\partial \dot{x}_j} \right) \delta x_j\, dz \right\} dt \quad (7.4.7)$$

$$\int_0^\theta \int_0^L \left(\frac{\partial H}{\partial(\ddot{x}_j)} \frac{\partial^2(\delta x_j)}{\partial z^2} \right) dz\, dt = \int_0^\theta \left\{ \left[\frac{\partial H}{\partial(\ddot{x}_j)} \frac{\partial(\delta x_j)}{\partial z} \right]_0^L - \left[\frac{\partial}{\partial z} \left(\frac{\partial H}{\partial \ddot{x}_j} \right) \delta x_j \right]_0^L + \int_0^L \frac{\partial^2 \left(\frac{\partial H}{\partial \ddot{x}_j} \right)}{\partial z^2} \delta x_j\, dz \right\} dt \quad (7.4.8)$$

$$\int_0^\theta \int_0^L \lambda_i \left[A_{ij} \frac{\partial(\delta x_j)}{\partial t} \right] dt\, dz = \int_0^L \left\{ \left[\lambda_i A_{ij}\, \delta x_j \right]_0^\theta - \int_0^\theta \frac{\partial(\lambda_i A_{ij})}{\partial t} \delta x_j\, dt \right\} dz \quad (7.4.9)$$

then Eq. 7.4.5 becomes

$$\delta I = \int_0^\theta \int_0^L \left\{ \left[\frac{\partial H}{\partial x_k} - \frac{\partial \left(\frac{\partial H}{\partial \dot{x}_k}\right)}{\partial z} + \frac{\partial^2 \left(\frac{\partial H}{\partial \ddot{x}_k}\right)}{\partial z^2} + \frac{\partial(\lambda_i A_{ik})}{\partial t} \right] \delta x_k + \left[\frac{\partial H}{\partial u_i} \right] \delta u_i \right\} dt\, dz$$

$$+ \int_0^\theta \left\{ \left[\frac{\partial \bar{G}_2}{\partial x_k(L, t)} + \frac{\partial H}{\partial \dot{x}_k} - \frac{\partial \left(\frac{\partial H}{\partial \ddot{x}_k}\right)}{\partial z} \right] \delta x_k(L, t) + \frac{\partial \bar{G}_2}{\partial y_j} \delta y_j(t) \right.$$

$$+ \frac{\partial \bar{G}_2}{\partial v_j(t)} \delta v_j(t) + \left[\frac{\partial H}{\partial(\ddot{x}_j)} \frac{\partial(\delta x_j)}{\partial z} \right]_0^L$$

$$\left. + \left[\frac{\partial \bar{G}_2}{\partial x_j(0, t)} - \frac{\partial H}{\partial \dot{x}_j} + \frac{\partial}{\partial z}\left(\frac{\partial H}{\partial \ddot{x}_j}\right) \right] \delta x_j(0, t) \right\} dt$$

$$+ \int_0^L \left\{ \left[\frac{\partial \bar{G}_1}{\partial x_k(\theta, z)} - \lambda_i A_{ik} \right] \delta x_k(z, \theta) + \left[\frac{\partial \bar{G}_1}{\partial w_k(z)} + \lambda_i A_{ik} \right] \delta w_k(z) \right\} dz \tag{7.4.10}$$

To remove the explicit dependence of δI on $\delta \mathbf{x}(z, t)$, let us define the adjoint variables, $\lambda_i(z, t)$, by

$$\boxed{\frac{\partial(\lambda_i A_{ik})}{\partial t} = -\left[\frac{\partial H}{\partial x_k} - \frac{\partial(\partial H/\partial \dot{x}_k)}{\partial z} + \frac{\partial^2(\partial H/\partial \ddot{x}_k)}{\partial z^2} \right] \qquad k = 1, 2, \ldots, n} \tag{7.4.11}$$

which causes the first term in Eq. 7.4.10 to vanish.

Now let us consider the three separate cases that can arise from the boundary conditions (Eqs. 7.3.3 and 7.3.4).

CASE 1. For those state variables having boundary conditions Eqs. 7.3.3a and 7.3.4a the boundary condition variations become:

$$\frac{\partial(\delta x_i(0, t))}{\partial z} = \left[\frac{\partial \bar{g}_i}{\partial x_j(0, t)} \delta x_j(0, t) + \frac{\partial \bar{g}_i}{\partial v_j(t)} \delta v_j(t) \right] \tag{7.4.12}$$

$$\frac{\partial(\delta x_i(L, t))}{\partial z} = \left[\frac{\partial \bar{h}_i}{\partial x_j(L, t)} \delta x_j(L, t) + \frac{\partial \bar{h}_i}{\partial v_j(t)} \delta v_j(t) \right] \tag{7.4.13}$$

CASE 2. For those state variables with boundary conditions of the form of Eqs. 7.3.3b and 7.3.4b, the variations

$$\left.\frac{\partial(\delta x_i)}{\partial z}\right]_0^L$$

are free and the variations

$$\delta x_i(0, t), \qquad \delta x_i(L, t)$$

vanish.

CASE 3. For those state variables with boundary conditions Eqs. 7.3.3c, and 7.3.4c, the variations

$$\left.\frac{\partial(\delta x_i)}{\partial z}\right]_0^L$$

are free and

$$\delta x_i(0, t) = \delta v_i(t), \qquad \delta x_i(L, t) = \delta y_i(t) \tag{7.4.14}$$

If we denote the state variables in Case 1 by index s, use index r for those in Case 2, and index p for those in Case 3, we can rewrite Eq. 7.4.10 as

$$\begin{aligned}
\delta I = &\int_0^\theta \int_0^L \frac{\partial H}{\partial u_i}\,\delta u_i\,dt\,dz \\
&+ \int_0^L \left\{\left[\frac{\partial \bar{G}_1}{\partial x_k(z,\theta)} - \lambda_i A_{ik}\right]\delta x_k(z,\theta) + \left[\frac{\partial \bar{G}_1}{\partial w_k(z)} + \lambda_i A_{ik}\right]\delta w_k(z)\right\} dz \\
&+ \int_0^\theta \left\{\frac{\partial H_2}{\partial v_s}\,\delta v_s(t) + \frac{\partial H_3}{\partial y_s}\,\delta y_s + \left[\frac{\partial H_2}{\partial x_s(0,t)} - \frac{\partial H(0,t)}{\partial \dot{x}_s} + \frac{\partial}{\partial z}\left(\frac{\partial H}{\partial \ddot{x}_s}\right)\right]\delta x_s(0,t)\right. \\
&+ \left.\left[\frac{\partial H_3}{\partial x_s(L,t)} + \frac{\partial H(L,t)}{\partial \dot{x}_s} - \frac{\partial}{\partial z}\left(\frac{\partial H}{\partial \ddot{x}_s}\right)\right]\delta x_s(L,t)\right\} dt \\
&+ \int_0^\theta \left\{\left[\frac{\partial \bar{G}_2}{\partial x_r(0,t)} - \frac{\partial H}{\partial \dot{x}_r} + \left(\frac{\partial H}{\partial \ddot{x}_r}\right)\right]\delta x_r(0,t)\right. \\
&+ \left[\frac{\partial G_2}{\partial x_r(L,t)} + \frac{\partial H}{\partial \dot{x}_r} - \frac{\partial(\partial H/\partial \ddot{x}_r)}{\partial z}\right]\delta x_r(L,t) \\
&+ \left.\left[\frac{\partial H}{\partial \ddot{x}_r}\frac{\partial(\delta x_r)}{\partial z}\right]_0^L\right\} dt + \int_0^\theta \left\{\left[\frac{\partial \bar{G}_2}{\partial v_p} - \frac{\partial H}{\partial \dot{x}_p} + \frac{\partial}{\partial z}\left(\frac{\partial H}{\partial \ddot{x}_p}\right)\right]\delta v_p(t)\right. \\
&+ \left.\left[\frac{\partial \bar{G}_2}{\partial y_p} + \frac{\partial H}{\partial \dot{x}_p} - \frac{\partial(\partial H/\partial \ddot{x}_p)}{\partial z}\right]\delta y_p(t) + \left[\frac{\partial H}{\partial(\ddot{x}_p)}\frac{\partial(\delta x_p)}{\partial z}\right]_0^L\right\} dt
\end{aligned} \tag{7.4.15}$$

where we have defined additional *Hamiltonians* as

$$H_1 \equiv \bar{G}_1 + \lambda_i A_{ik} w_k \tag{7.4.16}$$

$$H_2 \equiv \bar{G}_2 - \frac{\partial H}{\partial \ddot{x}_i}(0, t)\bar{g}_i \tag{7.4.17}$$

$$H_3 \equiv \bar{G}_2 + \frac{\partial H(L, t)}{\partial \ddot{x}_i}\bar{h}_i \tag{7.4.18}$$

Now to cause the coefficients of the arbitrary variations

$$\delta x_s(0, t), \quad \delta x_s(L, t), \quad \frac{\partial(\delta x_r)}{\partial z}\bigg]_0^L, \quad \frac{\partial(\delta x_p)}{\partial z}\bigg]_0^L$$

to vanish, we must specify the following boundary conditions on the adjoint variables.

For Case 1 boundary conditions:

$$\left\{\frac{\partial H_2}{\partial x_s(0, t)} - \frac{\partial H(0, t)}{\partial \dot{x}_s} + \frac{\partial}{\partial z}\left[\frac{\partial H(0, t)}{\partial \ddot{x}_s}\right]\right\} = 0 \tag{7.4.19}$$

$$\left\{\frac{\partial H_3}{\partial x_s(L, t)} - \frac{\partial H(L, t)}{\partial \dot{x}_s} + \frac{\partial}{\partial z}\left[\frac{\partial H(L, t)}{\partial \ddot{x}_s}\right]\right\} = 0 \tag{7.4.20}$$

For Cases 2 and 3 boundary conditions:

$$\frac{\partial H}{\partial \ddot{x}_r}\bigg]_0^L = \frac{\partial H}{\partial \ddot{x}_p}\bigg]_0^L = 0 \tag{7.4.21}$$

In addition if the terminal state, $\mathbf{x}(z, \theta)$, is completely unspecified, the terminal conditions on $\boldsymbol{\lambda}$ become

$$\lambda_i(z, \theta)\, A_{ik} = \frac{\partial \bar{G}_1}{\partial x_k(z, \theta)} \qquad k = 1, 2, \ldots, n \tag{7.4.22}$$

It should be noted that if the partial differential equations are not second order in some of the state variables, $x_q(z, t)$, then $\partial H/\partial \ddot{x}_q \equiv 0$, and Case 2 or 3 boundary conditions are possible only at one side. If, for example, $x_q(L, t)$ was unspecified, then the coefficient of $\delta x_q(L, t)$ in Eq. 7.4.15 must vanish. The boundary condition on $\lambda_q(L, t)$ would then be

$$\frac{\partial G_2}{\partial x_q(L, t)} + \frac{\partial H}{\partial \dot{x}_q} = 0 \tag{7.4.23}$$

Thus these results apply to both first- and second-order partial differential equations. Applying these results reduces the variation in I to

$$\begin{aligned}
\delta I = & \int_0^\theta \int_0^L \frac{\partial H}{\partial u_i}\, \delta u_i\, dx\, dt + \int_0^\theta \left[\frac{\partial H_2}{\partial v_s(t)}\, \delta v_s(t) + \frac{\partial H_3}{\partial y_s(t)}\, \delta y_s(t)\right] dt \\
& + \int_0^\theta \left\{\left[\frac{\partial \bar{G}_2}{\partial v_p} - \frac{\partial H}{\partial \dot{x}_p} + \frac{\partial}{\partial z}\left(\frac{\partial H}{\partial \ddot{x}_p}\right)\right] \delta v_p(t)\right. \\
& \left. + \left[\frac{\partial \bar{G}_2}{\partial y_p} + \frac{\partial H}{\partial \dot{x}_p} - \frac{\partial(\partial H/\partial \ddot{x}_p)}{\partial z}\right] \delta y_p(t)\right\} dt \\
& + \int_0^L \left[\frac{\partial H_1}{\partial w_i(z)}\, \delta w_i(z)\right] dz
\end{aligned} \tag{7.4.24}$$

where the influence of the variation $\delta\mathbf{u}$, $\delta\mathbf{v}$, $\delta\mathbf{y}$, $\delta\mathbf{w}$ on the objective δI is now clear. Since the variations δu_i, δv_j, δy_k, δw_i are all arbitrary, a necessary condition for $\delta I \leq 0$ and the nominal policies $\bar{\mathbf{u}}$, $\bar{\mathbf{v}}$, $\bar{\mathbf{y}}$, $\bar{\mathbf{w}}$ to be optimal is that the coefficients of the variations vanish. Thus we can collect our results into the following weak maximum principle:

THEOREM. In order for the control trajectories $\bar{\mathbf{u}}$, $\bar{\mathbf{v}}$, $\bar{\mathbf{y}}$, $\bar{\mathbf{w}}$ to be optimal for the problem defined by Eqs. 7.3.1 to 7.3.5 and subject to the upper- and lower-bound constraints

$$\begin{aligned} u_{i*} &\leq u_i \leq u_i^* \\ v_{j*} &\leq v_j \leq v_j^* \\ y_{k*} &\leq y_k \leq y_k^* \\ w_{l*} &\leq w_l \leq w_l^* \end{aligned} \tag{7.4.25}$$

it is necessary that

$$\frac{\partial H}{\partial u_i} = 0 \tag{7.4.26}$$

for $u_i(z, t)$ unconstrained and H be a maximum when $u_i(z, t)$ is constrained. If u_i is only a function of z, then

$$\int_0^\theta \frac{\partial H}{\partial u_i}\, dt = 0 \tag{7.4.27}$$

must hold for unconstrained $u_i(z)$ and $\int_0^\theta H\, dt$ be maximized with respect to constrained $u_i(z)$. Similarly, if u_i is only a function of t, then

$$\int_0^L \frac{\partial H}{\partial u_i}\, dz = 0 \tag{7.4.28}$$

must hold for unconstrained $u_i(t)$ and $\int_0^L H\, dz$ must be a maximum with respect to constrained $u_i(t)$.

Furthermore, it is necessary that

$$\frac{\partial H_1}{\partial w_l} = 0 \tag{7.4.29}$$

$$\frac{\partial H_2}{\partial v_s} = 0 \tag{7.4.30}$$

$$\frac{\partial \bar{G}_2}{\partial v_p} - \frac{\partial H(0, t)}{\partial \dot{x}_p} + \frac{\partial}{\partial z}\left(\frac{\partial H(0, t)}{\partial \ddot{x}_p}\right) = 0 \tag{7.4.31}$$

$$\frac{\partial H_3}{\partial y_s} = 0 \tag{7.4.32}$$

$$\frac{\partial \bar{G}_2}{\partial y_p} + \frac{\partial H(L, t)}{\partial \dot{x}_p} - \frac{\partial(\partial H(L, t)/\partial \ddot{x}_p)}{\partial z} = 0 \tag{7.4.33}$$

must hold for unconstrained $w_l(z)$, $v_s(t)$, $v_p(t)$, $y_s(t)$, $y_p(t)$, respectively, and these quantities must be nonnegative at the upper bounds on the controls and nonpositive at the lower bounds. If any of the w_l, v_s, v_p, y_s, y_p, are unconstrained constant parameters, then the necessary conditions become

$$\int_0^L \frac{\partial H_1}{\partial w_l}\, dz = 0 \tag{7.4.34}$$

$$\int_0^\theta \frac{\partial H_2}{\partial v_s}\, dt = 0 \tag{7.4.35}$$

$$\int_0^\theta \left[\frac{\partial \bar{G}_2}{\partial v_p} - \frac{\partial H(0,t)}{\partial \dot{x}_p} + \frac{\partial}{\partial z}\left(\frac{\partial H(0,t)}{\partial \ddot{x}_p}\right)\right] dt = 0 \tag{7.4.36}$$

$$\int_0^\theta \frac{\partial H_3}{\partial y_s}\, dt = 0 \tag{7.4.37}$$

$$\int_0^\theta \left(\frac{\partial \bar{G}_2}{\partial y_p} + \frac{\partial H(L,t)}{\partial \dot{x}_p} - \frac{\partial(\partial H(L,t)/\partial \ddot{x}_p)}{\partial z}\right) dt = 0 \tag{7.4.38}$$

The adjoint variables $\lambda_i(z,t)$ are defined by Eqs. 7.4.11, and Eqs. 7.4.19 to 7.4.23 and H, H_1, H_2, H_3 by Eqs. 7.4.6, and 7.4.16 to 7.4.18.

We hope that the reader was not unduly intimidated by the apparent complexity of the theorem given in Eqs. 7.4.25 to 7.4.38. The rather involved nature of these expressions is caused by the fact that we wish to present a fairly general statement of the necessary conditions for optimality for the system described by Eq. 7.3.1. The hope is that the reader can apply the results of the theorem directly to many real problems and will have to *derive* the necessary conditions only for very unusual problems not falling within this framework.

If the reader is still not reassured by these remarks, he is urged to review Sections 6.1 to 6.3, which should lead to the realization that the lumped parameter results developed in Chapter 6 represent but a special case of the more general and, therefore, more involved treatment that is given here.

In order to illustrate the application of these general results to a particular problem, we shall produce the necessary conditions for optimality for the slab-heating problem discussed in Section 7.2.

EXAMPLE 7.5.1. From the general formulation, produce the necessary conditions for optimality of the heat flux program, $v(t)$, for the optimization problem described by Eqs. 7.2.1 to 7.2.6. Let us assume for the moment that the coefficients α, β are constant.

Solution. First we shall define the needed Hamiltonians:

$$H = [T - T_s(z)]^2 + \lambda(z, t)\left(\frac{\alpha}{\beta}\right)\frac{\partial^2 T}{\partial z^2}$$

$$H_2 = -\left(\frac{\alpha}{\beta}\right)\lambda(0, t)v(t)$$

$$H_3 = 0$$

Then the necessary condition (from Eq. 7.4.30) for $v(t)$ to be optimal is that

$$v(t) = \begin{cases} v^* & \text{for } \dfrac{\alpha}{\beta}\lambda(0, t) > 0 \\ v_* \le v \le v^* & \text{for } \dfrac{\alpha}{\beta}\lambda(0, t) = 0 \\ v_* & \text{for } \dfrac{\alpha}{\beta}\lambda(0, t) < 0 \end{cases}$$

where the adjoint equation (from Eq. 7.4.11) is

$$\frac{\partial \lambda(z, t)}{\partial t} = -\left[2(T - T_s) + \frac{\alpha}{\beta}\frac{\partial^2 \lambda(z, t)}{\partial z^2}\right]$$

Clearly the terminal state $\lambda(z, \theta)$ is unspecified and the boundary conditions are Case 1, so that the boundary conditions on λ (from Eqs. 7.4.19, 7.4.20, and 7.4.22) become

$$\frac{\alpha}{\beta}\frac{\partial}{\partial z}[\lambda(0, t)] = 0$$

$$\frac{\alpha}{\beta}\frac{\partial}{\partial z}[\lambda(L, t)] = 0$$

$$\lambda(z, \theta) = 0$$

Thus we have specified the necessary conditions for $v(t)$ to be optimal by simply plugging into the general equations given in the theorem. We note the fact, which is of considerable practical interest, that the *optimal heat flux must either correspond to the upper bound* (the maximum allowable value) or the optimal flux is zero. The only exception to this stipulation is the case when $\lambda(0, t) = 0$. Thus we have learned the form of the optimal program without performing any calculations. One could readily test likely candidates for the optimal program $v(t)$ by solving the given adjoint partial differential equations and examining the behavior of $\lambda(0, t)$.

7.5 SOME COMPUTATIONAL PROCEDURES

Just as Pontryagin's maximum principle formed the basis of computational approaches to the solution of lumped parameter trajectory optimization problems in Chapter 6, the distributed maximum principle of the last section forms the basis of a number of computational procedures for distributed parameter trajectory optimization problems. The most commonly applied method is:

The Control Vector Iteration Technique

This procedure is very similar to the one described in Chapter 6 and makes use of the fact that if the initial estimates $\bar{\mathbf{u}}$, $\bar{\mathbf{v}}$, $\bar{\mathbf{y}}$, $\bar{\mathbf{w}}$ are nonoptimal, then a gradient correction

$$\delta u_i(z, t) = \epsilon_0 \frac{\partial H}{\partial u_i} \tag{7.5.1}$$

$$\delta w_l(z) = \epsilon_1 \frac{\partial H_1}{\partial w_l} \tag{7.5.2}$$

$$\delta v_s(t) = \epsilon_2 \frac{\partial H_2}{\partial v_s} \tag{7.5.3}$$

$$\delta v_p(t) = \epsilon_3 \left[\frac{\partial \bar{G}_2}{\partial v_p} - \frac{\partial H}{\partial \dot{x}_p} + \frac{\partial}{\partial z}\left(\frac{\partial H}{\partial \ddot{x}_p}\right)\right] \tag{7.5.4}$$

$$\delta y_s(t) = \epsilon_4 \frac{\partial H_3}{\partial y_s} \tag{7.5.5}$$

$$\delta y_p(t) = \epsilon_5 \left[\frac{\partial \bar{G}_2}{\partial y_p} + \frac{\partial H}{\partial \dot{x}_p} - \frac{\partial(\partial H/\partial \ddot{x}_p)}{\partial z}\right] \tag{7.5.6}$$

will show the greatest local improvement in δI (cf. Eq. 7.4.21) for sufficiently small positive ϵ_0, ϵ_1, ϵ_2, ϵ_3, ϵ_4, ϵ_5. The detailed algorithm then is

(i) Guess $u_i(z, t)$, $v_j(t)$, $y_k(t)$, $w_l(z)$, $0 \leq t \leq \theta$, $0 \leq z \leq L$.

(ii) Solve the state Eq. 7.3.1 together with the boundary conditions Eqs. 7.3.2, 7.3.3, and 7.3.4. Compute I from Eq. 7.3.5.

(iii) Solve the adjoint Eqs. 7.4.11 together with the boundary condition (Eqs. 7.4.19 to 7.4.23).

(iv) Correct $u_i(z, t)$, $v_j(t)$, $y_k(t)$, $w_l(z)$ by Eqs. 7.5.1 to 7.5.6, where the ϵ_i are so chosen as to maximize I. A multivariable search may be used, or alternatively we may assume, $\epsilon_i = a_i\epsilon_0$ $i = 1, 2, \ldots, 5$ and perform an initial scaling of the a_i followed by a single variable search on ϵ_0 at each iteration.

(v) Return to step (ii) and iterate.

Just as in the lumped parameter trajectory optimization problems, these procedures progress very rapidly in the initial stages, but slow down considerably as the optimum is approached. Thus efforts are being made to extend second-order ascent procedures as well as conjugate gradient methods to these problems.

From a practical standpoint, computational difficulties would arise (caused largely by inadequate computer memory) if we were to tackle problems in several dimensions and with a large number of control and state variables by using this technique. We note, however, that it is quite feasible to carry out the optimization of systems modeled by partial differential equations and having a number of state and control variables. Indeed, a host of such problems have been tackled by chemical and control engineers; some references will be made to such work in subsequent sections of this chapter.

For practical reasons we shall restrict ourselves in the illustrative examples to be presented, to systems described by partial differential equations with relatively few state and control variables.

To demonstrate this control vector iteration procedure we shall determine the optimal inlet temperature program for a train of packed bed reactors whose catalyst is subject to deactivation (cf. [3–4] for the treatment of similar problems).

EXAMPLE 7.5.1. Let us consider the problem of disposing of exhaust gases from a smelting or other ore-processing operation. One solution which has been employed to avoid the air pollution resulting from SO_2 and other noxious components in the stack gases is to oxidize the material (e.g., SO_2 to SO_3 for the production of sulphuric acid). Let us consider, furthermore, that this oxidation is to be carried out over some catalyst (e.g., V_2O_5 for SO_2 oxidation [5]) which is subject to deactivation with time. Because the reaction is exothermic and is assumed to be reversible, a number of adiabatic stages are employed with interstage cooling as shown in Fig. 7.2. We assume that species A is the reactant and B is the oxidation product. Thus the reaction

$$A \underset{k_2}{\rightleftharpoons} B$$

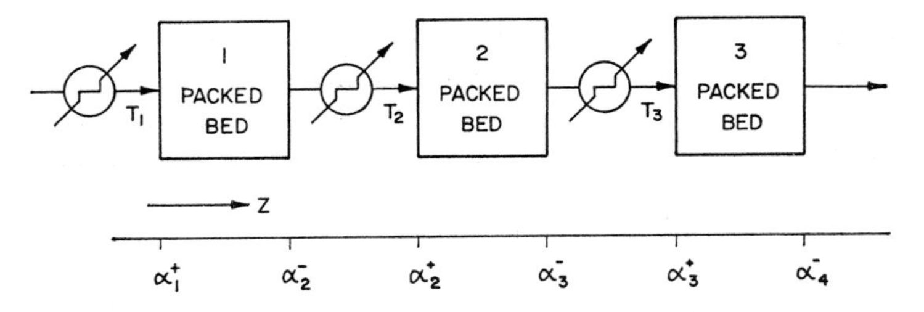

Fig. 7.2 Optimization of the reactors used for pollutant oxidation.

is to be carried out in the three adiabatic packed bed reactors sketched in Fig. 7.2. The modeling equations are given as

$$u \frac{\partial C_B}{\partial z'}(z', t') = \psi(z', t')[k_1(T)(C_T - C_B) - k_2(T)C_B];$$

$$\text{(Mass balance on the product)} \qquad \begin{matrix} 0 \le t' \le \theta \\ 0 \le z' \le L \end{matrix} \qquad (7.5.7)$$

$$\rho C_p u \frac{\partial T(z', t')}{\partial z'} = (-\Delta H)\psi(z', t')[k_1(T)(C_T - C_B) - k_2(T)C_B];$$

$$\text{(Heat balance)} \qquad \begin{matrix} 0 \le t' \le \theta \\ 0 \le z' \le L \end{matrix} \qquad (7.5.8)$$

$$T(\alpha_1'^+, t') = T_1; \quad T(\alpha_2'^+, t') = T_2, \quad T(\alpha_3'^+, t') = T_3, \quad C_B(0, t') = C_{Bf} \qquad (7.5.9)$$

which represent steady-state material and heat balances in the reactor train. The quantities α_i' denote the points of separation between the beds, C_T is the total feed concentration, and C_{Bf} is the feed concentration of B. The total catalyst lifetime is θ, and the total reactor length is L. The reaction rate constants are given by $k_i = A_{i0} \exp(-E_i/RT)$, $i = 0, 1, 2$. The decline in catalyst activity $\psi(z', t')$ at each point in the bed, can be described by

$$\frac{\partial \psi(z', t')}{\partial t'} = -k_0(T)\psi^2; \qquad \begin{matrix} 0 \le t' \le \theta \\ 0 \le z' \le L \end{matrix} \qquad (7.5.10)$$

where the initial activity is taken to be unity for fresh catalyst, that is,

$$\psi(z', 0) = 1.0 \qquad (7.5.11)$$

Thus if the time scale for catalyst decay is much longer than the time scale for the dynamics of the reactor, then Eqs. 7.5.7 to 7.5.11 are the modeling equations for the system.

Let us suppose that we wish to control the interstage coolers (i.e., the inlet temperatures T_1, T_2, T_3) so as to maximize the conversion of A over the catalyst lifetime, θ. However, due to heat exchange constraints, it is assumed that the possible inlet temperatures are bounded by $T_* \leq T_i \leq T^*$. This is a practical optimization problem because, by raising the inlet temperature, we both increase the conversion of A from Eqs. 7.5.7 and 7.5.8, and hasten the deactivation of the catalyst through Eq. 7.5.10. Thus there is an optimal inlet temperature progression $T_1(t')$, $T_2(t')$, $T_3(t')$ which must be determined, and we do this by applying the control vector iteration technique to the problem.

Solution. Let us first recognize that Eqs. 7.5.7 and 7.5.8 are not independent, but can be related by the transformation

$$T(z', t') = T_i + \left(\frac{-\Delta H}{\rho C_p}\right)[C_B(z, t) - C_B(\alpha_i', t)] \qquad i = 1, 2, 3 \quad (7.5.12)$$

because of the adiabatic operation.

Now we define the new variables

$$x_1(z, t) = \frac{C_B}{C_T}, \quad x_2(z, t) = \psi(z, t), \quad u_1 = \frac{RT_1}{E_1}, \quad u_2 = \frac{RT_2}{E_1},$$
$$u_3 = \frac{RT_3}{E_1}, \quad p = \frac{E_1}{E_0}, \quad p_1 = \frac{E_2}{E_1} \quad (7.5.13)$$

$$\tau_k = \frac{RT}{E_1}, \quad \beta_i = \frac{A_{i0} L C_T}{u} \quad i = 1, 2; \quad \rho = A_0 \theta, \quad z = \frac{z'}{L}, \quad t = \frac{t'}{\theta}$$

$$\alpha_i = \frac{\alpha_i'}{L}, \quad x_{1f} = \frac{C_{Bf}}{C_T}, \quad J = \frac{(-\Delta H) R C_T}{\rho C_p E_1}, \quad u_{k^*} = \frac{RT_*}{E_1}, \quad u_k{}^* = \frac{RT^*}{E_1}$$

so that the modeling equations become

$$0 = -\frac{\partial x_1(z, t)}{\partial z} + x_2(z, t)\left[\beta_1 \exp\left(-\frac{1}{\tau_k}\right)(1 - x_1) - \beta_2 \exp\left(-\frac{p_1}{\tau_k}\right) x_1\right]$$
$$0 \leq t \leq 1, \quad 0 \leq z \leq 1 \quad (7.5.14)$$

or

$$0 = -\frac{\partial x_1(z, t)}{\partial z} + \hat{f}(x_1, x_2, u_k)$$

which describes the reactor conversion. The catalyst activity can be determined from

$$\frac{\partial x_2(z, t)}{\partial t} = -\rho (x_2)^2 \exp\left[-(p\tau_k)^{-1}\right] = \hat{g}(x_1, x_2, u_k) \qquad 0 \leq t \leq 1, \quad 0 \leq z \leq 1 \quad (7.5.15)$$

where

$$x_1(0, t) = x_{1f}, \qquad x_2(z, 0) = 1.0 \tag{7.5.16}$$

and Eq. 7.5.12 becomes

$$\tau_k(z, t) = u_k(t) + J[x_1(z, t) - x_1(\alpha_k, t)] \qquad k = 1, 2, 3 \tag{7.5.17}$$

The *objective functional*, which is the cumulative conversion of A over a catalyst lifetime, now becomes

$$I = \int_0^1 x_1(1, t)\, dt \tag{7.5.18}$$

The Hamiltonians of interest H, H_3 become

$$H = \lambda_1(z, t)\left[-\frac{\partial x_1(z, t)}{\partial z} + \hat{f}(x_1, x_2, u_k)\right] + \lambda_2(z, t)\hat{g}(x_1, x_2, u_k) \tag{7.5.19}$$

$$H_3 = x_1(1, t) \tag{7.5.20}$$

where the adjoint variables are given (cf. Eq. 7.4.11) by

$$0 = -\left[\frac{\partial \lambda_1(z, t)}{\partial z} + \lambda_1 \frac{\partial \hat{f}}{\partial x_1} + \lambda_2 \frac{\partial \hat{g}}{\partial x_1}\right] \tag{7.5.21}$$

$$\frac{\partial \lambda_2(z, t)}{\partial t} = -\lambda_1 \frac{\partial \hat{f}}{\partial x_2} - \lambda_2 \frac{\partial \hat{g}}{\partial x_2} \tag{7.5.22}$$

with boundary conditions (cf. Eqs. 7.4.22 and 7.4.23)

$$\lambda_1(1, t) = 1 \tag{7.5.23}$$

$$\lambda_2(z, 1) = 0 \tag{7.5.24}$$

The computational procedure then is as follows:

(i) Guess $u_k(t)$ $0 \leq t \leq 1$, $k = 1, 2, 3$.

(ii) Solve the state Eqs. (7.5.14 and 7.5.15) forward in z, t using the method of characteristics (or finite differences); compute I.

(iii) Solve the adjoint Eqs. 7.5.21 and 7.5.22 backward in z, t.

(iv) Correct the controls $u_k(t)$ by

$$\overset{\text{new}}{u_k(t)} = \overset{\text{old}}{u_k(t)} + \epsilon_0 \int_{\alpha_k}^{\alpha_{k+1}} \left(\frac{\partial H}{\partial u_k}\right) dz \tag{7.5.25}$$

where $k = 1, 2, 3$, $\alpha_1 = 0$, $\alpha_2 = \frac{1}{3}$, $\alpha_3 = \frac{2}{3}$, $\alpha_4 = 1$ and ϵ_0 is determined by a one-dimensional search.

(v) Return to step (ii) and iterate.

It is important to note that because there are three beds, control u_1 only applies over $0 \leq z < \frac{1}{3}$, u_2 over $\frac{1}{3} \leq z < \frac{2}{3}$, and u_3 over $\frac{2}{3} \leq z < 1$.

This explains the limits on the integral in Eq. 7.5.25. This computational algorithm was applied for the set of parameters $\beta_1 = 5.244 \times 10^5$, $\beta_2 = 2.28 \times 10^9$, $\rho = 1300$, $u_{k*} = 0.070$, $u_k^* = 0.080$, $p = 1.648$, $p_1 = 1.666$, $J = 0.005$, $x_{if} = 0$, and the result after five iterations is shown in Fig. 7.3. The convergence of the algorithm was checked by successfully producing the same optimal temperature progression from two starting points. The inlet

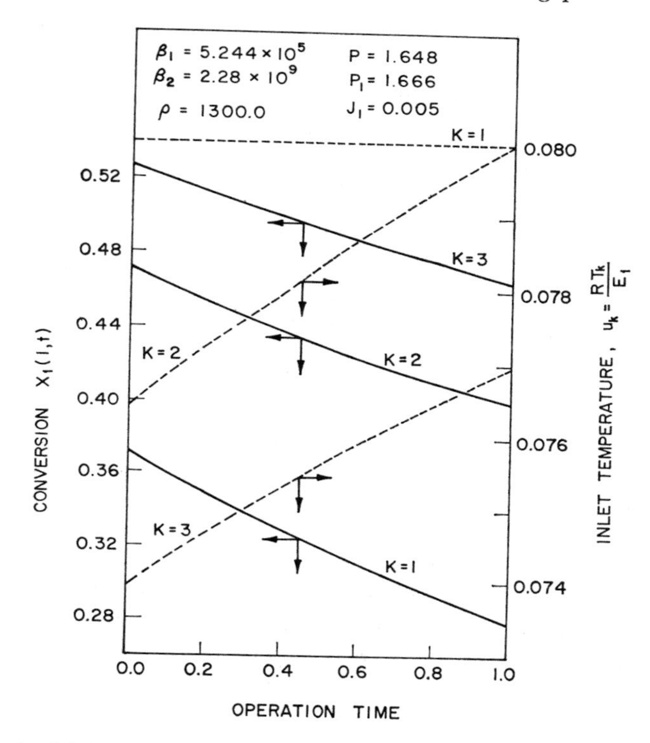

Fig. 7.3 Optimal inlet temperature progression for Example 7.5.1, an oxidation reaction.

temperature is seen to be at the upper bound always for the first bed, and it rises to compensate for catalyst decay in the other beds. The low conversion ($\sim$50%) of A is due to the fact that the reaction is equilibrium limiting, which even the falling temperature from bed to bed cannot overcome. The optimal temperature progressions shown in Fig. 7.3 produce about a 10% improvement in the objective over a constant inlet temperature policy of $u_k = 0.080$, $k = 1, 2, 3$.

The fact that this optimal progression was found from two starting points in about five minutes of computing time (IBM 360/75) illustrates the practicality of this control vector iteration procedure for complex problems. More detailed descriptions of this approach may be found in Refs. 3 and 4.

Another computational approach which has definite advantages in these problems is:

The Control Vector Parameterization Method

This technique has essentially the same form as for lumped parameter problems, which were discussed in Section 6.7.

One represents the controls in terms of trial functions

$$u_i(z, t) = \sum_{j=1}^{s_1} a_{ij} \phi_{ij}(z, t) \tag{7.5.26}$$

$$v_j(t) = \sum_{k=1}^{s_2} b_{jk} \eta_{jk}(t) \tag{7.5.27}$$

$$y_j(t) = \sum_{j=1}^{s_3} c_{jk} \xi_{jk}(t) \tag{7.5.28}$$

$$w_i(z) = \sum_{k=1}^{s_4} e_{ik} \nu_{ik}(z) \tag{7.5.29}$$

and then parameter optimization techniques are used to determine the optimal values for the coefficients a_{ij}, b_{jk}, c_{jk}, e_{ik}. We note that parameterization in terms of state variables may be done as well in order to develop a feedback control law.

There is very little computational experience with this approach, but one might expect that it would be reasonably efficient. An important metallurgical system is optimized using this approach in Chapter 8.

There have been a number of other computational techniques suggested for the solution of distributed parameter trajectory optimization problems. Sage [1] gives several examples in which the partial differential equations have been discretized in the spatial direction and the resulting set of ordinary differential equations then treated in the standard way.

Zahradnik et al. [6] and Bosarge [7] suggest the use of approximate methods in which all the variables $\mathbf{x}(z, t)$, $\boldsymbol{\lambda}(z, t)$, $\mathbf{u}$, $\mathbf{v}$, $\mathbf{y}$, $\mathbf{w}$ are expanded in trial functions, and then the method of weighted residuals is used to evaluate the coefficients; however, there has been little computational experience to date on nonlinear problems.

In principle a direct substitution approach similar to that discussed in Chapter 6 could be used for these problems. However, the uncoupling problem of explicitly representing the optimal control in terms of the state and adjoint variables can rarely be done in practice; thus direct substitution methods usually cannot be applied.

7.6 A BATCH PACKED BED REACTOR

As a further illustration of distributed parameter optimization problems, let us consider a batch packed bed reactor of depth L which is insulated on the sides and bottom and is subject to a radiating source from above. The scheme shown in Fig. 7.4 is used for the batchwise reduction of various pelletized materials.

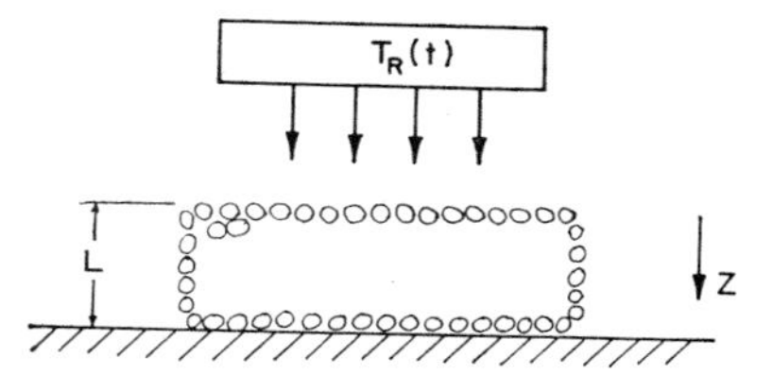

Fig. 7.4 A batch packed bed reactor.

The modeling equations become

$$\frac{\partial T}{\partial t} = \frac{1}{\beta(T)} \frac{\partial[\alpha(T)\, \partial T/\partial z]}{\partial z} - Jr(T, s); \qquad 0 \leq t \leq \theta, \quad 0 \leq z \leq L \qquad (7.6.1)$$

$$\frac{\partial s}{\partial t} = -r(s, T) \qquad 0 \leq t \leq \theta, \quad 0 \leq z \leq L \qquad (7.6.2)$$

$$\frac{\partial T}{\partial z}(0, t) = v(t) \qquad (7.6.3)$$

$$\frac{\partial T}{\partial z}(L, t) = 0 \qquad (7.6.4)$$

$$T(z, 0) = T_0(z) \qquad (7.6.5)$$

$$s(z, 0) = s_0(z) \qquad (7.6.6)$$

where $T(z, t)$ is the temperature in the bed, and $s(z, t)$ is the fraction of the solid remaining unreduced at any point in the bed at time t. J is the heat of reaction so that the quantity, $Jr(T, s)$ represents the rate of heat adsorption due to the endothermic reaction. The nonlinear term $r(T, s)$ is the reaction rate at any point in the bed, and $v(t)$ is the heat flux received at the upper surface of the bed which is to be controlled optimally between the limits

$$v_* \leq v(t) \leq v^* \qquad (7.6.7)$$

The optimization problem arises because it is necessary to minimize the time required to heat the bed throughout so as to attain complete conversion of the pellets and then to cool the bed back to a temperature, T_c. Once the temperature T_c is reached the products may be removed from the reactor while avoiding the possibility of reoxidation. This latter cooling is facilitated by the circulation of a gas across the top of the bed producing the lowest (negative) heat flux v_* given in Eq. 7.6.7.

An objective functional which should produce the optimum discussed above is

$$I[v(t)] = \int_0^L s(z, \theta)^2 \, dz + c \int_0^L [T(z, \theta) - T_c]^3 \, dz \tag{7.6.8}$$

which if minimized reflects the wish to achieve complete conversion and yet force the $T(z, \theta)$ below T_c. The powers in Eq. 7.6.8 are chosen to emphasize that large deviations from the desired values of s and T are much more detrimental than small deviations.

An additional requirement for the problem is that the solid material should not melt; thus there is a state variable constraint

$$T(0, t) \leq T_m \tag{7.6.9}$$

which must be satisfied by the optimal control. The optimization problem then is to choose $v(t)$ such that I in Eq. 7.6.8 is minimized while satisfying the constraints (Eqs. 7.6.1 to 7.6.7 and Eq. 7.6.9).

The problem described above is characteristic of ferroalloy production, by a variety of processes. A particular example, the Simplex process will be discussed in some detail in Section 8.3 of the following chapter.

7.7 A BROADER VIEW

Now that the reader has been introduced to the basic concepts of distributed parameter trajectory optimization, let us take a brief overlook of the entire field so as to provide material for further reading.

There are several excellent recent reviews of the literature in the field. Brogan [8] has a very comprehensive discussion of both integral and partial differential equations and the maximum principles for each. In this work a number of examples are given and several computational algorithms are presented for optimal control synthesis. Butkovskii et al. [9] give a rather complete review of the Soviet work in this area, most of which was contributed by the authors of the review. However, no general purpose computational algorithms are discussed for synthesis.

Seinfeld and Lapidus [10] have a very readable review of the literature through 1967 in which they present some computational results of their own for problems of engineering interest. A recent symposium [11] has a great number of papers dealing with these problems. Sage [1] and Denn [2] have a very good general discussion of distributed parameter optimization in their monographs.

The state of the art at present is that the theory has been highly developed [8, 9, 12] and good computational algorithms are available for many linear systems [8, 9, 13–20]. However, work has been reported only recently on computational results for nonlinear systems. Because nonlinear problems are of great interest to metallurgists, several case studies will be cited.

Denn et al. [21], Seinfeld and Lapidus [10], and Paynter et al. [22] have treated in detail the optimal control of a tubular reactor with radial temperature and concentration gradients. Paynter et al. [22] propose an approximate method of control synthesis using radially averaged values of the adjoint variable in the control vector iteration step. Denn [23], Jackson [24], Chang and Bankoff [25], Bertran and Chang [26], and Ogunye and Ray [27–29] report results of having used gradient methods with the distributed maximum principle with moderate to good success. The convergence of the gradient methods seemed to depend strongly on the problem being considered.

Some very early work on the optimization of nonlinear distributed parameter systems was carried out by Ostrovskii and Volin [30–32]. However, they have reported very little computational experience with their suggested algorithms.

7.8 CONCLUDING REMARKS

The subject of this chapter is the most sophisticated and advanced of the book. The theory as well as computational studies on distributed parameter optimization problems are the subjects of current research, so that new, more efficient computational procedures should appear in the next few years. Hence, the topics of this chapter carry the reader to the fringes of research in the area.

"Why has this degree of sophistication been necessary?" one might ask. The answer is that a large number of metallurgical and chemical processes are nonlinear and distributed in nature. Thus any successful optimization of these systems requires, at the very least, a knowledge of the introductory material presented here. It is hoped that the recognition of this practical necessity will inspire a diligent study of this chapter.

PROBLEMS

The problems given here are simple ones designed to help the reader apply the necessary conditions of optimality. For more substantial posed problems, see Section 8.5.

1. Let us consider unsteady state, one-dimensional heat conduction in a slab, with spatially uniform internal heat generation, the intensity of which can be made time dependent. One face of the slab is insulated and the other surface receives thermal radiation from a temperature source, which can also be controlled. Problems of this type could occur in systems where radiant heating is augmented by induction heating. The modeling equations are then given as:

$$\frac{\partial x}{\partial t} = \alpha \frac{\partial^2 x}{\partial y^2} + u(t)$$
$$0 \leq t \leq 1$$
$$0 \leq y \leq 1$$

with boundary conditions

$$\frac{\partial x}{\partial y} = 0, \quad y = 0; \quad \frac{\partial x}{\partial y} = \beta(v^4 - x^4), \quad y = 1$$

where the radiant source temperature $v(t)$ can be controlled as a function of time. It is desired to take the slab from an initially uniform temperature $x_0 = 0$ to as close as possible to a desired temperature distribution $x_s(y) = 1 + 2y$ by choosing $v(t)$, $u(t)$ constrained by

$$0 \leq u \leq 3$$
$$0 \leq v \leq 6$$

such that the objective

$$I = \int_0^1 \int_0^1 \{[x(y, t) - x_s(y)]^2 + \mu_1[u(t)]^2\}\, dt\, dy + \mu_2 \int_0^1 v^4\, dt$$

is minimized.

(a) Determine the necessary conditions for optimality of $v(t)$, $u(t)$ for this problem.

(b) Suggest a computational scheme for calculating the optimal control variables $u(t)$, $v(t)$.

(c) Repeat parts (a) and (b) when $\mu_1 \equiv 0$ in the objective functional.

2. Consider a tubular, packed bed reactor in which the following single, irreversible reaction is being carried out:

$$A \xrightarrow{k_1} B$$

The reaction is catalytic and the activity of the catalyst decays if overheating occurs in the presence of B. Let us designate the rate constant for this decay by k_2. On assuming plug flow, the dimensionless form of the heat and material balances is given as:

$$\frac{\partial x_1}{\partial t} + \frac{\partial x_1}{\partial z} = -x_2 u x_1 \qquad \begin{array}{l} x_1(0, t) = 1 \\ x_1(z, 0) = 1 \end{array}$$

$$\begin{cases} 0 \leq t \leq 1 \\ 0 \leq z \leq 1 \end{cases}$$

$$\frac{\partial x_2}{\partial t} = -(x_2)^2 \beta u^p (1 - x_1) \qquad x_2(z, 0) = 1$$

where x_1 is the reactant concentration, x_2 the catalyst activity, and $u(z, t) = k_1$ represents the dimensionless bed temperature. β is a parameter and p is the ratio of activation energies of k_1 and k_2. We note that in contrast to Example 7.5.1, here we assume perfect control over the bed temperature at each position and time; in addition, the decay is supposed rapid enough that $\partial x_1/\partial t$ cannot be neglected. The optimization problem is to choose $u(z, t)$ $(u_* \leq u \leq u^*)$ such that the total yield of B is maximized; that is, maximize the functional

$$I = \int_0^1 [1 - x_1(1, t)]\, dt$$

(a) Write down the necessary conditions for optimality of $u(z, t)$.
(b) Suggest a computational scheme for calculating the optimal trajectories $u(z, t)$.

REFERENCES

1. A. P. Sage, *Optimum Systems Control*, Prentice-Hall, 1968.
2. M. M. Denn, *Optimization by Variational Methods*, McGraw-Hill, 1970.
3. A. F. Ogunye and W. H. Ray, *AIChE J.* **17,** 43, 365 (1971).
4. A. F. Ogunye and W. H. Ray, *I and EC Process Des. Dev.* **9,** 619 (1970); **10,** 410, 416 (1971).
5. C. M. Crowe et al., *Chemical Plant Simulation*, Prentice-Hall, 1971.
6. R. L. Zahradnik and L. L. Lynn, *Proc.* 1970 *JACC*, paper 22E.
7. E. Bosarge, *Proceedings IFAC Symposium on the Control of Distributed Parameter Systems*, Banff, Canada, June 1971.
8. W. L. Brogan, *Adv. Control Syst.* **6,** 222 (1968), C. T. Leondes, Ed.
9. A. G. Butkovskii, A. I. Egorov, and K. A. Lurie, *SIAM J. Control* **6,** 437 (1968).
10. J. H. Seinfeld and L. Lapidus, *Chem. Eng. Sci.* **23,** 1461 (1968).
11. *Proceedings of IFAC Symposium on the Control of Distributed Parameter Systems*, Banff, Canada, June 1971.

12. R. Jackson, *Int. J. Control* **4,** 127, 585 (1966).

13. M. M. Denn, *I and EC Fund.* **7,** 410 (1968).

14. B. M. Raspopov, *Automotika i Telemekh.* February 1968, p. 185.

15. L. B. Koppel, Y. P. Shih, and D. R. Coughanowr, *I and EC Fund.* **7,** 286 (1968).

16. L. B. Koppel and Y. P. Shih, *I and EC Fund.* **7,** 414 (1968).

17. Y. Yavin and Y. Rasis, *Int. J. Control* **11,** 153 (1970).

18. M. A. Hassan and K. D. Solberg, *Automatica* **6,** 409 (1970).

19. F. L. Alvarado and R. Mukundan, *Int. J. Control* **9,** 665 (1969).

20. M. Kim and S. H. Gajwani, *IEEE Trans. Auto. Control* **AC-13,** 191 (1968).

21. M. M. Denn, R. D. Gray, and J. R. Ferron, *I and EC Fund.* **5,** 59 (1966).

22. J. D. Paynter, J. S. Dranoff, and S. G. Bankoff, *Ind Eng Chem. Proc. Des.* **9,** 303 (1970).

23. M. M. Denn, *Int. J. Control* **4,** 167 (1966).

24. R. Jackson, *Trans. Inst. Chem. Engr.* **45,** T160 (1967).

25. K. S. Chang and S. G. Bankoff, *AIChE J.* **15,** 410, 414 (1969).

26. D. R. Bertran and K. S. Chang, *AIChE J.* **16,** 897 (1970).

27. A. F. Ogunye and W. H. Ray, *AIChE J.* **17,** 43, 365 (1971).

28. A. F. Ogunye and W. H. Ray, *I and EC Process Des.* **9,** 619 (1970).

29. Ibid., **10** 410, 416 (1971).

30. Y. M. Volin and G. M. Ostrovskii, *Automatika i Telemeckh.* **25,** 1197 (1964).

31. G. M. Ostrovskii and Yu. M. Volin, *Methods of Optimizing Chemical Reactors,* Izdatel'stvo "Khimiya", Moscow, 1967.

32. G. M. Ostrovskii and Yu. M. Volin, *Methods of Optimizing Complex Chemical-Technological Systems,* Izdatel'stvo "Khimya" Moscow, 1970.

8 More Complex Optimization Problems

8.1 INTRODUCTION

In the preceding Chapters 1 to 7 we dealt with problems where the governing equations describing the system were established, or at least taken for granted, and our main concern was to apply the various optimization techniques.

Our primary purpose in these examples was illustration; hence conciseness, brevity, and simplicity were necessary constraints which had to be observed in the selection of the majority of these problems. While this approach is necessary from the pedagogical viewpoint, it could leave the reader with a misleading impression, as far as the practical world is concerned.

In many practical optimization problems, be they concerned with the establishment of the optimum operating conditions in an existing unit or with the development of a new optimal design, a suitable mathematical model of the system may not be available. This lack of mathematical models is thought to apply particularly in the case of metallurgical systems where little is available in the way of "standard models" even for the common "unit operations."

Our objective in this chapter is to illustrate, through case studies, the application of optimization techniques to systems where the development of process models has to be undertaken as part of the project.

For this reason, this chapter is dedicated to metallurgical engineers, and the examples to be presented will be drawn from the field of process metallurgy. Chemical engineers have a somewhat easier task, in that mathematical models have been developed for the majority of "standard components" such as chemical reactors, distillation columns, heat exchangers, etc. from which chemical plants are composed. This fact has been taken for granted in the presentation of some of the more detailed chemical examples in the earlier parts of the text (cf. Sections 4.9 and 5.6). It is thought, that in spite of the strong metallurgical flavor, the case studies described here will be of interest to chemical engineers, particularly as metals processing companies are now beginning to employ chemical engineers in their operations.

It follows from the foregoing, that in the majority of metallurgical systems a great deal of preparatory (or preliminary) work may be needed before the "optimization proper" can begin. As in many real-life situations, "foresight" is helpful and an awareness of the techniques one is planning to use for the optimization step may have a significant bearing on the modeling equations that are to be developed.

In a practical situation the optimization process has to be developed in three stages.

(i) The identification of the problem and the establishment of the criteria for optimization in a physical sense.

(ii) The development of the process model.

(iii) The optimization calculations.

It is hoped that the following examples will help the reader develop the ability to identify practical optimization problems—initially perhaps through analogies to the situations that are described here.

Regarding the organization of the Chapter, Sections 2 to 4 deal with fully worked, substantial examples, whereas Section 5 contains a number of problems that are posed, but not solved. In each instance we shall start with a statement of the physical nature of the problem and stress, wherever possible, the inherent physical characteristics of these systems that show a promise of success from the viewpoint of optimization.

8.2 THE OPTIMAL TEMPERATURE PROGRESSION FOR THE REDUCTION OF A METAL OXIDE SPHERE WHERE SINTERING OF THE METAL PHASE MAY OCCUR†

The Physical Problem

In general, the rate at which metal oxide pellets react with a reducing gas (say H_2, CO) tends to increase with increasing temperature. This behavior is usually caused by

(i) The fact that the reaction rate constant increases with increasing temperature.
(ii) The diffusion coefficients increase with increasing temperature, although not as rapidly as the reaction rate constant.
(iii) The equilibrium conditions may become more favorable at higher temperatures.

There exist, however, a number of systems where, upon increasing the temperature beyond a certain level, the rate of reaction is actually reduced and may indeed diminish to negligibly low values. The reduction of nickel oxide with hydrogen above about 750°C is a typical example for such systems, and there is reason to believe that similar phenomena could be of importance in some direct reduction processes [2].

The physical explanation for this behavior is that at high temperatures the porous metal shell formed on reduction sinters, and this sintered shell offers a greatly increased diffusional resistance.

For such systems there may exist a clearcut optimum reaction temperature, where the temperature is high enough for the reaction and the diffusional processes to occur as rapidly as possible, but where no significant sintering can take place.

Intuitively, such a physical picture may also suggest the existence of an *optimal temperature progression*. The example that follows deals with a particular case of this rather broad class of problems.

† This example, reported in more detail in Ref. [1], is used with the permission of Pergamon Press Ltd.

Formulation

Let us consider the reduction of a metal oxide sphere through its reaction with a reducing gas, according to the following scheme:

$$\underset{\text{Gas}}{A} + \underset{\text{Solid}}{S_{\text{I}}} \rightleftarrows \underset{\text{Gas}}{B} + \underset{\text{Solid}}{S_{\text{II}}}$$

Let us consider the reaction reversible and assume that the heat of reaction is small.

Let us assume, furthermore, that diffusion through the reacted (reduced) shell contributes a significant resistance, so that the "shrinking core" or topochemical model is appropriate.

Accordingly, after some reaction has taken place, a pellet of initial radius R_0 will consist of *an unreacted core*, in the region $0 \leq R \leq R_\alpha$ and of a *reacted*

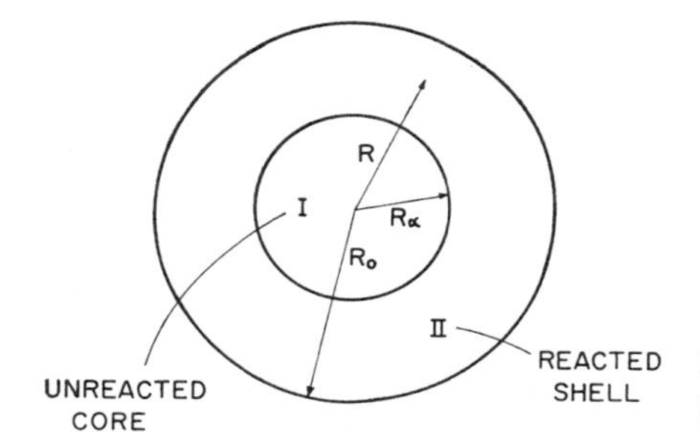

Fig. 8.1 Schematic representation of the gas-solid reaction system.

shell in the region $R_\alpha \leq R \leq R_0$, as sketched in Fig. 8.1. The overall reaction scheme will involve the following sequential steps:

(i) Mass transfer of reactant A to the surface of the sphere.

(ii) Diffusion of A through the reacted shell; since we allow the shell to sinter, the diffusion coefficient will, in general, depend on both position and time.

(iii) Heterogeneous chemical reaction at the interface separating the reacted and unreacted regions, that is, at $R = R_\alpha$.

(iv) Diffusion of the gaseous product B across the reacted shell, again with a time- and position-dependent diffusion coefficient.

(v) Mass transfer of product B from the outer surface of the sphere to the surrounding gas.

If we assume further, that the chemical reaction step (iii) is of first order, and that there is no gaseous product B present in the gas surrounding the pellet, then the problem may be readily stated in mathematical terms [3].

Before proceeding with this formulation, it is convenient to list the principal symbols that will be used:

R = radial coordinate within the pellet
R_0 = pellet radius
R_α = radius of "reaction front"
C = gas concentration
$D(R, t)$ = effective diffusivity
h = external mass transfer coefficient
C_{A0} = reactant gas concentration in bulk gas stream
ρ_m = true molecular density of reactant solid
P = initial porosity of reactant solid
k = first-order reaction rate constant
K_E = equilibrium constant
t = time
E_{AR} = activation energy for sintering
τ_0 = pre-exponential constant for reaction
τ = characteristic time for sintering
$\mathscr{R}$ = gas constant
T = absolute temperature
D_0 = initial diffusivity
D_∞ = minimum value of the diffusivity

On noting that the diffusion coefficients are not constant, the conservation of the gaseous reactant and product in the reacted shell may be written as:

$$\frac{\partial}{\partial R}\left(D_A R^2 \frac{\partial C_A}{\partial R}\right) = 0 \qquad R_\alpha < R < R_0 \tag{8.2.1}$$

$$\frac{\partial}{\partial R}\left(D_B R^2 \frac{\partial C_B}{\partial R}\right) = 0 \qquad R_\alpha < R < R_0 \tag{8.2.2}$$

The functional relationship between D_A, D_B, t, and R is related to the advancement of the reaction front, and will be stated subsequently.

The conservation of the gaseous species A and B at the outer surface of the pellet is given as:

$$D_A \frac{\partial C_A}{\partial R} = h(C_{A0} - C_A) \qquad R = R_0 \tag{8.2.3}$$

$$D_B \frac{\partial C_B}{\partial R} = -hC_B \qquad R = R_0 \tag{8.2.4}$$

Finally, a mass balance at the reaction front, or the rate of advancement of the reaction front, may be written as

$$-\rho_m(1 - P)\frac{dR_\alpha}{dt} = k\left(C_A - \frac{C_B}{K_E}\right) = D_A\frac{\partial C_A}{\partial R} \qquad R = R_\alpha \quad (8.2.5)$$

Equations 8.2.1 to 8.2.5 together with the appropriate relationships defining $D(R, t)$ represent a complete statement of the kinetic problem.

What follows is a convenient algebraic manipulation, which allows us to transform the problem of three simultaneous differential equations to a single first-order differential equation containing two integral expressions. However, we note that strictly speaking, we do not have three independent differential equations because Eqs. 8.2.1 and 8.2.2 are related through the overall mass balance on the system. From Eq. 8.2.5 we have

$$C_A = -\frac{\rho_m}{k}(1 - P)\frac{dR_\alpha}{dt} + \frac{C_B}{K_E} \qquad \text{at} \quad R = R_\alpha \qquad (8.2.6)$$

Upon integrating Eq. 8.2.1 twice we obtain

$$C_A - C_A|_{R=R_\alpha} = A\int_{R_\alpha}^{R}\frac{dR}{D_A R^2} \qquad (8.2.7)$$

where A is an integration constant:

$$A = D_A R^2\frac{\partial C_A}{\partial R} = -D_B R^2\frac{\partial C_B}{\partial R} \qquad (8.2.8)$$

But from Eq. 8.2.5 we also have

$$D_A R_\alpha^{\ 2}\frac{\partial C_A}{\partial R} = -R_\alpha^{\ 2}\frac{dR_\alpha}{dt}\rho_m(1 - P) \qquad \text{at} \quad R = R_\alpha \qquad (8.2.9)$$

which is just an expression of the continuity of A. Thus using Eq. 8.2.8 we may write

$$-\rho_m(1 - P)\frac{dR_\alpha}{dt} = \frac{A}{R_\alpha^{\ 2}} \qquad (8.2.10)$$

On substituting into Eq. 8.2.6 we have

$$C_A = \frac{A}{kR_\alpha^{\ 2}} + \frac{C_B}{K_E}; \qquad R = R_\alpha \qquad (8.2.11)$$

Equation 8.2.2 may now be integrated twice to obtain

$$C_B - C_B|_{R=R_\alpha} = -A\int_{R_\alpha}^{R}\frac{dR}{D_B R^2} \qquad (8.2.12)$$

Let

$$I_A(R_\alpha, t) = \int_{R_\alpha}^{R_0} \frac{dR}{D_A R^2}, \qquad I_B(R_\alpha, t) = \int_{R_\alpha}^{R_0} \frac{dR}{D_B R^2} \tag{8.2.13}$$

Thus from Eq. 8.2.12 we have

$$C_B\big|_{R=R_0} + A I_B = C_B\big|_{R=R_\alpha} \tag{8.2.14}$$

On substituting into Eq. 8.2.11 we obtain

$$C_A\big|_{R=R_\alpha} = \frac{A}{kR_\alpha^2} + \frac{C_B\big|_{R=R_0} + A I_B}{K_E} \tag{8.2.15}$$

but from Eq. 8.2.4 we have that

$$R^2 D_B \frac{\partial C_B}{\partial R} = -A = -R_0^2 h C_B\big|_{R=R_0} \tag{8.2.16}$$

and therefore

$$C_B\big|_{R=R_\alpha} = \frac{A}{hR_0^2} + A I_B \tag{8.2.17}$$

Substituting in Eq. 8.2.15, we have

$$C_A\big|_{R=R_\alpha} = \frac{A}{kR_\alpha^2} + \frac{A I_B + A/R_0^2 h}{K_E} \tag{8.2.18}$$

Substituting Eq. 8.2.18 in Eq. 8.2.7 at $R = R_0$

$$C_A\big|_{R=R_0} = A I_A + \frac{A}{kR_\alpha^2} + \frac{A I_B + A/R_0^2 h}{K_E} \tag{8.2.19}$$

Substituting in Eq. 8.2.3 and making use of Eq. 8.2.8 we have

$$h\left(C_{A0} - A I_A - \frac{A}{kR_\alpha^2} - \frac{A I_B - A/R_0^2 h}{K_E}\right) R_0^2 = A \tag{8.2.20}$$

Thus

$$hC_{A0} = \frac{A}{R_0^2} + hAI_A + \frac{hA}{kR_\alpha^2} + \frac{hAI_B}{K_E} + \frac{A}{K_E R_0^2} \tag{8.2.21}$$

or, on rearranging

$$A = \frac{hC_{A0}}{[1/R_0^2 + hI_A + h/kR_\alpha^2 + hI_B/K_E + 1/K_E R_0^2]} \tag{8.2.22}$$

Finally, on recalling Eq. 8.2.5 we obtain the desired relationship:

$$\frac{dR_\alpha}{dt} = \frac{-R_0^2 h C_{A0}}{R_\alpha^2 \rho_m (1 - P)[1 + 1/K_E + hR_0^2(I_A + I_B/K_E + 1/kR_\alpha^2)]} \tag{8.2.23}$$

with the initial condition

$$R_\alpha = R_0, \qquad \text{at} \quad t = 0 \tag{8.2.24}$$

Equation 8.2.23 relates the rate of advancement of the reaction front to the various known property values of the system. The only unknown quantities in this equation are the two integrals, I_A and I_B.

Had the diffusion coefficients been constant, I_A and I_B could have been evaluated immediately and upon combining these values with the definition of A, viz. Eq. 8.2.22, Eq. 8.2.23 would have reduced to the well-known expression for the "shrinking core" model.

For a variable diffusivity, as is the case here, we need a relationship between the diffusion coefficient, time and position, for the various temperatures.

Let us consider that the temperature dependence of the reaction rate constant is given by

$$k = k_0 e^{-E_{AR}/\mathscr{R}T} \tag{8.2.25}$$

In order to relate the effective diffusion coefficient to position and time, let us consider that the net effect of *sintering* within the reacted zone will be to cause the decrease of D, according to a first-order kinetic expression, that is,

$$\frac{d}{dt}[D(R,t)] = -\frac{D}{\tau} \qquad \text{for} \quad R_\alpha < R \leq R_0 \tag{8.2.26a}$$

with

$$D = D_0, \qquad \text{for} \quad 0 \leq R \leq R_\alpha \tag{8.2.26b}$$

where

$$\tau = \tau_0 \exp\left(\frac{E_{AS}}{\mathscr{R}T}\right) \tag{8.2.27}$$

Equations 8.2.23 to 8.2.27 together with Eq. 8.2.13 provide a complete statement of the problem.

Equations 8.2.26 and 8.2.27 would cause the diffusion coefficient to tend to zero for large values of t/τ which may not be realistic. In order to overcome this difficulty it was decided to work in terms of an effective diffusion coefficient, defined as:

$$D_{\text{eff}} = D + D_\infty \tag{8.2.28}$$

where D_∞ is the diffusion coefficient in the fully sintered medium.

These equations may be solved numerically. In order to avoid dealing with an excessive number of parameters and without loss of generality, we shall assume that $K_E \rightarrow \infty$ and $D_A = D_B$ throughout.

A typical set of computed results is shown in Fig. 8.2 on a plot of the time taken to achieve 95% reaction against the reaction temperature, which was kept constant, for any particular run.

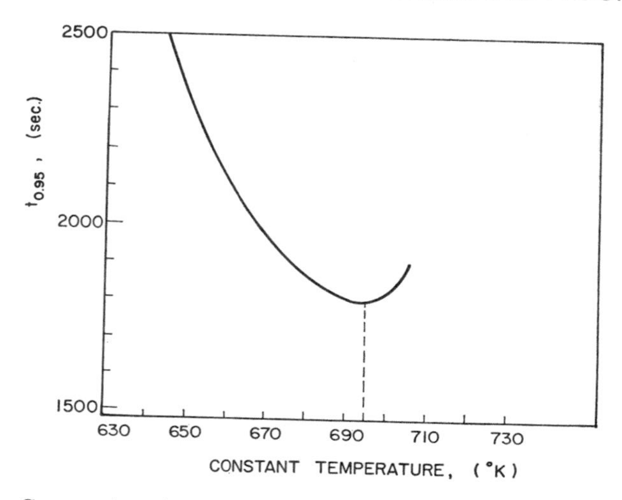

Fig. 8.2 Computed results, showing a plot of the time required to attain 95% conversion against the operating temperature—operation at fixed temperatures.

The other parameters were kept constant, and are given below in Table 8.2.1

TABLE 8.2.1 NUMERICAL VALUES OF THE PARAMETERS CHOSEN FOR THE CALCULATION

$$R_0 = 1 \text{ cm}$$
$$h = 10 \text{ cm/sec}$$
$$C_{A0} = 10^{-5} \text{ gm/cm}^3$$
$$D_0 = 3.0 \text{ cm}^2/\text{sec}$$
$$D_\infty = 0$$
$$k = 4.0 \times 10^{-5} \exp(-15{,}000/\mathscr{R}T)$$

An inspection of Fig. 8.2 shows that for these fixed kinetic parameters, and for various sintering characteristics, as reflected by E_{AS} and τ_0, there may exist an optimal fixed reaction temperature, which would lead to a minimum in the time required to achieve a given extent of reaction—95% in the present case.

The question is whether this performance could be improved by considering a temperature progression, rather than a fixed temperature. Thus we shall determine the optimal temperature progression $T(t)$ so as to minimize the time required for a given degree of conversion.

Solution. The modeling Eqs. 8.2.13, and 8.2.23 to 8.2.28 are a set of partial differential and ordinary differential equations with moving boundaries which can be written in the form:

$$\frac{\partial I(R,t)}{\partial R} = \frac{H(R - R_\alpha)}{D_{\text{eff}} R^2} \qquad I(0,t) = 0 \tag{8.2.29}$$

$$\frac{\partial D(R,t)}{\partial t} = -\frac{D}{\tau} H(R - R_\alpha) \qquad D(R,0) = D_0 \tag{8.2.30}$$

where $H(x)$ is the Heaviside function

$$H(x) = \begin{cases} 1 & x > 0 \\ 0 & x < 0 \end{cases} \tag{8.2.31}$$

Equations 8.2.23 and 8.2.28 complete the description of the system. Our objective then is to find $T(t)$, such that the total time required to attain a given degree of conversion is minimized.

These systems could be treated by any of the standard methods of distributed parameter system optimization discussed in Chapter 7, if one makes special provision for the generalized function $H(x)$. However, in this problem we shall use the method of control vector parameterization because, as discussed in Chapters 6 and 7, this technique is convenient to use in conjunction with existing models of the system [2, 3].

For this problem, the control vector parameterization method consists of expanding temperature as a function of time

$$T(t) = \sum_{i=1}^{n} \phi_i(a_i, t) \tag{8.2.32}$$

(where the ϕ_i are a set of known functions of t with undetermined coefficients) and then finding the optimal set of coefficients a_i. In this work a number of such trial functions were tested, including a fourth-order power series in t_i; however, the functional form

$$T(t) = a_2 + (a_1 - a_2)e^{-t/a_3} \tag{8.2.33}$$

seemed to offer the most promise, allowing constant temperature ($a_3 \to \infty$), a rising temperature profile [$(a_1 - a_2) < 0$], a falling temperature profile [$(a_1 - a_2) > 0$], a sharp change in temperature (a_3 small), a gradual change in temperature (a_3 large), and so on.

Physically the parameters a_i are

a_1 = initial temperature
a_2 = final temperature
a_3 = time constant for temperature change

The actual optimization was carried out by using the following procedure:

(i) An initial guess was made for a_1, a_2, a_3.

(ii) The modeling equations were solved with the temperature progression given in Eq. 8.2.33 until the specified conversion was obtained.

(iii) A new set of a_1, a_2, a_3 were then chosen, using the conjugate direction method of Powell [4] to minimize the time to achieve the specified conversion.

(iv) Step (ii) was repeated with subsequent iterations until the optimal set of a_1, a_2, a_3 (and thus the optimal temperature progression) was found.

The results of the optimization are given in the following.

Computed Results

In view of the large number of parameters, the effect of which could be investigated, the selection of the physical constants for the computation is necessarily somewhat arbitrary. Nonetheless, it is thought that the values actually chosen, a range of which corresponds to the NiO/H_2 system, could be regarded as representative.

The chemical and physical parameters used in the computation were those given in Table 8.2.1, which were common to all the cases considered, with the exception that nonzero values were also considered for D_∞.

The additional property values used, and some of the computed results are summarized in Table 8.2.2.

TABLE 8.2.2 COMPUTED RESULTS FOR THE THREE CASES CONSIDERED

	CASE I		CASE II		CASE III	
D_∞/D_0	Optimal temperature progression time (sec)	Optimal constant temperature time (sec)	Optimal temperature progression time (sec)	Optimal constant temperature time (sec)	Optimal temperature progression time (sec)	Optimal constant temperature time (sec)
0	1752	1798	2163	2218	[a]	—
0.1	1508	1552	2148	2204	2564	2625
0.2	1258	1270	2003	2044	2356	2395

Case I: $\tau_0 = 1.8$, $E_{AS} = 7500$, 95% conversion.
Case II: $\tau_0 = 5 \times 10^{-8}$, $E_{AS} = 30{,}000$, 95% conversion.
Case III: $\tau_0 = 5 \times 10^{-8}$, $E_{AS} = 30{,}000$, 98% conversion.

[a] Unreasonably high reaction time.

The principal parameters that were varied included the conversion, D_∞ and the activation energy in the rate expression for sintering. Three cases were considered:

CASE I. Here the activation energy for sintering was half that of the chemical reaction and the conversion was set at 95%.

CASE II. Here the activation energy for sintering was twice that for the chemical reaction (and the value of τ_0 was appropriately adjusted to give an initial "rate constant" comparable to that of Case I) and conversion again was set at 95%.

CASE III. This was identical to Case II, but with a higher conversion, at 98%. The actual results of the computation are summarized in Table 8.2.2 and further details on the behavior of these systems are provided in Figs. 8.3 to 8.7.

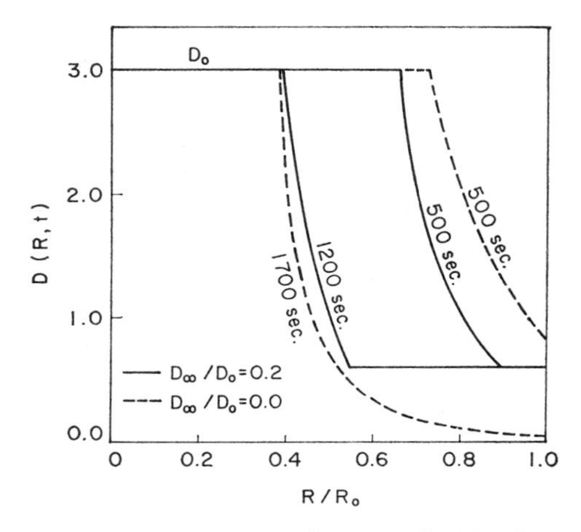

Fig. 8.3 Typical spatial distribution of the effective diffusivity for various times and $D\infty/D_0$ ratios.

Commenting on these graphs first, Fig. 8.3 shows the spatial distribution of the effective diffusion coefficient at three different times, for a particular temperature progression. The curves clearly show the progressive decrease in the value of the diffusion coefficient, with increased time of exposure, within the reacted regions. This behavior is, of course, consistent with the model adopted for representing the system.

Figure 8.4 shows the optimal temperature and conversion trajectories for Case I, with $D_\infty = 0$. It is seen that the optimal temperature progression is a monotonously decreasing temperature from an initially high value. Inspection of the graph also shows that the initial value of the temperature

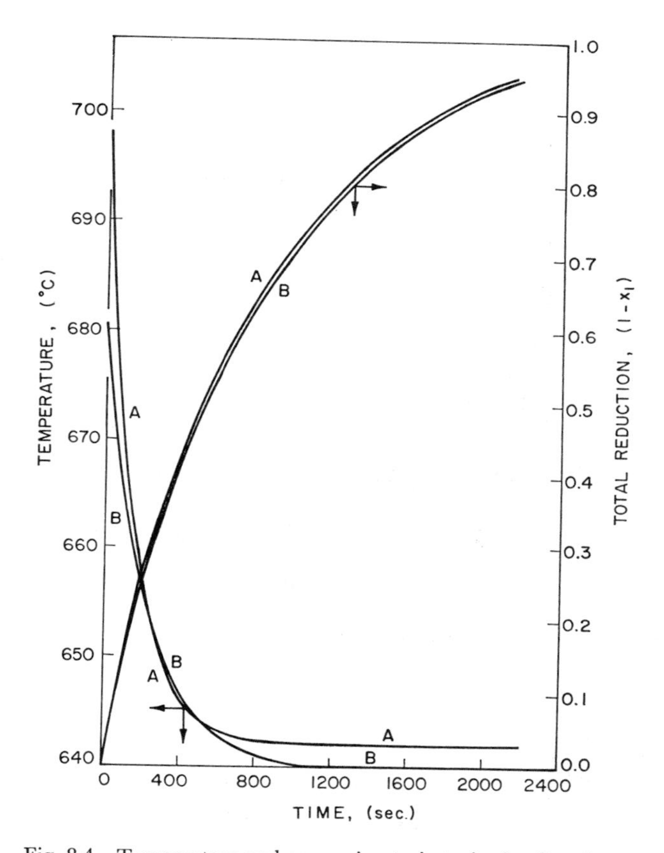

Fig. 8.4 Temperature and conversion trajectories for Case I.

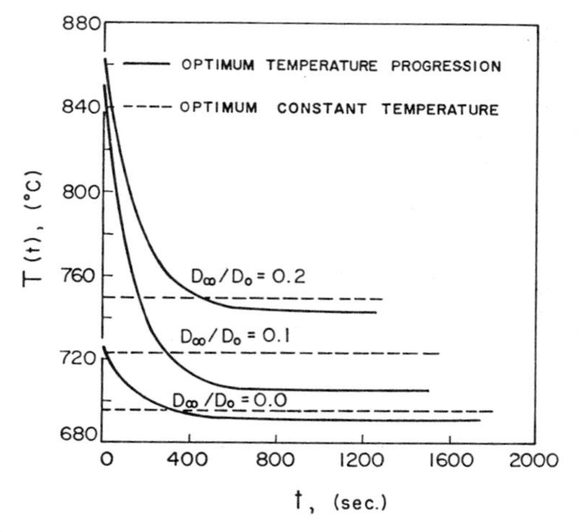

Fig. 8.5 Optimal temperature progession for Case I; also shown are the optimal constant temperatures.

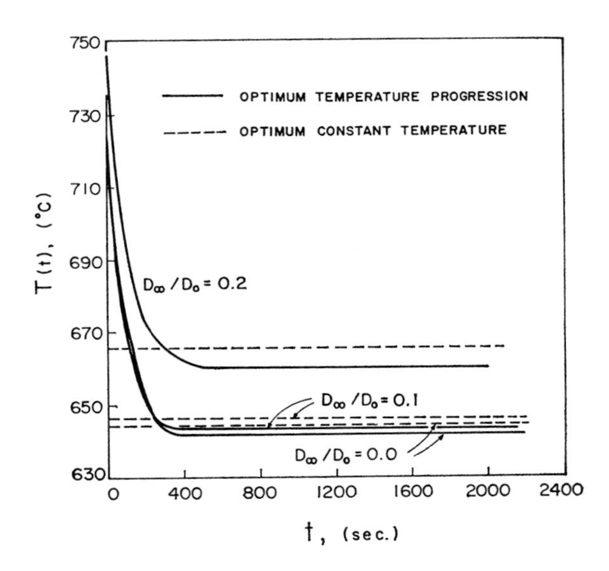

Fig. 8.6 Optimal temperature progression for Case II; also shown are the optimal constant temperatures.

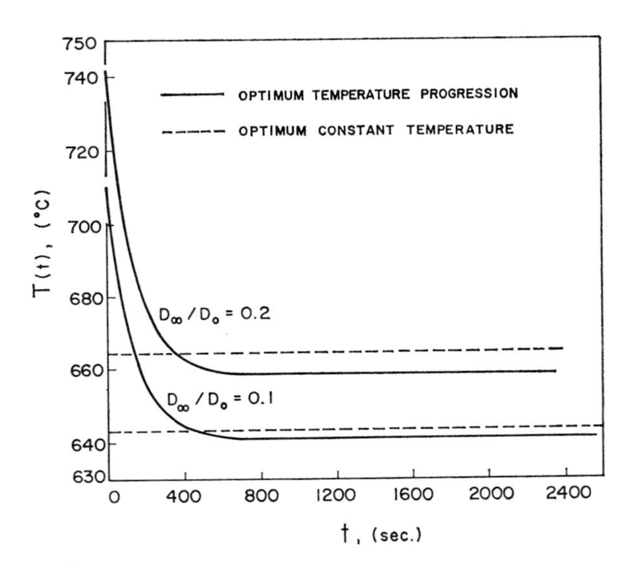

Fig. 8.7 Optimal temperature progressions for Case III; also shown are the constant optimal temperatures.

has a relatively minor effect on the time required to attain a given degree of conversion. The two curves shown in Fig. 8.4 were typical of the optimization results from two different starting values of a_1, a_2, a_3. It is clear that the conversion trajectories are nearly identical.

Figures 8.5 to 8.7 show plots of the optimal temperature progression, for various values of the D_∞/D_0 ratio, for the three cases given in Table 8.2.2.

As expected, the lower the D_∞/D_0 ratio, the smaller are both the initial and the final values of the optimal temperature.

Inspection of Table 8.2.2 shows that for all the cases considered, an optimal temperature progression offers very little advantage over operation at a constant optimal temperature.

Before concluding this example it may be worthwhile to comment on our findings and on their consequences.

On purely physical grounds the use of an optimal temperature progression appeared promising for the systems considered, because of the apparent conflict brought about by the changing temperature, i.e., the conflict between the increased rate of reaction and the increased rate of sintering which occurs simultaneously, upon raising the temperature.

The computed results have shown, however, that for a fairly wide range of parameters, little can be gained by employing a temperature progression, rather than a constant optimal temperature.

We may explain this result, through the use of hindsight, by noting that the effect of the total diffusional resistances I_A and I_B on the reaction rate (Eq. 8.2.23) is an integral of the pointwise value of D. Therefore, the reaction rate is only sensitive to the $D(R, t)$ profile through the fact that the gas flux varies as R^2. This explains both the form of the optimal policy, which allows higher temperatures initially (more sintering near the surface where R^2 is large), and the fact that the temperature program [affecting the $D(R, t)$ profile] improves matters only slightly. It should be emphasized that even this failure to improve matters through an *optimal temperature progression* has important practical implications. Inspection of Table 8.2.2 readily shows that operation at the easily implemented optimal constant temperature may offer quite marked advantages over operation at any other (fixed) temperature level. For the case shown, a temperature displaced by 20°C from the optimum would cause a 10 to 15% increase in the reaction time. Thus for the system considered, appreciable improvements are possible, through optimization with respect to a constant reaction temperature.

A few additional practical comments may be in order at the conclusion of this example.

The system described here had an inherent conflict between chemical kinetics and sintering, the rate of both of which was increased by increasing temperature. Problems of this type were found to occur in the reduction of nickel oxide, but there are indications that similar problems may arise in the reduction of some iron oxide pellets in the blast furnace.

In many oxide reduction processes there are no such conflicts, and here the optimum operation would correspond to the highest temperature that may be allowed by material considerations, such as the mechanical integrity of the reactant, refractory problems, etc. Under these conditions the optimum would occur at the boundary defined by the appropriate constraint.

Finally, we may envision situations, where the conflict is brought about by the high temperatures favoring kinetics, but not equilibria. Systems of this type might benefit from an optimal temperature progression.

In conclusion, the reader is asked to recall Example 6.5.1, where superficial model development lead us to propose an optimal *temperature progression*, rather than a constant, optimal temperature for an essentially identical problem. This fact underlines the importance of using the correct model for the purpose of optimization.

8.3 OPTIMIZATION OF A BATCH PACKED BED REACTOR, HEATED BY RADIATION

In this section we shall solve a particular example of a class of problems first discussed in Section 7.6—the optimal operation of a batch packed bed reactor, heated by radiation.

In the production of ferroalloys, certain processes rely on the reaction between a metal carbide and a metal oxide to produce a metal (ferroalloy) and carbon monoxide, e.g.

$$MeO + MeC = 2Me + CO$$

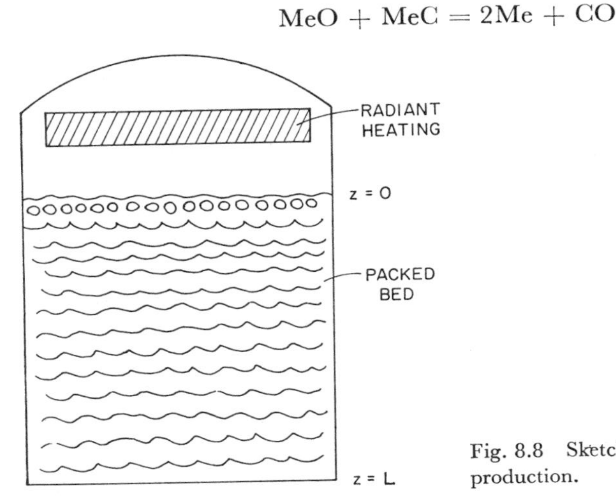

Fig. 8.8 Sketch of the process for ferroalloy production.

Such reactions would generally occur at a high temperature and since a reducing atmosphere is mandatory, one possible scheme of effecting the reaction is sketched in Fig. 8.8.

Pellets made of the solid reactant are placed in a packed bed and then the system is heated by radiation from a resistance heated element.

Within the bed, heat is transferred both by conduction and radiation; once the bed temperature is raised to a sufficiently high value so that the reaction does proceed at a finite rate, a reaction zone is formed, which then proceeds downward. For an endothermic reaction, the rate at which the reaction front propagates depends on both the chemical kinetics and on the rate of heat transfer to the reaction zone.

A sketch of the expected temperature profile at some intermediate time is shown in Fig. 8.9. After some time the bottom of the bed, i.e., the $z = L$

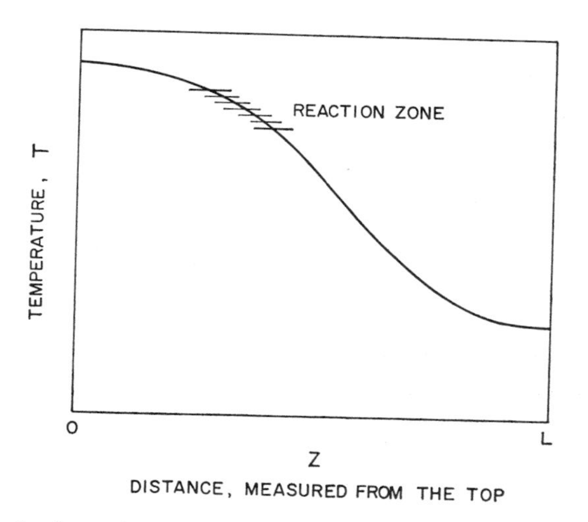

Fig. 8.9 Sketch of a typical temperature profile, for intermediate times, in Example 8.3.

plane reaches a high enough temperature, so that the reaction can proceed there at a reasonable rate. Finally, when the reaction is completed even in the $z = L$ plane, the radiative heat source is turned off, and the system is allowed to cool.

Once a certain critical upper limit of the temperature is reached, further cooling is effected by circulating an inert gas, and then the solid product is discharged.

If we denote the time taken to heat and complete the reaction by t_h, the time required for cooling by t_c, and the time required for charging and discharging of the solids by t_{ro}, then in a physical sense our objective would be to maximize the production rate, that is,

$$\text{Max}\left(\frac{L}{t_h + t_c + t_{r_o}}\right) \tag{8.3.1}$$

Still, purely physical reasoning suggests the following possible avenues:

(i) We might consider varying the bed depth. Deep beds would reduce the fraction of the time spent loading and unloading, but in contrast the time required to heat and cool would be correspondingly increased. On the other hand, very shallow beds would take a shorter time to cool, but here the "dead time" of loading and unloading could become significant.

(ii) Another approach, possibly pursued in a complementary fashion, would be to stop the heating of the upper surface some time before the reaction is completed at the lower end of the furnace. If it were possible to attain a 100% conversion by this procedure, *both the heating and cooling times would be reduced*, thus resulting in an increased production rate.

A variation on the same theme would be to examine the possible benefits that may be gained from employing a temperature progression, when switching from the heating to the cooling cycle.

(iii) Finally, it could be worthwhile to consider the role played by the size of the granular solid reactant; as heat is transferred by radiation, within limits, the larger the particles the larger the "mean free path," and hence, the greater the effective (radiative) conductivity. On the other hand, for larger particles the diffusional resistance within the solid could become the rate limiting factor.

Having established the possibility of optimizing the system on purely physical grounds, let us proceed with the formulation of the problem.

The conservation of thermal energy within the bed may be written as

$$\rho C_p \frac{\partial T}{\partial t} = \frac{\partial}{\partial y}\left(\tilde{k}\frac{\partial T}{\partial y}\right) - R(x, T)\,\Delta H \rho_0 \phi \tag{8.3.2}$$

where $\tilde{k}$ is the effective thermal conductivity

$R(x, T)$ is the rate of the endothermic chemical reaction, thus

$-R(x, T)\,\Delta H \rho_0 \phi$ corresponds to rate of heat absorption within the system and

x is the mole fraction of the reactant solid

After Downing [5] the effective conductivity of the packed bed system may be expressed by the following empirical equation, valid for $T > 500°\text{R}$:

$$\tilde{k} = \frac{(1-\phi)k_s kT^3}{k_s + kT^3} + \phi kT^3 \tag{8.3.3}$$

where k_s = thermal conductivity of the solids

ϕ = void fraction

$$k = \frac{0.692\epsilon d_p}{10^8}\ (\text{Btu/hr ft }°\text{F}) \tag{8.3.4}$$

d_p = particle diameter (in.)

ϵ = emissivity

Before we can proceed further, we have to obtain an expression for $R(x, T)$, and we have to state the boundary conditions, again on the basis of physical considerations.

Let us consider, that for a particular system under consideration, the reaction term may be approximated by

$$R(x, t) = \frac{4.57 \times 10^{19} e^{1.23\times 10^5/T}(1-x)}{0.45 + d_p^{\,2}} \tag{8.3.5}$$

Regarding the initial and boundary conditions, let us specify the initial values of the bed temperature and of the reactant concentration, that is,

$$T = T_i, \qquad t = 0 \tag{8.3.6}$$

$$x = 1, \qquad t = 0 \tag{8.3.7}$$

Let us consider, furthermore, that during the heating period, the upper surface of the bed receives thermal radiation from a source at a temperature T_E, $T_a \leq T_E \leq T_{\text{Max}}$ while the bottom layer is insulated.

During the cooling period we shall assume that the uppermost layer of the bed is maintained at some low temperature T_a while the bottom layer is insulated.

Thus we have

$$\tilde{k}\frac{\partial T}{\partial y} = \epsilon\sigma(T_E^{\,4} - T^4) \qquad 0 \leq t \leq t_h; \qquad y = 0 \tag{8.3.8a}$$

and

$$T = T_a, \quad y = 0 \qquad t_h \leq t \leq t_c \tag{8.3.8b}$$

and

$$\frac{\partial T}{\partial y} = 0, \qquad y = L \tag{8.3.9}$$

The system of Eqs. 8.3.1–8.3.9 is a nonlinear two-point boundary value problem, which was solved numerically, on the CDC 6400 digital computer.

The following property values were used in the computation.

$$\begin{aligned} \Delta H &= 500 \text{ Btu/lb} \\ T_i &= 38°\text{C} \\ T_a &= 115°\text{C} \\ T_{E_{\text{Max}}} &= 1500°\text{C} \\ k_s &= 20 \text{ Btu/hr ft°F} \\ \rho &= 150 \text{ lb/ft}^3 \\ d_p &= 0.0625 \text{ ft} \\ \epsilon &= 0.7 \\ \phi &= 0.62 \end{aligned}$$

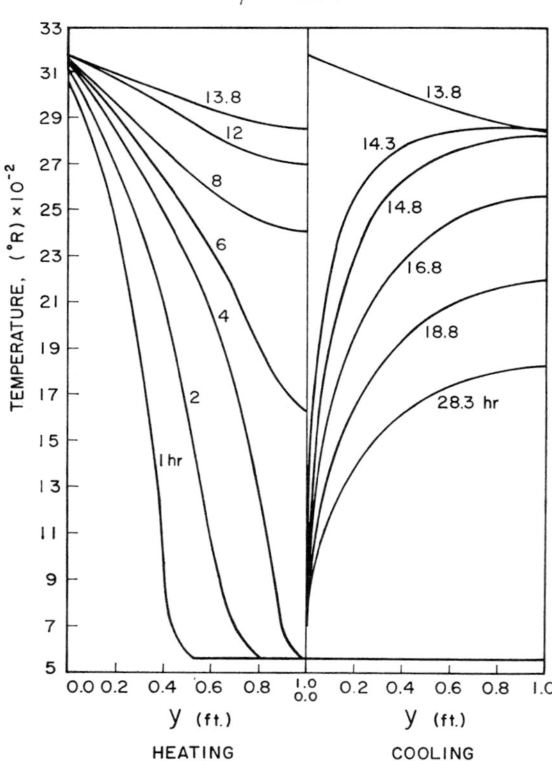

Fig. 8.10 Typical computed temperature profiles in the bed in Example 8.3.

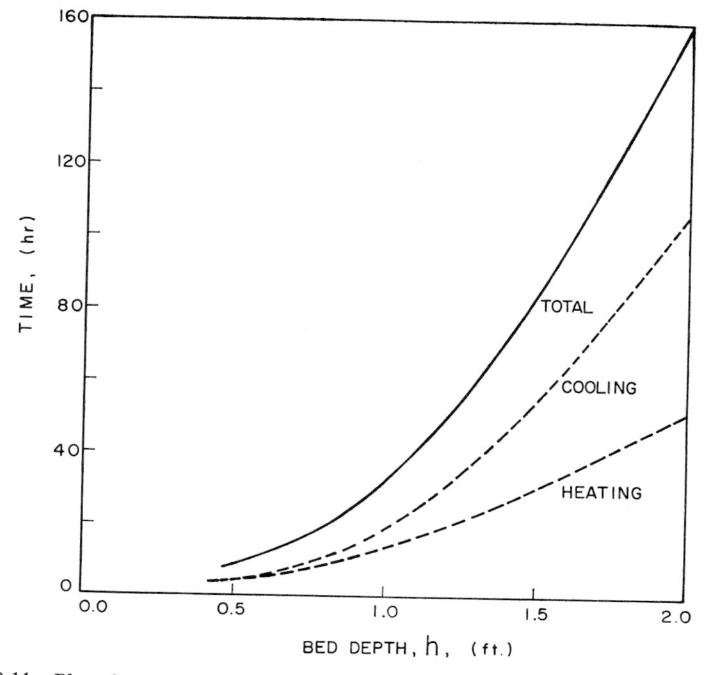

Fig. 8.11 Plot of t_c, t_h, and $(t_c + t_h)$ against the bed depth in Example 8.3.

Some computed results are shown in Figs. 8.10 to 8.12.

Figure 8.10 shows a set of typical temperature profiles in the bed at various times, and illustrates the development of the temperature field within the system. In carrying out the actual optimization process, let us examine the various facets of the problem separately.

The Effect of the Bed Depth

In assessing the effect of the bed depth, Eqs. 8.3.1 to 8.3.9 were solved numerically, the particle size was fixed at 0.0625 ft and the switch was made from heating to cooling when the conversion reached 99% of the bottom layer, that is, when

$$x < 0.01, \qquad \text{at } y = L \tag{8.3.10}$$

Figure 8.11 shows a plot of t_h, t_c, and $(t_c + t_h)$ against the bed depth. Since these were monotonic curves, there was no need to use any single variable elimination methods (cf. Chapter 3) to obtain these plots in the most economical manner.

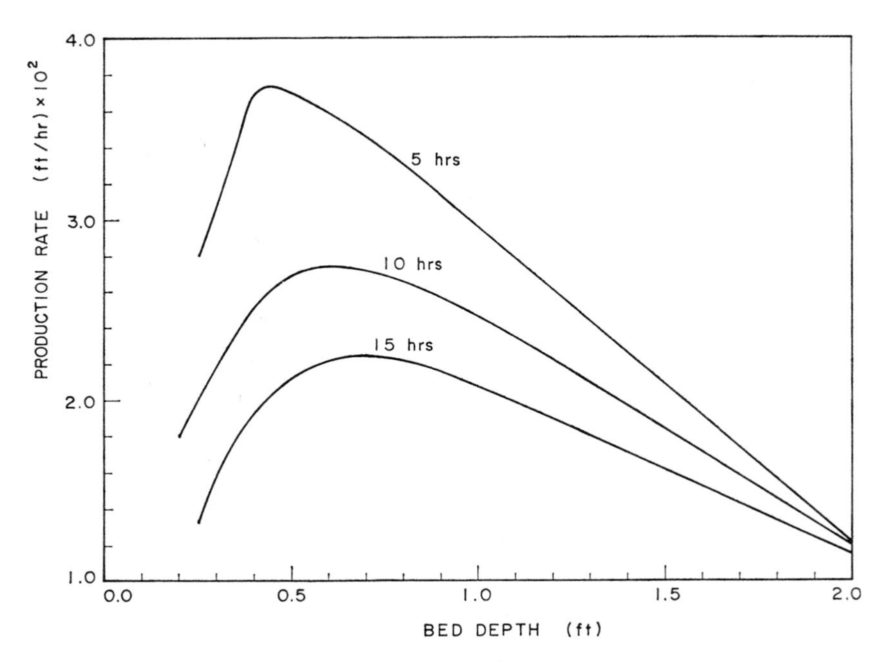

Fig. 8.12 Plot of the production rate against the bed depth with the turn around time, t_{r0} as a parameter.

As seen in Fig. 8.11, the quantity $(t_h + t_c)$ increases quite rapidly with an increasing bed depth. This behavior is to be expected, at least qualitatively, since for a case of pure conduction $(t_h + t_c)$ is proportional to L^2.

However, the quantity of interest is the production rate, which may be defined as

$$\text{production rate} = \frac{L}{t_c + t_h + t_{ro}} \tag{8.3.11}$$

Figure 8.12 shows a plot of the production rate as a function of the bed depth with t_{ro}, the turn around time, as a parameter.

It is seen that the production rate shows relatively sharp maxima which correspond to progressively shallower bed depths as the turn around time is decreased.

The plots shown in Fig. 8.12 were readily constructed by performing simple slide rule calculations, using the information in Fig. 8.11. Had the construction of Fig. 8.12 been more complex, then a single variable elimination method, e.g., the use of the golden section, could have been used to find the maxima of these curves.

Optimal Heating Program

As noted earlier in the statement of the problem, the production rate could be improved by the implementation of an *optimal heating program*, the essential feature of which would be to manipulate the source temperature in a particular time-dependent way so that after heating is complete, the upper surface could be cooled, while the lower regions of the bed are still reacting to the desired degree of conversion.

We can use the theory developed in Chapter 7 to determine the structure of the optimal heating policy. The *Hamiltonian* at the surface $y = 0$ is

$$H_1 = A(t) + B(t)\,T_E^{\,4} \tag{8.3.12}$$

where $A(t)$ and $B(t)$ depend on the temperature $T(0, t)$ and the adjoint variable for the problem, $\lambda(0, t)$. (As an exercise the reader may wish to determine A and B explicitly.) Thus the distributed maximum principle tells us that to minimize H_1, we must have

$$T_E = \begin{cases} T_{E_{\text{Max}}} & \text{for} \quad B < 0 \\ T_a < T_E < T_{E_{\text{Max}}} & \text{for} \quad B = 0 \text{ (singular arc)} \\ T_a & \text{for} \quad B > 0 \end{cases} \tag{8.3.13}$$

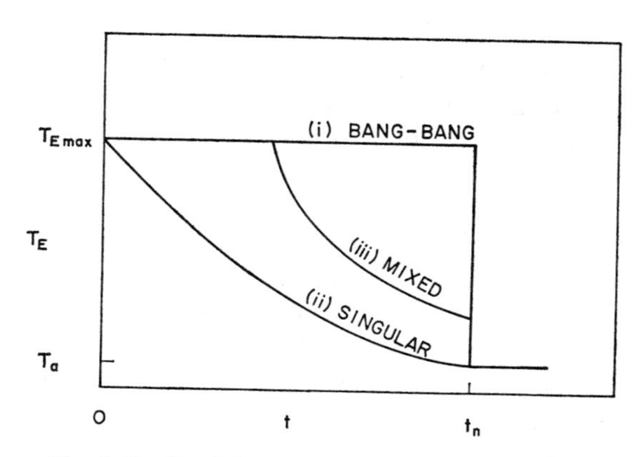

Fig. 8.13 Possible optimal heating programs.

Thus the optimal heating program must take one of the three forms shown in Fig. 8.13:

(i) Bang-bang policy, where T_E switches from maximum heating $T_{E_{\text{Max}}}$ to T_a (i.e., cooling) in a stepwise fashion.

(ii) Singular policy, where $B(t) = 0$, $0 \leq t \leq t_h$ and T_E takes on intermediate values.

(iii) Mixed policy. A mixture of the bang-bang and singular arcs.

Here we shall confine our attention to the "bang-bang" policy and only a brief mention will be made of the other alternatives.

Step Switch

This possibility was explored by determining, by trial and error, the switching time which gave exactly 99% conversion at the bottom of the bed. This was termed the "optimal" switching time, for this complex minimum time optimization problem. The results are summarized in Table 8.3.1.

TABLE 8.3.1 COMPUTED DATA FOR OPTIMAL SWITCH OF HEATING TO COOLING

Bed depth	Time of switch (hr)	Total time required $t_h + t_c$ optimal switch (hr)	Total time required $t_h + t_c$ switch on completion of the reaction (hr)
1 ft	13.0	30	32.2
2 ft	48.0	149.5	158.9

Inspection of Table 8.3.1 shows that the "optimal switch" would allow a 6 to 8% improvement in the production rate, compared with the procedure where the switch occurs only after the completion of the reaction to 99% conversion.

Temperature Progression

As noted earlier, we could also explore the possibility of an optimum temperature progression, which as sketched in Fig. 8.13 could correspond either to a singular arc, or to a "mixed policy."

One possible way of exploring such an optimal temperature progression would be to assign some trial function, such as

$$T_s = a_0 + a_1 t + a_2 t^2$$

or

$$T_s = a_0 + a_1 e^{-a_2 t} \tag{8.3.12}$$

to the dependence of T_s, the temperature of the top surface, and time, and then search for the optimal values of the coefficients a_0, a_1, and a_2, e.g. by the methods described in Chapter 3. It could be reasoned on physical grounds that a temperature progression—as opposed to the bang-bang policy—is unlikely to improve matters. Indeed, some preliminary calculations, which are not reproduced here, seem to support this contention.

The Effect of the Particle Size

Finally, let us consider the possibility of improving the performance of the system by specifying an optimal vertical distribution of particle sizes.

The reader will recall that all the preceding calculations were performed using a particular pellet size.

Upon considering the definition of the effective bed conductivity, Eqs. 8.3.3 to 8.3.4 and the expression used for describing the rate of the reaction, viz. Eq. 8.3.2, a potential conflict brought about by changing the particle sizes becomes immediately apparent. One may reason that the process depends critically on the rate at which heat may be transferred through the bed and therefore it would be desirable to use as large a pellet as practical, since this arrangement will lead to a high effective conductivity.

In contrast, Eq. 8.3.5 shows that the rate of reaction is inversely proportional to the particle size. It follows that it may be profitable to search for an optimum particle size, $d_{p\ \mathrm{opt}}$, or possibly for an optimal distribution of sizes along the reactor, that is, $d_p(y)$.

Such a search was indeed carried out and it was found that the effects of heat transfer predominate, which would lead us to choose the largest practical pellet size.

To summarize, we examined the behavior of a packed bed reactor where an endothermic reaction was carried out in a batchwise manner and where thermal radiation was the predominant mode of heat transfer.

Through the establishment of an appropriate model it was found that there exists a clearcut optimum bed depth and that the process may be improved by discontinuing the heating before all of the bed has reacted.

Neither of these ideas was immediately apparent before the construction of the model (with an optimization in mind), and it is possible, indeed likely, that their implementation would bring about a substantial improvement in the operation.

8.4 THE OPTIMIZATION OF THE PRIMARY END OF AN INTEGRATED STEEL PLANT—A NONLINEAR PROBLEM

Let us consider the optimization of the primary end of an integrated steel plant, sketched in Fig. 8.14. As can be seen, the system consists of two blast

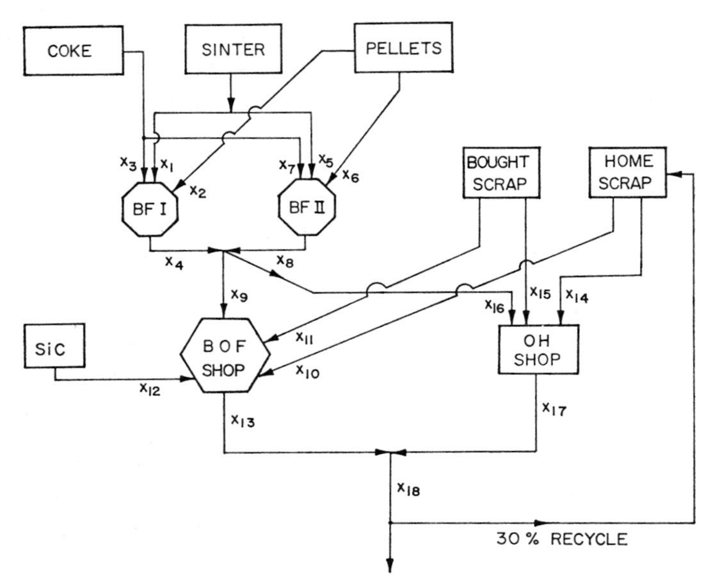

Fig. 8.14 Schematic layout of the primary end of an integrated steel plant in Example 8.4.

furnaces, a basic oxygen steelmaking shop and an open hearth shop. The blast furnaces may be fed with sinter and with pellets, and the rate at which coke is to be supplied to these units depends on both the feed composition and the production rate.

The hot metal from the blast furnaces may be fed either to the BOF shop or to the open hearth shop. Both these steelmaking facilities are capable of utilizing scrap. The open hearth shop may be run on a 100% scrap charge, whereas the BOF shop has a fixed ratio of (scrap/hot metal). This ratio may, however, be increased if silicon carbide is being fed to the system as a supplemental fuel. It is noted, furthermore, that the scrap which is being supplied to the BOF and the open hearth shop consists of "home scrap" that results from the subsequent processing of the crude steel, and of "bought scrap." For a given situation the home scrap is a fixed proportion of the crude steel output, whereas there exists a clear option to purchase additional amounts of scrap from outside sources.

The problem is, given the quantitative relationships between the variables mentioned, to find the optimal operating conditions for producing crude steel at a given production rate.

The reader will recall that a somewhat similar problem was discussed in Chapter 4 (Example 4.4.4) where linear programming was used for handling the problem. It is noted that linear programming mandates that both the constraints and the objective functions are linear, or at least may be linearized over finite intervals. Here we shall treat the more general problem where this restriction need not apply. For the majority of practical problems one would encounter (e.g., where the modeling equations would be deduced empirically by data fitting) the modeling equations used will be nonlinear.

The Modeling Equations

Let us define the following principal symbols:

x_1	sinter input, into Blast Furnace I (BF-I) (tons/annum)
x_2	pellet input into BF-I (tons/annum)
x_3	coke input into BF-I (tons/annum)
x_4	hot metal output from BF-I
x_5	sinter input into BF-II
x_6	pellet input into BF-II
x_7	coke input into BF-II
x_8	hot metal output from BF-II
x_9	hot metal to the basic oxygen shop (BOF)
x_{10}	home scrap into BOF
x_{11}	bought scrap into BOF
x_{12}	silicon carbide into BOF
x_{13}	crude steel output from BOF
x_{14}	home scrap into open-hearth shop (OH)
x_{15}	bought scrap into OH
x_{16}	hot metal input into OH (from the blast furnaces)
x_{17}	crude steel output from OH
x_{18}	total crude steel output
x_{19}	total home scrap
x_{20}	total bought scrap
R_1	coke rate for BF-I, that is, x_3/x_4
R_2	coke rate for BF-II, that is, x_7/x_8
c_1	cost of sinter ($/ton)
c_2	cost of pellets, ($/ton)
c_3	cost of coke ($/ton)
c_4	prorated production cost in BF-I ($/ton)
c_8	prorated production cost in BF-II ($/ton)
c_{11}	cost of bought scrap
c_{12}	cost of SiC ($/ton)
c_{13}	prorated production cost in BOF ($/ton)
c_{17}	prorated production cost in OH ($/ton)
F_1	fixed cost of BF-I ($/annum)
F_2	fixed cost of BF-II ($/annum)
F_3	fixed cost of BOF ($/annum)

F_4 fixed cost of OH (\$/annum)
T_4 total cost of BF-I production (\$/annum)
T_8 total cost of BF-II production (\$/annum)
T_{13} total cost of BOF production (\$/annum)
T_{17} total cost of OH production (\$/annum)
$d_4 = T_4/x_4$ unit cost of hot metal from BF-I (\$/ton)
$d_8 = T_8/x_8$ unit cost of hot metal from BF-II (\$/ton)
$d_9 = (T_4 + T_8)/(x_4 + x_8)$ unit cost of hot metal (\$/ton)
$d_{13} = T_{13}/x_{13}$ unit cost of crude steel from BOF (\$/ton)
$d_{17} = T_{17}/x_{17}$ unit cost of crude steel from OH (\$/ton)
$d_{18} = (T_{13} + T_{18})/x_{18}$ unit cost of crude steel output (\$/ton)

Let us now proceed with the statement of the modeling equations.

Blast Furnace I

Operating Range: 0.7 to 1.6 M tons/annum
1.4 tons of sinter produce 1 ton of hot metal
1.1 tons of pellets produce 1 ton of hot metal
Maximum capacity with sinter: 1.2 M tons of hot metal/annum
The coke rate R_1 is related to the production rate and to the feed composition by the following empirical expression:

$$R_1 = 0.4(1 + 0.5e^{-0.71x_2/x_1}) + 0.3(1 - x_4)^2 \tag{8.4.1}$$

Equation 8.4.1 expresses the fact that the coke rate R_1 depends on the fraction of pellets fed to the system [6] and also on the production rate. An increase in the fraction of pellets tends to reduce the coke rate—to a limiting value of 0.4 in the present case. The second term on the right-hand side indicates that for a fixed pellet/sinter ratio the coke rate shows a minimum at a particular production rate—1 M tons/annum in the present case. The following additional relationships are available:

$$x_4 = 0.715x_1 + 0.91x_2 \tag{8.4.2}$$

$$x_1 \leq 1.7 \tag{8.4.3}$$

$$0.7 \leq x_4 \leq 1.6 \tag{8.4.4}$$

(Capacity constraints)

$$x_3 = R_1 x_4 \tag{8.4.5}$$

(Expression for total coke input rate)

$$\underset{\substack{\text{Total cost}\\ \$/\text{annum}}}{T_4} = \underset{\text{Cost of sinter}}{c_1x_1} + \underset{\text{Cost of pellets}}{c_2x_2} + \underset{\text{Cost of coke}}{c_3x_3} + \underset{\substack{\text{Prorated}\\ \text{production cost}}}{c_4x_4} + \underset{\text{Fixed cost}}{F_1} \tag{8.4.6}$$

Finally,

$$d_4 = \frac{T_4}{x_4}$$

$$\text{Unit cost} \quad \frac{\text{total cost}}{\text{production rate}}$$

Blast Furnace II

Operating range: 0.4–0.8 M tons/annum
1.4 tons of sinter produce 1 ton of hot metal
1.1 tons of pellets produce 1 ton of hot metal
Maximum capacity with sinter: 0.6 M tons/annum
Empirical relationship between coke rate, production rate and feed composition:

$$R_2 = 0.5(1 + 0.4e^{-0.7x_6/x_5}) + 0.35(0.5 - x_2)^2 \tag{8.4.7}$$

Additional relationships:

$$x_8 = 0.715x_5 + 0.91x_6 \tag{8.4.8}$$

$$0.4 \leq x_8 \leq 0.8 \tag{8.4.9}$$

(Capacity constraint)

$$x_5 \leq 0.84 \tag{8.4.10}$$

(Limit on sinter input)

$$x_7 = R_2 x_8 \tag{8.4.11}$$

Coke input = (Coke rate) × (Hot metal production)

$$T_8 = c_1x_5 + c_2x_6 + c_3x_7 + c_8x_8 + F_2 \tag{8.4.12}$$

Total cost ($/annum) = Cost of sinter + Cost of pellets + Cost of coke + Prorated production cost + Fixed cost

$$d_8 = \frac{T_8}{x_8}$$

$$\text{Unit cost} \quad \frac{\text{Total cost}}{\text{Production rate}}$$

Basic Oxygen Furnace

Maximum capacity: 3.5 M tons/annum of crude steel production
Yield: 90% of both scrap and hot metal feed
Hot metal and scrap have to be fed at a minimum ratio of 4:1

The use of SiC allows an increase in the proportion of the scrap that may be melted; 1 ton of SiC allows the charging of an additional 12 tons of scrap. The maximum amount of SiC that may be charged is $\frac{1}{24}$th of the hot metal charge.

The above constraints may be put in the following algebraic form:

$$\underset{\text{Crude steel output}}{x_{13}} = \underset{\text{Hot metal} \times \text{yield}}{0.9x_9} \quad \underset{\text{Scrap} \times \text{yield}}{0.9(x_{10} + x_{11})} \tag{8.4.13}$$

$$x_9 \geq 4(x_{10} + x_{11} - 12x_{12}) \tag{8.4.14}$$

Fixed ratio: $\dfrac{\text{Hot metal}}{\text{Scrap}}$

$$x_{13} \leq 3.5 \tag{8.4.15}$$

(Capacity constraint)

$$x_{12} \leq 24x_9 \tag{8.4.16}$$

(Limit on the use of SiC)

The total annual cost for the BOF is then given as:

$$\underset{\text{Annual cost}}{T_{13}} = \underset{\text{Fixed cost}}{F_3} + \underset{\text{Cost of hot metal}}{x_9\, d_9} + \underset{\text{Cost of bought scrap}}{c_{11}x_{11}} + \underset{\text{Cost of SiC}}{c_{12}x_{12}}$$

$$+ \underset{\text{Prorated production cost}}{c_{13}x_{13}} + \underset{\text{Cost of home scrap}}{(c_{10}x_{10})^{\dagger}} \tag{8.4.17}$$

The unit cost is then given as:

$$d_{13} = \frac{T_{13}}{x_{13}} \tag{8.4.18}$$

The Open Hearth Shop

Maximum capacity: 1.5 M tons/annum with a 100% scrap charge
2.0 M tons/annum with a 100% hot metal charge

Yield: 92%

The following relationships apply

$$\underset{\text{Total output}}{x_{17}} = \underset{\text{yield} \times \text{(home scrap + bought scrap + hot metal)}}{0.92(x_{14} + x_{15} + x_{16})} \tag{8.4.19}$$

$$2.0 \geq x_{16} + 1.33(x_{14} + x_{15}) \tag{8.4.20}$$

(Capacity constraint)

† Was taken as zero.

The total cost is given as:

$$\begin{array}{ccccccccccc} T_{17} & = & F_4 & + & c_{17}x_{17} & + & c_{10}x_{14} & + & c_{11}x_{15} & + & d_9x_{16} \\ \text{Total annual cost} & & \text{Fixed cost} & & \text{Prorated operating cost} & & \text{Cost of home scrap} & & \text{Cost of bought scrap} & & \text{Cost of hot metal} \end{array} \qquad (8.4.21)$$

Unit cost:

$$d_{17} = \frac{T_{17}}{x_{17}}$$

$$\begin{array}{ccc} (x_{10} + x_{14}) & = & 0.3x_{18} \\ \text{Total home scrap} & & 0.3 \times \text{total production} \end{array} \qquad (8.4.22)$$

$$\begin{array}{ccccc} x_{18} & = & x_{17} & + & x_{13} \\ \text{Total crude steel produced} & & \text{Crude steel from OH} & & \text{Crude steel from BOF} \end{array} \qquad (8.4.23)$$

$$\begin{array}{ccccc} (x_4 + x_8) & = & x_9 & + & x_{16} \\ \text{Hot metal output} & & \text{Hot metal to BOF} & & \text{Hot metal to OH} \end{array} \qquad (8.4.24)$$

$$d_9 = \frac{T_4 + T_8}{x_4 + x_8} \qquad (8.4.25)$$

Unit cost of hot metal

The parameters to be used for this problem are:

Production rate: 3 M tons/annum
SiC available: 0.1 M tons/annum
} other raw materials unlimited

$$c_1 = \$21 \qquad c_{11} = 0 \qquad F_1 = 7.2 \times 10^6$$

$$c_2 = \$30 \qquad c_{12} = \$180 \qquad F_2 = 5.0 \times 10^6$$

$$c_3 = \$25 \qquad c_{13} = \$15 \qquad F_3 = 10.5 \times 10^6$$

$$c_4 = c_8 = 0 \qquad c_{17} = \$26 \qquad F_4 = 3.0 \times 10^6$$

The Optimization Technique

It would be possible to apply a number of different techniques in order to optimize this complex, nonlinear structure. One could treat the system as a large nonlinear program and apply one of the nonlinear programming algorithms of Chapter 4. Alternatively, one could choose to decompose the problem by using dynamic programming, the discrete maximum principle, or other methods and optimize the subsystems iteratively. The best choice depends on the particular problem considered.

Because the problem is largely linear with only two nonlinear processes (BF-I and BF-II), we shall choose a strategy which treats the whole plant by linear programming (LP) and applies nonlinear programming for the iterative treatment of the blast furnaces only. The algorithm, developed in more detail in Ref. 7, is as follows:

1. Select an initial guess of the values of R_1 and R_2.
2. Use linear programming to minimize the total cost of steel production T_{17}, subject to the linear constraints Eqs. 8.4.2 to 8.4.5, 8.4.8 to 8.4.11, 8.4.13 to 8.4.16, 8.4.19 to 8.4.20, and 8.4.22 to 8.4.24.
3. For hot metal production fixed by the LP, use the gradient projection algorithm to minimize the costs of production of the blast furnaces (i.e., for BF-I, minimize T_4 subject to constraints Eqs. 8.4.1 to 8.4.5; for BF-II, minimize T_8 subject to constraints Eqs. 8.4.7 to 8.4.11).
4. Calculate new values for R_1 and R_2 by the formula

$$\begin{aligned} R_1 &= (1-\epsilon)\bar{R}_1 + \hat{R}_1 \\ R_2 &= (1-\epsilon)\bar{R}_2 + \hat{R}_2 \end{aligned} \qquad 0 \leq \epsilon \leq 1 \qquad (8.4.27)$$

where $\bar{R}_i$ is the value from step 2, and $\hat{R}_i$ the value from step 3. Then re-solve the LP as in step 2.

(i) If the sinter, coke, pellets, and hot metal values are the same as the subproblem optimization in step 3 predicts, then the optimum has been found.
(ii) If the hot metal production changes from steps 2 and 3, then go to step 3 and reoptimize the nonlinear subsystems for the new hot metal production rates.
(iii) If the hot metal production or R_1 and R_2 remain fixed, but sinter, coke, or pellet requirements change, then the algorithm cycles—go to step 5.

5. Calculate new R_1 and R_2 values from the results of the *LP* solution in step 4 and use these to recompute the *LP* solution. Repeat this procedure until the hot metal production changes, then go to step 3.

It should be noted that this procedure is a *feasible one* in that the hot metal production is maintained constant during subsystem optimization; however, the technique produces nonfeasible solutions from the *LP* because R_1 and R_2 are not correct until the optimum is reached. For further discussions of convergence, special problems, etc. see Ref. 7.

Optimization Results

The detailed results arising from applying this procedure to the problem described are given in Tables 8.4.1 to 8.4.7. The values of R_1 and R_2 were

TABLE 8.4.1 LINEAR PROGRAMMING RESULTS

Unit	Coke	Sinter	Pellets	SiC	Bought scrap	Home scrap	Hot metal	Steel
BFI	0.608	1.7	0				1.216	
BFII	0.3	0.84	0				0.601	
BOF				0.079	1.408		1.816	2.902
OH					0.218	1.285		1.383
Total	0.908	2.54	0	0.079	1.626	1.285	1.816	4.285

Total cost: \$236.19 $\times 10^6$.
Unit cost: \$55.12/ton.
Iteration number: 1.
$R_1 = 0.5$, $R_2 = 0.5$.

initially estimated as 0.5 and the LP described in step 2 carried out. The results, given in Table 8.4.1, show how the iron should be made with coke rates $R_1 = R_2 = 0.5$. The resulting hot metal values were used to optimize the subproblems as described in step 3 of the algorithm. The results of this optimization are shown in Table 8.4.2. The R_1 and R_2 values were then used in the LP again as in step 4; ϵ was chosen as 1.0. The results, given in Table 8.4.3, show that hot metal changes in BF-I, and the pellets, sinter, and coke mix change for both subsystems; thus we must go to step 3. In step 3 the subsystem optimization program produced the results in Table 8.4.4. Notice that there is no optimization needed for BF-II because the hot metal production had not changed. The coke rates R_1 and R_2 from Table 8.4.4 are unfortunately the same as from Table 8.4.2, so no improvement can be made, and the iterative scheme cycles; thus we go to step 5. We compute R_1 and R_2 from Table 8.4.3 and use these to produce Table 8.4.5 (see page 311) in which the hot metal from both BF-I and BF-II changes. We can now compute R_1 and R_2 from Eq. 8.4.26 with $\epsilon = 0.8$ (we used $\epsilon = 1$ with no success). These results are shown in Table 8.4.6 (see page 312). These hot metal values were used in step 3 to produce the subsystem optima shown in Table 8.4.7 (see page 313). The closeness of the results in Tables 8.4.6 and 8.4.7 indicates that the technique has finally converged and thus Table 8.4.6 shows the optimal policy.

The shadow prices (cf. Section 4.4) are given in Table 8.4.6 and show that the limitation of SiC is a bottleneck in the operation so that one could afford to pay up to $205.24/ton to purchase additional supplies. Again we must stress to the reader that this particular conclusion is necessarily the result of the particular price structure assumed for the purpose of computation. Higher scrap prices or lower pellet prices would quite likely alter this picture.

We hope, however, that through this example the reader can visualize how nonlinear processes or processes having distributed parameters could be treated within the basic framework of linear programming.

TABLE 8.4.2 SUBSYSTEM OPTIMIZATION RESULTS

	BFI	BFII
Hot metal input	1.216	0.601
Output		
Coke	0.793	0.351
Sinter	1.424	0.598
Pellets	0	0.200
R_1, R_2	0.661	0.584

Iteration number: 1.

TABLE 8.4.3 LINEAR PROGRAMMING RESULTS

Unit	Coke	Sinter	Pellets	SiC	Bought scrap	Home scrap	Hot metal	Steel
BFI	0.673	1.424	0				1.018	
BFII	0.351	0.84	0				0.601	
BOF				0.1	1.605		1.619	2.902
OH					0.218	1.285		1.383
Total	1.024	2.264	0	0.1	1.823	1.285	1.619	4.285

Total cost: $\$241.92 \times 10^6$.
Unit cost: $56.46/ton.
Iteration number: 2.
$R_1 = 0.661$, $R_2 = 0.584$.

TABLE 8.4.4 SUBSYSTEM OPTIMIZATION RESULTS

	BFI	BFII
Input hot metal	1.018	0.601
Output		
Coke	0.673	0.351
Sinter	1.424	0.598
Pellets	0	0.200
R_1, R_2	0.661	0.584

Iteration number: 2.

8.5 SOME FURTHER OPTIMIZATION PROBLEMS

The reader will have noted from the worked examples presented in the preceding sections of this chapter, that it may require considerable computational effort to tackle these more realistic problems.

Since the full statement of these problems and the development of their solution poses appreciable demands on space, we shall conclude this chapter and the text by posing a number of additional, realistic optimization problems. We shall emphasize the basic problems, the *conflicts* or *trade-offs*, which indicate that optimization is required. In addition, we shall outline the possible techniques available for solution, without giving any actual results.

These examples should help the reader to identify additional practical problems which may benefit from this approach.

EXAMPLE 8.5.1. We wish to optimize the performance of a process in which a high chrome, low carbon steel is being produced. The process operates in the following manner: molten iron containing chromium and carbon is introduced into a vessel, where it is bottom blown with a mixture of argon and oxygen. The carbon contained in the melt reacts with the oxygen according to

$$C_{metal} + O_{metal} = CO_{gas} \tag{8.5.1}$$

In addition, some of the chromium will also be oxidized, which is regarded as the undesirable side reaction, according to

$$3Cr_{metal} + 4O_{metal} = Cr_3O_4 \tag{8.5.2}$$

TABLE 8.4.5 LINEAR PROGRAMMING RESULTS

Unit	Coke	Sinter	Pellets	SiC	Bought scrap	Home scrap	Hot metal	Steel
BFI	0.694	1.5	0				1.051	
BFII	0.405	0.80	0				0.568	
BOF				0.1	1.605		1.619	2.902
OH					0.218			1.383
Total	1.099	2.30	0	0.1	1.823		1.619	4.285

Total cost: $244.21 × 10^6$.
Unit cost: $56.99/ton.
Iteration number: 3.
$R_1 = 0.661$, $R_2 = 0.704$.

TABLE 8.4.6 LINEAR PROGRAMMING RESULTS

Unit	Coke	Sinter	Pellets	SiC	Bought scrap	Home scrap	Hot metal	Steel
BFI	0.804	1.7	0				1.215	
BFII	0.274	0.564	0				0.403	
BOF				0.1	1.605		1.618	2.902
OH					0.218	1.285		1.383
Total	1.078	2.264	0	0.1	1.823	1.285	1.618	4.285
Shadow price				25.24[a] $205.24[b]			46.37[b]	61.77[b]

Total cost: 243.27×10^6.
Unit cost: $56.77/ton.
Iteration number: 4.
$R_1 = 0.661$, $R_2 = 0.680$.
[a] Capacity constraint.
[b] Outside purchase option.

TABLE 8.4.7 SUBSYSTEM OPTIMIZATION RESULTS

	BFI	BFII
Input hot metal	1.215	0.403
Output		
Coke	0.804	0.274
Sinter	1.700	0.564
Pellets	0	0
R_1, R_2	0.661	0.681

Iteration number: 4.

The chemical reactions described by Eqs. 8.5.1 and 8.5.2 will be the only ones considered in this formulation, although in practical systems some other components may also react, in particular, FeO will certainly form. Another practical consideration which will not be pursued here is to provide suitable additions so as to promote the formation of slag with the required consistency and composition.

Nonetheless, it is thought that the oxidation of carbon and of chromium constitute the essential part of the problem, so that our simplification will not distort the principal features of the system.

One of our objectives in running this process in an optimal fashion is to reduce the initial carbon content to the desired level, while minimizing the oxidation (i.e., loss) of chromium.

For a fixed vessel size, charge composition, and product specification, the control variables are the oxygen and argon blowing rates, both of which could be made time dependent.

The conflicts, needed in any optimization procedure, are provided by the following inherent features of the system.

(i) At high temperatures the oxidation of carbon is favored over the oxidation of chromium.

(ii) At high temperatures the erosion of the refractory lining of the vessel becomes more pronounced, which will require more frequent relining.

(iii) The higher the oxygen flow rate the more rapid is the temperature rise, which would favor carbon oxidation; however, carbon oxidation may also be favored by blowing an argon rich oxygen-argon mixture, which would reduce the partial pressure of carbon monoxide in the product gas at the expense of using argon.

In the following we shall construct a simplified mathematical model for the process, which will then be used in the subsequent calculations for finding the optimal trajectories of the oxygen and argon flow rates.

The Mathematical Model

Let us consider a vessel, containing an iron-carbon-chromium melt, which is blown with an argon + oxygen mixture.

The mathematical model for the process will consist of a heat balance, a mass balance, combined with a suitable expression for the equilibrium relationships governing the system; these latter two in combination with the heat balance, will allow us to calculate the relative rates at which chromium and the carbon are oxidized.

Heat Balance

Let us assume that the initial molten metal charge is 100 tons and that the heat capacity of the system, comprised of the slag and of the metal, is not altered significantly during the blow. Then the heat balance may be written as

$$MC_p \frac{dT}{dt} = q_{\mathrm{C}} + q_{\mathrm{Cr}} - q_L - q_{\mathrm{Ar}} \tag{8.5.3}$$

where $M = 2 \times 10^5$ lb is the mass of the melt charged to the unit.

$C_p = 0.2$ Btu/lb °R the effective heat capacity of the melt and the slag that is formed subsequently.

$q_{\mathrm{C}} =$ is the rate of heat generation due to the oxidation of carbon.

$q_{\mathrm{Cr}} =$ is the rate of heat generation due to the oxidation of the chromium.

$q_L =$ represents the rate of heat losses from the system including the sensible heat carried away by the CO gas.

$q_{\mathrm{Ar}} =$ represents the amount of heat removed from the system by the argon gas.

By using available thermodynamic information [8] q_{C}, q_{Cr}, and q_{Ar} may be written as:

$$q_{\mathrm{Cr}} = -M \frac{dw_{\mathrm{Cr}}}{dt} \times 5.3 \times 10^3 \quad \text{(Btu/min)} \tag{8.5.4}$$

$$q_{\mathrm{C}} = -M \frac{dw_{\mathrm{C}}}{dt} \times 3.15 \times 10^3 \quad \text{(Btu/min)} \tag{8.5.5}$$

where t is time in minutes and w_{Cr} and w_{C} denote the mass fraction of chromium and carbon in the melt, respectively.

$$q_{\mathrm{Ar}} = 0.23(T_B - 600)R_{\mathrm{Ar}}^{\cdot} \tag{8.4.6}$$

where T_B is the bath temperature in °F and
R_{Ar} is the mass flow rate of the argon, in lb/minute and 600°F is a reference temperature

Finally, the heat loss from the system of 100-ton capacity may be estimated as a first approximation from data reported in the literature [9].

Mass Balance

The overall mass balance on oxygen may be written on the assumption that all the oxygen fed to the system will be used up by oxidizing chromium and carbon. Thus we have

$$R_{O_2} = -1.33M\frac{dw_C}{dt} - 0.415M\frac{dw_{Cr}}{dt} \tag{8.5.7}$$

In order to assess the relative rates at which carbon and chromium are oxidized, reference must be made to the available equilibrium data.

By considering the chemical equilibrium for the reactions (Eq. 8.5.1 and Eq. 8.5.2) and on assuming unit activity for Cr_3O_4 and one atmosphere CO partial pressure, a family of curves may be constructed with the temperature as a parameter showing the amount of carbon and chromium that may coexist in molten iron, under these conditions [10]. Such a set of curves is shown in Fig. 8.15.

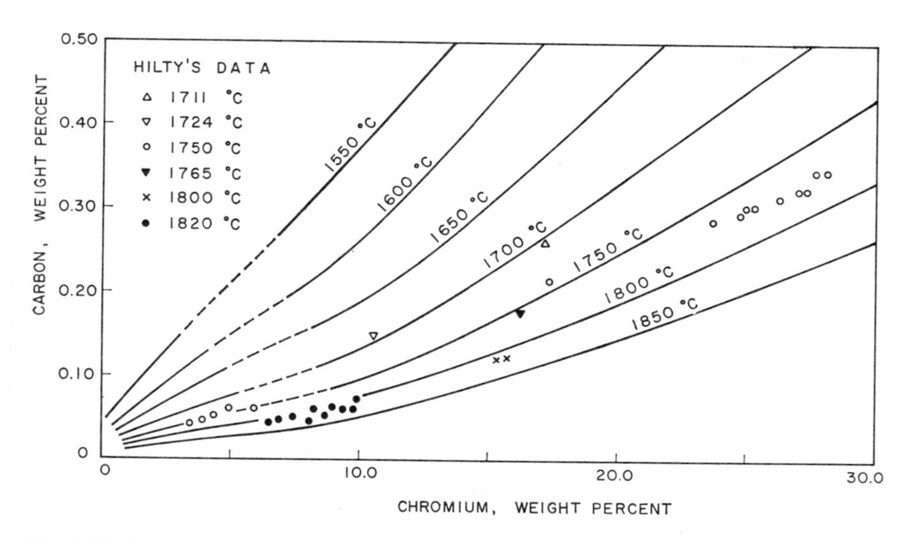

Fig. 8.15 Equilibrium relationship for the iron-chromium-carbon system at 1 atm CO pressure, after Ref. 8.

This plot may be readily used for calculating the equilibrium compositions for CO pressures other than one atmosphere by multiplying the ordinate by the CO partial pressure, and thus changing the scale. Thus for a melt with 20% chromium at 0.1 atmosphere CO partial pressure and at 1800°C the equilibrium carbon content is about 0.02 wt %.

The family of curves shown in Fig. 8.15 may be represented by a functional relationship of the following type:

$$w_{\mathrm{C}}\Big|_{eq} = f_1(w_{\mathrm{Cr}})\Big|_T \qquad \text{or} \qquad w_{\mathrm{Cr}}\Big|_{eq} = f_2(w_{\mathrm{C}})\Big|_T \tag{8.5.8}$$

In other words, at a fixed temperature we can express the amount of chromium that would be in equilibrium with a given carbon content, or alternatively, we may calculate the amount of carbon that would be in equilibrium with a given chromium content.

In order to illustrate how these equilibrium relationships may be used to estimate the relative rates at which carbon and chromium are oxidized, let us examine Fig. 8.16 showing $w_{\mathrm{C}_{eq}} = f_1(w_{\mathrm{Cr}})$ corresponding to two different temperatures.

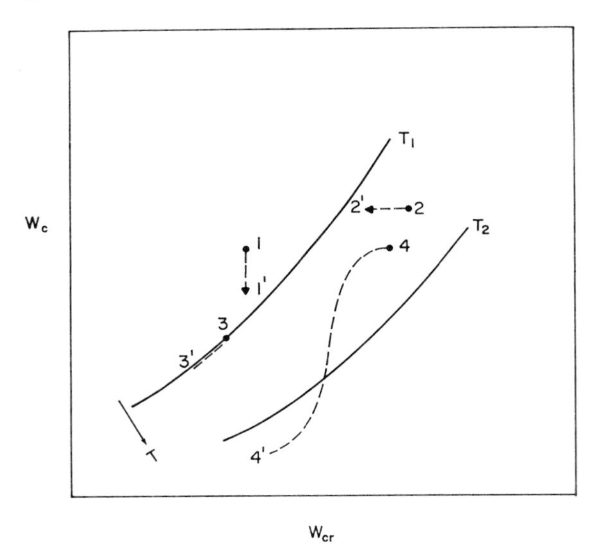

Fig. 8.16 Possible paths for the simultaneous oxidation of oxygen and carbon.

Let us consider an initial bath composition corresponding to point 1 on the graph, at a temperature T_1. Here the carbon content is in excess of what is in equilibrium with the chromium present, so we shall assume that, under these conditions, nearly all the oxygen blown will be used to oxidize carbon, as indicated by the path between points 1 and 1′.

An alternative possibility is that the initial bath composition corresponds to point 2, at a temperature T_1. Here the chromium is in excess of that in equilibrium with the carbon, so that the oxygen will be used to oxidize chromium almost exclusively, as indicated by the horizontal line drawn between 2 and 2′.

Finally, we may consider a situation where the carbon and chromium content of the bath are in equilibrium; under these conditions there will be a simultaneous reduction in the carbon and chromium content, as we move along the equilibrium line between the points 3 and 3′ as shown in the graph.

The path followed by the carbon and the chromium contents during a given blow, over which the temperature will change significantly, will be made up of a number of these elementary steps. One such possible path, covering a range of temperatures is also sketched in Fig. 8.16 connecting the points 4 and 4′.

The Objective Functional

The objective functional to be minimized is the total cost of production, which may be expressed as

$$\text{Min}\left\{ \begin{aligned} I = {} & \underbrace{8.0 \times 10^4 (w_{\text{Cr}}|_{t=0} - w_{\text{Cr}}|_{t=t_f})}_{\text{Cost of chromium loss into the slag}} \\ & + \underbrace{a(w_{\text{C}}|_{tf} - w_{\text{C}}|_{sp})}_{\text{Penalty for exceeding the specified maximum carbon content}} \\ & + \underbrace{C_1 \int_0^{t_f} R_{O_2}\, dt}_{\text{Total cost of oxygen blown}} + \underbrace{C_2 \int_0^{t_f} R_{\text{Ar}}\, dt}_{\text{Total cost of argon blown}} \\ & + \underbrace{C_3 \int_0^{t_f} dt}_{\text{Capital and operating charges}} + \underbrace{C_4 \int_0^{t_f} F\, dt,}_{\text{Cost of refractory erosion}} \end{aligned} \right\}$$

$$F = 0; \qquad T < 3050°\text{F}$$

$$F = (T - 3050)^2, \qquad T > 3050°\text{F} \tag{8.5.9}$$

One would then proceed by assigning numerical values to the cost factors, C_1–C_4, and by expressing the equilibrium relationship, appearing in Fig. 8.15 in a numerical form, e.g. by curve fitting. The modeling equations, established on the basis of heat and mass balance considerations would then take the following general form:

$$\frac{dw_c}{dt} = f_1(T_1, w_c, w_c\big|_{eq}, R_{O_2}, R_{Ar}) \tag{8.5.10}$$

$$\frac{dw_{cr}}{dt} = f_2(T_1, w_{cr}, w_{cr}\big|_{eq}, R_{O_2}, R_{Ar}) \tag{8.5.11}$$

$$\frac{dT}{dt} = \frac{1}{MC_p}[q_c(t) + q_{cr}(t) - q_L - q_{Ar}] \tag{8.5.3}$$

By using these modeling equations we can calculate the instantaneous bath composition and bath temperature. A set of nonoptimal results, showing the

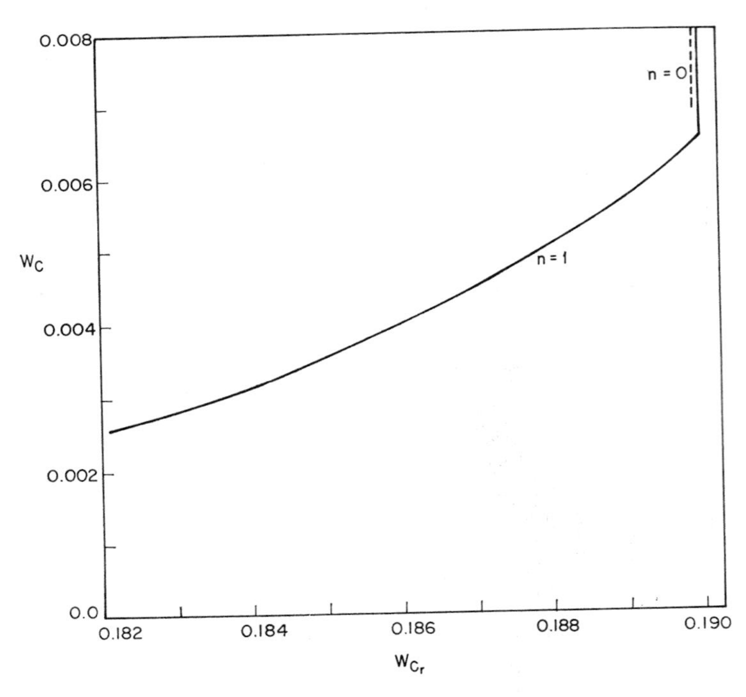

Fig. 8.17 A typical computed trajectory of carbon and chromium concentrations for non-optimal conditions in Example 8.5.1.

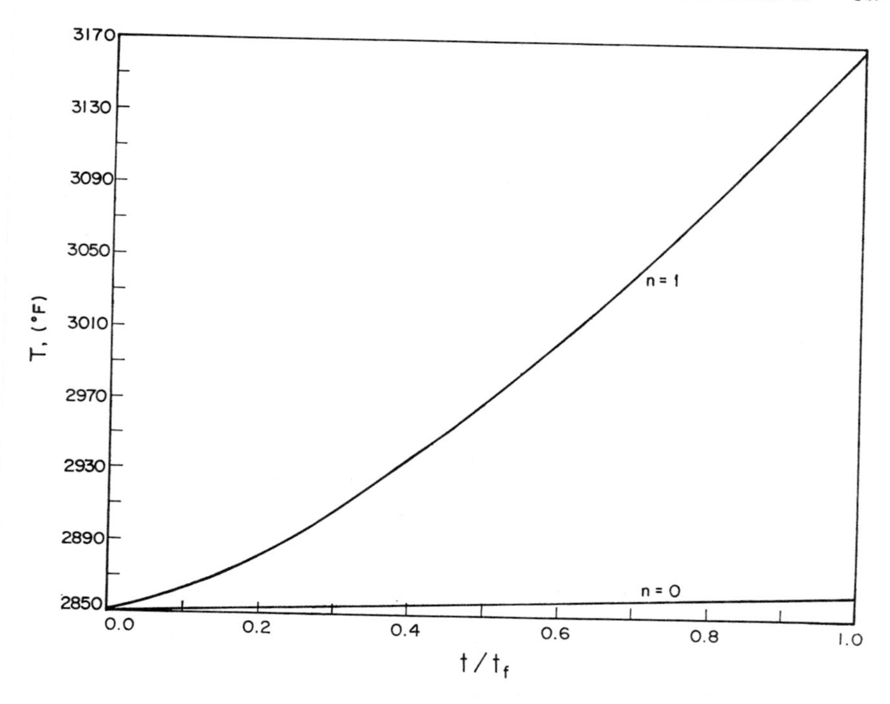

Fig. 8.18 A typical computed temperature progression in Example 8.5.1.

time dependence of the bath composition and the bath temperature are given in Figs. 8.17 and 8.18, respectively.

The general behavior shown by these two graphs appears to be consistent with the experimental observations reported by Choulet et al. [11], whose article is recommended for a good review of recent work on the decarburization of ferrochrome. Using the modeling equations given above, the optimization can now be performed by employing one of the techniques described in Chapter 6, namely control vector iteration or control vector parameterization. In some ways, the latter technique might be more convenient, as physical intuition suggests that the optimal blowing arrangement could involve a progressively increasing argon/oxygen ratio.

EXAMPLE 8.5.2. Let us consider the solidification of a ferroalloy or a fused cast refractory material.

Ideally, one would wish to extract heat at the maximum rate from the outer solid surface.

However, here the limitation is provided by the need to avoid excessive thermal stresses.

The problem may be formulated as

$$\rho C_p \frac{\partial T}{\partial t} = \frac{\partial}{\partial y}\left(k \frac{\partial T}{\partial y}\right) \qquad 0 \leq y \leq Y \tag{8.5.12}$$

(Heat balance in the solidified shell)

$$t = 0; \qquad Y = 0 \tag{8.5.13}$$

$$y = Y, \qquad T = T_{mp} \tag{8.5.14}$$

$$y = Y; \qquad L\rho \frac{dY}{dt} = k \frac{\partial T}{\partial y} \tag{8.5.15}$$

and

$$y = 0; \qquad k \frac{\partial T}{\partial y} = f(T, T_E) \tag{8.5.16}$$

Here

Y	position of the solidification front
L	latent heat of solidification
T_E	coolant temperature at outer solid surface

Our objective may now be expressed in a mathematical form as follows:

$$\mathrm{Min}(t_f) \tag{8.5.17}$$

that is, to minimize the time required for completely solidifying the slab, while satisfying the constraint

$$\frac{\partial T}{\partial y}(y, t) \leq \frac{\partial T}{\partial y}(y)\Big|_{\mathrm{Max}} \tag{8.5.18}$$

i.e., the temperature gradient within the slab must not exceed a given critical value.

Physical reasoning will tell us that for a slab, this maximum temperature gradient, i.e., flux, will occur at the outer surface of the slab; thus the condition (Eq. 8.5.18) specifies a maximum heat flux which may not be exceeded.

It follows, that from the viewpoint of optimization, this problem is a trivial one.

It is noted that the considerations would be drastically different for cylindrical or spherical geometries, where appreciable thermal stresses may occur in the region of the center of the specimen. Under these conditions the problem would be a more difficult one, which could be tackled by the methods discussed in Chapter 7.

Regarding the preferred computational procedure, the moving boundary problem could be conveniently represented using a technique of Green's functions, which seems to allow considerable computational economy [12].

The actual procedure for optimization could involve a control vector iteration procedure, where the adjoint equations would have to have a special form due to the moving solidification boundary and constraint (Eq. 8.5.18). However, the control parameterization procedure discussed in the previous chapter seems to offer more promise because it would require no calculation of the adjoint variables and thus could be used directly with the Green's function method for state variable calculation. The constraint (Eq. 8.5.18) could be handled by means of a penalty added to the objective (Eq. 8.5.17).

We shall leave the actual optimization of this problem as an exercise for the reader.

EXAMPLE 8.5.3. We wish to optimize a process in which iron oxide pellets are reduced to iron, through contacting with a $CO + H_2$ mixture in a shaft furnace, according to the following reactions:

$$Fe_2O_3 + 3CO \rightleftarrows 2Fe + 3CO_2 \qquad -(\sim 100 \text{ Btu/lb iron}) \tag{8.5.19}$$

and

$$Fe_2O_3 + 3H_2 \rightleftarrows 2Fe + 3H_2O \qquad +(\sim 380 \text{ Btu/lb iron}) \tag{8.5.20}$$

A sketch showing the schematic flow diagram of the reaction system is shown in Fig. 8.19.

It is seen that the solid pellets are fed into the system at the top of the bed, at a molar flow rate of G_s.

Fresh gas enters the system at a rate of $G_{g,f}$ which is mixed with recycled gas $G_{g,r}$ prior to being introduced into the system, at the bottom of the bed.

In addition a side stream of gas $G_{g,s}$ is introduced into the bed at an intermediate level.

We note that the reaction of hematite with CO is exothermic, whereas the reduction process, if carried out with hydrogen, is endothermic. It is known, furthermore, that the reactions should be carried out within the temperature range 600 to 900°C, and that equilibrium conditions dictate that the exit gas must contain at least 55 to 60% $(CO + H_2)$ [13].

General Considerations

Before proceeding with the detailed formulation of the problem, let us consider, in a qualitative manner, the options that are available to us in optimizing the system, together with the possible trade-offs involved. Some of these, which may be deduced from inspection of Fig. 8.19 and from the physical nature of the problem, are listed below:

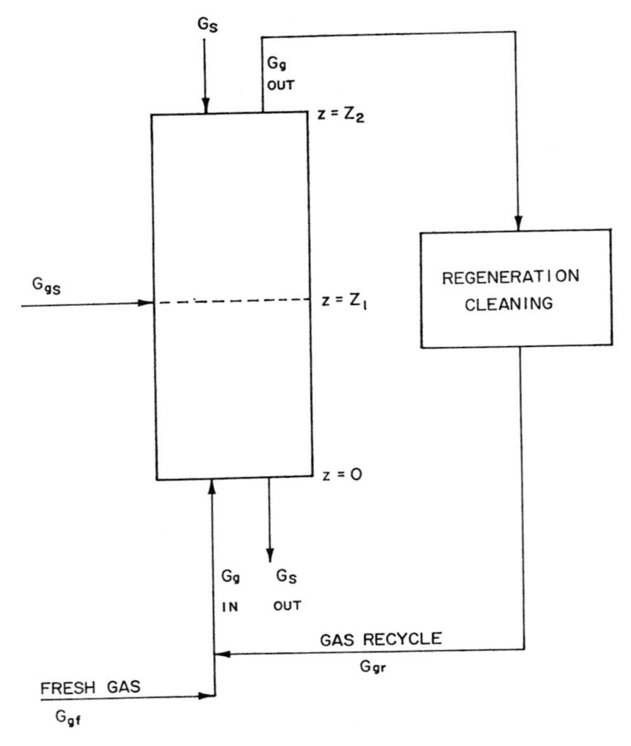

Fig. 8.19 Schematic representation of the flow sheet in Example 8.5.3.

Inlet Gas Composition

Since the reaction with CO is exothermic and the reaction with hydrogen is endothermic, there may be an optimal inlet gas composition, which would result in a uniform bed temperature throughout; whether it would be worthwhile to use such a $CO + H_2$ mixture will have to be determined in the light of the economic factors.

Exit Gas Composition

The obvious trade-offs available with regard to the exit gas composition are the following: if the exit gas composition is close to equilibrium a very deep bed is required, but the load on the regenerative system (removal of H_2O and CO_2) is correspondingly reduced. Conversely, if the exit gas is far from equilibrium, the increased effective driving force will allow a shorter bed to be used—at the expense of a greater load on the regeneration facility.

Use of the Side Stream

Based on past experience, the use of a side stream may be attractive under certain conditions; if the reaction is predominately endothermic, then it may be advantageous to introduce a relatively hot side stream at an intermediate point, so as to obviate the need for indirect heating of the system.

The optimal quantity and location of the side stream could again form the objective of an optimization study.

Additional Factors

Additional factors that could be examined include the inlet gas temperature, the degree of regeneration, and the temperature of the side stream.

Formulation

In order to state the modeling equations, let us assume that the kinetics of the reaction are known, and are expressible as:

$r_{H_2} = r(X_s, X_{H_2}, X_{CO}, T, h)$
$r_{CO} = r(X_s, X_{H_2}, X_{CO}, T, h)$
z_1 = axial position of side stream entrance
z_2 = top of the bed
z = axial distance measured from the gas inlet
G_g, G_s = molar velocities of the gaseous and solid streams, respectively
X_{H_2}, X_{CO} = mole fraction of H_2 and CO in the gas streams
X_s = mole fraction of solid reactant in the solid stream
T_g, T_s = gas and solid temperatures
C_{pg}, C_{ps} = specific heat of the gaseous and the solid streams
h_D, h = heat and mass transfer coefficients
a = surface area per unit volume
$\Delta H_{H_2}, \Delta H_{CO}$ = heat of reactions of H_2 and CO, respectively
$r(X_s, X_{H_2}, X_{CO}, T_s, h_D)$ = the rate of reaction, which depends on the gas and the solid composition, on the temperature (as a result of both equilibrium and kinetic considerations) and on the mass transfer coefficient
G_{gf}, G_{gs}, G_{gr} = molar velocities of the fresh gas streams, the side stream, and the recycle stream

Subscripts

gf	fresh gas stream
gs	side gas stream
gr	recycle gas stream
i	refer to inlet

$$\frac{d}{dz}(G_g X_{H_2}) = r_{H_2}(X_s, X_{H_2}, X_{CO}, T_s, h_D) \tag{8.5.21}$$

Mass balance on the hydrogen

$$\frac{d}{dz}(G_g X_{CO}) = r_{CO}(X_s, X_{H_2}, X_{CO}, T_s, h_D) \tag{8.5.22}$$

Mass balance on the CO

$$\frac{d}{dz}(G_s X_s) = r_{CO} + r_{H_2} \tag{8.5.23}$$

Mass balance on the solids

$$\frac{d}{dz}(T_g G_g C_{pg}) = ha(T_g - T_s) \tag{8.5.24}$$

Heat balance on the gas

$$\frac{d}{dz}(T_s G_s C_{ps}) = ha(T_s - T_g) + \Delta H_{H_2} r_{H_2} + \Delta H_{CO} r_{CO} \tag{8.5.25}$$

Heat balance on the solids

A set of equations of this type will apply to the domains $0 \leq z \leq z_1$ and $z_1 \leq z \leq z_2$.

Thus the boundary conditions may be written

at z $= 0$

$$T_g = \frac{(G_{gf} T_i) C_{pgf} + (G_{gr} T_g) C_{pgr}}{G_{gf} C_{pgf} + G_{gr} C_{pgr}} \tag{8.5.26}$$

$$X_{CO} = \frac{G_{gf} X_{gfCO} + G_{gr} X_{grCO}}{G_{gf} + G_{gr}} \tag{8.5.27}$$

with a similar expression for H_2

$$G_{ig} = G_{gf} + G_{gr} \tag{8.5.28}$$

at z $= z_1$

$$G_g = G_{gf} + G_{gr} + G_{gs} \tag{8.5.29}$$

$$T_g = \frac{G_g T_g C_{pg}\big|_{z=(z_1-dz)} + (G_{gs} T_{gs} C_{pgs})}{G_g C_{pg} + G_{gs} C_{pgs}} \tag{8.5.30}$$

$$X_{CO} = \frac{G_g X_{gCO}\big|_{z=(z_1-dz)} + G_{gs} X_{gsCO}}{G_g + G_{gs}} \tag{8.5.31}$$

where, again, a similar expression would apply for H_2
at $z = z_2$

$$X_s = X_{si} \tag{8.5.32}$$

$$T_s = T_{si} \tag{8.5.33}$$

Equations 8.5.21 to 8.5.33 together with the information on r_{H_2} and r_{CO} and the appropriate correlations for h and h_D, represent the mathematical model of the system.

The optimization problem could then be solved through the use of the techniques in Chapters 3 and 4. The state equations are a rather complex two-point boundary value problem due to the countercurrent flow, the side stream of gas, and the recycle of the exit gas stream. Thus an iterative technique will have to be used to satisfy all of these boundary conditions. However, this would only require one additional computational step in the optimization algorithm. Because the parameters to be chosen (e.g., inlet and side stream gas compositions, bed length, inlet and side-stream gas temperatures, etc.) are independent of the spatial variable, then a multivariable search algorithm not requiring derivatives (e.g., Powell's method) could be used to determine the minimum cost (or maximum profit) from the system. Thus the optimization can be performed with a standard program (cf. Appendix C) immediately after a simulation of the modeling equations is available.

EXAMPLE 8.5.4.

The Optimization of Flotation Circuits

Flotation is a process whereby the grains of minerals contained in a pulp or a slurry are caused to rise selectively to the surface in a tank or a cell, by the action of air bubbles that are introduced into the system.

The grains are caught in the froth formed at the top and then withdrawn, while the undesirable impurities (gangue) are removed through the bottom of the cell.

Typically, the operation of a flotation circuit involves a large number of cells, and may be treated by the techniques developed in Chapter 5 for serially structured systems.

Problems in flotation could be particularly suited to optimization. Because information is now becoming available on both the static and the dynamic modeling of these systems, optimization of these systems is quite timely.

In this context the work of Bull [14], Woodburn and King [15], and Smith [16] would be particularly notable.

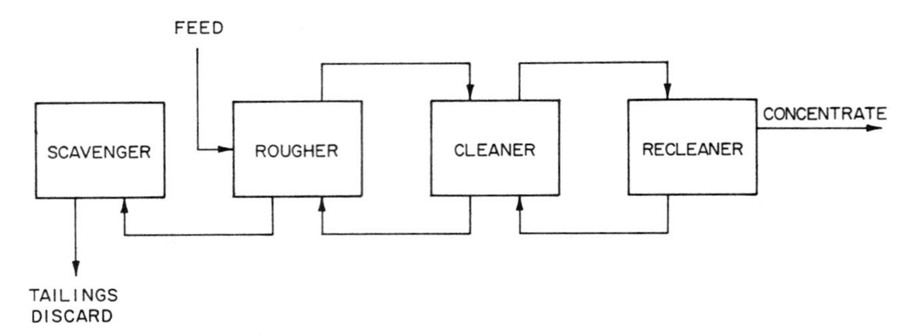

Fig. 8.20 Schematic representation of a flotation circuit.

Figure 8.20 shows a schematic representation of a typical flotation circuit. The feed is introduced into the *rougher cells*, the froth from which is passed to the *cleaner cells* and finally to the *recleaner cells*, which then yield the desired product, the concentrate, at a specified purity.

The underflow from each of these cells is returned to the preceding stage in order to effect the additional recovery of material. The underflow from the rougher is passed to the *scavenger cells*, the tailings from which are discarded.

On the assumption that the material is perfectly mixed in each of the cells (the blocks shown in Fig. 8.20 correspond to a battery of cells, rather than to individual units), and that the rate at which the concentrate is transferred from the pulp into the froth follows a first-order rate law, models have been developed for the description of these systems.

The rate constants or flotation constants for each of the units have to be determined empirically and will be affected by changes in the operating conditions. Thus we would need a correlation showing the changes in these velocity constants with operating conditions in order to carry out the optimization. This could be done in an adaptive mode so that new "predicted" optimal operations could be used to update the model constants and this new model reoptimized, and so on, until the optimal conditions are found.

To illustrate the nature of the modeling equations, we shall quote here the circuit models proposed by Smith and Bjerring [16] in connection with their dynamic modeling studies.

The following symbols are used:

V	volume of cell
α	ratio: solid mass/total mass in the froth
M	mass of solids in the cell pulp (slurry)
c	mass ratio: mineral/solids in the cell
F_n	tailings feed rate at stage n

F_{cn} concentrate feed rate at stage n
k_1, k_2 specific rate constant for the flotation of the mineral and the gangue, respectively
r_1, r_2 mass density of the mineral and of the gangue, respectively
R mass density of the solids in the feed pulp
n number of cells in the section

Tailings Flow

$$F_{n+1} = \frac{F_n M_n V_{n-1}}{V_n M_{n-1}} - \frac{1}{\alpha_n}\left[\left(\frac{k_1}{r_1} - \frac{k_2}{r_2}\right)C_n + \frac{k_2}{r_2}\right]\frac{M_n^{\,2}}{V_n} \tag{8.5.34}$$

Concentrate Flow

$$F_{c,n} = [(k_1 - k_2)C_n + k_2]M_n \tag{8.5.35}$$

Tailings Density

$$R_{n+1} = \frac{M_n}{V_n} \tag{8.5.36}$$

Concentration in Tailings

$$C_{i(n+1)} = C_i \tag{8.5.37}$$

Through the use of the modeling Eqs. 8.5.34 to 8.5.37, and when the flotation constants k_1 and k_2 are known (for each cell system), one can apply the techniques of Chapter 5 to determine the optimal flow and recirculation rates in the system.

8.6 CONCLUDING REMARKS

The material in Chapter 8 represents the application of the techniques described in the earlier chapters of the text to more realistic and therefore more complex problems. It was seen that in contrast to the simple examples used for the illustration of principles in the earlier portion of the text, the actual development of the mathematical model for the process required a significant, if not major part of the total effort.

In many respects the engineer concerned with the optimization of metallurgical operations faces far more difficult tasks than those involved in the optimization of chemical or purely physical processing units. Often the metallurgical plants involve large, complex batch or semibatch operations which have not yet been satisfactorily modeled; thus in the majority of cases any optimization procedure has to be preceded by the construction of a suitable mathematical model. In view of the possible complexities of these problems, a team effort, involving the collaboration of metallurgical and chemical (or electrical) engineers would be, perhaps, the most promising approach.

It would be unrealistic to suggest that, upon assimilating the major part of the material contained in this text, a metallurgical engineer would be ready to tackle unaided the very difficult tasks involved in optimizing integrated metallurgical operations. Nonetheless, the ideas developed here should provide a useful start, both in improving the communications with experts in the areas of operational reasearch and applied mathematics, and in representing a first step in developing one's expertise in process optimization. Of even greater importance is the possibility that this book will help to stimulate a new awareness of the potential in process optimization which could change quite drastically the general attitude to process improvement and process development in the metals processing field.

The tasks involved are difficult but the potential payoff is correspondingly great. It is our hope that after opening a crack in the door, this text will stimulate a great deal of further effort in this general area so that, in the years to come, process optimization will become a standard part of the metallurgical engineering curriculum.

Although this last chapter was dedicated to problems drawn from the metals processing field, the material should be of interest to chemical engineers as well. The actual methodology employed, emphasizing model building prior to actual optimization, could be useful in many chemical engineering systems, where process models are either unavailable, or are in a form which is not immediately compatible with optimization procedures. One cannot overemphasize the fact that awareness of the physical nature of the system to be optimized is the most important ingredient of any successful optimization project.

Finally we hope that the metallurgical examples given in this chapter may draw the attention of chemical engineers to the very interesting area of process metallurgy—to which their talents could be applied with a very good potential for success.

REFERENCES

1. J. W. Evans, J. Szekely, W. H. Ray, and Y. K. Chuang, *Chem. Eng. Sci.* **28,** (1973).
2. J. Szekely and J. W. Evans, *Metall. Trans.* **2,** 1691, 1699 (1971).
3. J. Szekely and J. W. Evans, *Chem. Eng. Sci.* **26,** 1901 (1971).
4. M. J. D. Powell, *Comput. J.* **7,** 155 (1964).
5. J. Downing, "Physical Chemistry of Ferroalloy Production," Vol. 26 *Proc. Electr. Steelmaking Conf.* **26,** (1968).

6. J. C. Agarwal, in *Blast Furnace Technology*, J. Szekely, Ed., Marcel Dekker, New York, 1972, p. 375.

7. W. H. Ray, J. Szekely, and M. Ajinkya, *Metall. Trans.* (in press).

8. J. F. Elliott and M. Gleiser, *Thermochemistry for Steelmaking*, Addison Wesley, 1960.

9. *Electric Furnace Steelmaking*, Clarence E. Sims, Ed., Interscience Publishers, New York, 1963, Chap. 19.

10. Ibid., Chap. 16.

11. R. J. Choulet, F. S. Death, and R. N. Dokken, *Can. Met. Q.* **10,** 9129 (1971).

12. Y. K. Chuang and J. Szekely, *Int. J. Heat Mass Transfer* **14,** 1285 (1971).

13. C. Bodsworth, *Physical Chemistry of Iron and Steel Manufacture*, Longmans, London, 1963, p. 303.

14. Rex Bull, Lecture Notes, Colorado School of Mines.

15. E. T. Woodburn, R. P. King, E. M. Buchalter, and S. E. Piper, *A Decade of Digital Computing*, A. Weiss, Ed., AIME, 1969.

16. H. W. Smith and A. K. Bjerring, *A Decade of Digital Computing*, A. Weiss, Ed., AIME, 1969.

Appendix A
Review of Elementary Matrix Algebra

While there exist a number of very good texts dealing with matrices, a partial list of which is given at the end of this section, it was thought worthwhile to present a brief, self-contained review of elementary matrix algebra.

This review covers the matrix manipulation steps that are used in the text, and should be consulted by the reader who feels unfamiliar with these techniques.

The array of numbers arranged in rows and columns, called a matrix, can be written with the notation, e.g.

$$\mathbf{A} = \begin{bmatrix} a_{11} & a_{12} & a_{13} \\ a_{21} & a_{22} & a_{23} \end{bmatrix} \tag{A.1}$$

where $\mathbf{A}$ is a 2×3 matrix (2 rows by 3 columns) in this instance. The elements $a_{11}, a_{12}, \ldots$, etc., have the notation that a_{ij} is the element of the ith row and jth column of $\mathbf{A}$.

As in the case of scalar quantities, matrices can be added, subtracted, multiplied, and so on, but special rules of algebra must be followed.

A.1 ADDITION AND SUBTRACTION

Only matrices having the same number of rows and columns can be added or subtracted. For example, given the matrices, **A**, **B**, **C**

$$\mathbf{A} = \begin{bmatrix} a_{11} & a_{12} & a_{13} \\ a_{21} & a_{22} & a_{23} \end{bmatrix}; \quad \mathbf{B} = \begin{bmatrix} b_{11} & b_{12} & b_{13} \\ b_{21} & b_{22} & b_{23} \\ b_{31} & b_{32} & b_{33} \end{bmatrix}; \quad \mathbf{C} = \begin{bmatrix} c_{11} & c_{12} & c_{13} \\ c_{21} & c_{22} & c_{23} \end{bmatrix} \tag{A.2}$$

only **A** and **C** can be added or subtracted; **B** cannot be added or subtracted from either **A** or **C** because it has a different number of rows. The sum of $\mathbf{A} + \mathbf{C} = \mathbf{D}$ can be written

$$\mathbf{D} = \mathbf{A} + \mathbf{C} = \begin{bmatrix} a_{11} & a_{12} & a_{13} \\ a_{21} & a_{22} & a_{23} \end{bmatrix} + \begin{bmatrix} c_{11} & c_{12} & c_{13} \\ c_{21} & c_{22} & c_{23} \end{bmatrix}$$

$$= \begin{bmatrix} a_{11} + c_{11} & a_{12} + c_{12} & a_{13} + c_{13} \\ a_{21} + c_{21} & a_{22} + c_{22} & a_{23} + c_{23} \end{bmatrix} \tag{A.3}$$

When a matrix is multiplied by a scalar, this is done by multiplying each element of the matrix by the scalar; for example,

$$k\mathbf{A} = k \begin{bmatrix} a_{11} & a_{12} & a_{13} \\ a_{21} & a_{22} & a_{23} \end{bmatrix} = \begin{bmatrix} ka_{11} & ka_{12} & ka_{13} \\ ka_{21} & ka_{22} & ka_{23} \end{bmatrix} \tag{A.4}$$

Thus we can multiply matrices by scalars before adding and subtracting them.

Some simple numerical examples of matrix addition and subtraction are given in the following:

$$\begin{bmatrix} 1 & 4 \\ 6 & 9 \end{bmatrix} + \begin{bmatrix} 2 & 3 \\ 9 & 5 \end{bmatrix} = \begin{bmatrix} 3 & 7 \\ 15 & 14 \end{bmatrix}$$

$$4\begin{bmatrix} 1 & 4 \\ 6 & 9 \end{bmatrix} - 3\begin{bmatrix} 2 & 3 \\ 9 & 5 \end{bmatrix} = \begin{bmatrix} -2 & 7 \\ -3 & 21 \end{bmatrix}$$

A.2 MULTIPLICATION

Matrix multiplication can be performed on two matrices **A** and **B** to form **AB** only if the number of columns of **A** is equal to the number of rows in **B**.

If this condition is satisfied then the matrices **A** and **B** are said to be *conformable* in the order **AB**.

Let us consider that the matrix **A** is an $m \times n$ matrix (m rows and n columns); then to form the product $\mathbf{P} = \mathbf{AB}$, **B** must be an $n \times r$ matrix (n rows and r columns). The resulting product, **P**, is then an $m \times r$ matrix (m rows and r columns).

If **A** and **B** are as defined by Eq. A.2, then **A** is a 3×2 matrix, **B** a 3×3 matrix and the product

$$\mathbf{P} = \begin{bmatrix} a_{11} & a_{12} & a_{13} \\ a_{21} & a_{22} & a_{23} \end{bmatrix} \begin{bmatrix} b_{11} & b_{12} & b_{13} \\ b_{21} & b_{22} & b_{23} \\ b_{31} & b_{32} & b_{33} \end{bmatrix} = \begin{bmatrix} p_{11} & p_{12} & p_{13} \\ p_{21} & p_{22} & p_{23} \end{bmatrix} \tag{A.5}$$

is a 2×3 matrix. The rule for determining the elements of P from the elements of A and B is given as:

$$p_{ij} = \sum_{k=1}^{n} a_{ik} b_{kj} \qquad i = 1, 2, \ldots, m, \quad j = 1, 2, \ldots, r \tag{A.6}$$

Thus the matrix **P** defined by Eq. A.5 can be written as:

$$\mathbf{P} = \begin{bmatrix} (a_{11}b_{11} + a_{12}b_{21} + a_{13}b_{31})(a_{11}b_{12} + a_{12}b_{22} + a_{13}b_{32})(a_{11}b_{13} + a_{12}b_{23} + a_{13}b_{33}) \\ (a_{21}b_{11} + a_{22}b_{21} + a_{23}b_{31})(a_{21}b_{12} + a_{22}b_{22} + a_{23}b_{32})(a_{21}b_{13} + a_{22}b_{23} + {}_{23}b_{33}) \end{bmatrix} \tag{A.7}$$

EXAMPLES

$$\begin{bmatrix} 3 & 4 \\ -2 & -1 \end{bmatrix} \begin{bmatrix} 1 & 2 \\ 2 & 5 \end{bmatrix}$$

$$= \begin{bmatrix} (3 \times 1 + 4 \times 2)(3 \times 2 + 4 \times 5) \\ (-2 \times 1 - 1 \times 2)(-2 \times 2 - 1 \times 5) \end{bmatrix} = \begin{bmatrix} 11 & 26 \\ -4 & -9 \end{bmatrix}$$

$$[x \quad y \quad z] \begin{bmatrix} a \\ b \\ c \end{bmatrix} = ax + by + cz \qquad \text{(scalar)}$$

$$\begin{bmatrix} x \\ y \\ z \end{bmatrix} [a \quad b \quad c] = \begin{bmatrix} ax & bx & cx \\ ay & by & cy \\ az & bz & cz \end{bmatrix}$$

It should be noted that only in the case of square matrices (number of rows = number of columns) can we form *both* the product **AB** and **BA**.

However, in general,

$$\mathbf{AB} \neq \mathbf{BA} \tag{A.8}$$

Those matrices for which $\mathbf{AB} = \mathbf{BA}$ are said to *commute* or to be *permutable*.

A.3 TRANSPOSE OF A MATRIX

The transpose of the matrix $\mathbf{A}$, denoted $\mathbf{A}^T$, is the matrix formed by interchanging the rows with the columns. For example, if

$$\mathbf{A} = \begin{bmatrix} a_{11} & a_{12} & a_{13} \\ a_{21} & a_{22} & a_{23} \end{bmatrix}$$

Then

$$\mathbf{A}^T = \begin{bmatrix} a_{11} & a_{21} \\ a_{12} & a_{22} \\ a_{13} & a_{23} \end{bmatrix}$$

Thus if $\mathbf{A}$ is an $m \times n$ matrix, then $\mathbf{A}^T$ is an $n \times m$ matrix. We note that

$$\mathbf{A}^T\mathbf{A} = \mathbf{A}\mathbf{A}^T \tag{A.9}$$

$$(\mathbf{AB})^T = \mathbf{B}^T\mathbf{A}^T \tag{A.10}$$

Also if $\mathbf{A}^T\mathbf{A} = \mathbf{I}$, where $\mathbf{I}$ is the identity matrix

$$\mathbf{I} = \begin{bmatrix} 1 & 0 & 0 & \cdot \\ 0 & 1 & 0 & \cdot \\ 0 & \cdot & 1 & \cdot \\ 0 & \cdot & \cdot & 1 \end{bmatrix} \tag{A.11}$$

then $\mathbf{A}$ is called an *orthogonal* matrix.

A.4 DETERMINANTS AND COFACTORS

The absolute value of a square matrix, denoted $|\mathbf{A}|$, is called the determinant of $\mathbf{A}$. For a simple 2×2 matrix,

$$\mathbf{A} = \begin{bmatrix} a_{11} & a_{12} \\ a_{21} & a_{22} \end{bmatrix} \tag{A.12}$$

we recall that

$$|\mathbf{A}| = a_{11}a_{22} - a_{21}a_{12} \tag{A.13}$$

For higher order matrices, the determinant is defined in terms of *cofactors*. The *cofactor* of the element a_{ij} is defined as the determinant of the matrix left when the ith row and jth column are removed, times the factor $(-1)^{i+j}$. For example, let us consider the 3×3 matrix

$$\mathbf{A} = \begin{bmatrix} a_{11} & a_{12} & a_{13} \\ a_{21} & a_{22} & a_{23} \\ a_{31} & a_{32} & a_{33} \end{bmatrix} \tag{A.14}$$

The *cofactor*, C_{12} of a_{12} is

$$C_{12} = (-1)^3 \begin{vmatrix} a_{21} & a_{23} \\ a_{31} & a_{33} \end{vmatrix} = -(a_{21}a_{33} - a_{31}a_{23}) \tag{A.15}$$

Similarly, the cofactor C_{23} is

$$C_{23} = (-1)^5 \begin{vmatrix} a_{11} & a_{12} \\ a_{31} & a_{32} \end{vmatrix} = -(a_{11}a_{32} - a_{31}a_{12}) \tag{A.16}$$

The determinant $|\mathbf{A}|$ of any $n \times n$ matrix can then be evaluated as the sum of the products of the elements of any row or column and their respective cofactors. That is

$$|\mathbf{A}| = \sum_{i=1}^{n} a_{ij}C_{ij} \qquad \text{(any column } j\text{)} \tag{A.17}$$

or

$$|\mathbf{A}| = \sum_{j=1}^{n} a_{ij}C_{ij} \qquad \text{(any column } i\text{)} \tag{A.18}$$

To illustrate, we see that the determinant of the 3×3 matrix given by Eq. A.14 can be found by expanding in the first row

$$\begin{vmatrix} a_{11} & a_{12} & a_{13} \\ a_{21} & a_{22} & a_{23} \\ a_{31} & a_{32} & a_{33} \end{vmatrix} = a_{11} \begin{vmatrix} a_{22} & a_{23} \\ a_{32} & a_{33} \end{vmatrix} - a_{12} \begin{vmatrix} a_{21} & a_{23} \\ a_{31} & a_{33} \end{vmatrix} + a_{13} \begin{vmatrix} a_{21} & a_{22} \\ a_{31} & a_{32} \end{vmatrix}$$

$$\begin{aligned} &= a_{11}a_{22}a_{33} - a_{11}a_{32}a_{23} + a_{12}a_{31}a_{23} - a_{12}a_{21}a_{33} \\ &\quad + a_{13}a_{21}a_{32} - a_{13}a_{31}a_{22} \end{aligned} \tag{A.19}$$

It is easily shown that the expansion by any other row or column yields the same result.

A.5 INVERSE OF A MATRIX

The matrix counterpart of division is the inverse operation. The inverse of the matrix **A**, denoted $\mathbf{A}^{-1}$ can be formed only if

(i) **A** is square
(ii) the determinant of **A**, (i.e., $|\mathbf{A}|$) is nonzero.†

The inverse of **A** is defined by

$$\mathbf{A}^{-1} = \frac{\mathbf{C}^T}{|\mathbf{A}|} \tag{A.20}$$

where **C** is the matrix of cofactors

$$\mathbf{C} = \begin{bmatrix} C_{11} & C_{12} & & & C_{1n} \\ C_{21} & C_{22} & & & \\ & & \cdot & & \\ & & & \cdot & \\ & & & & \cdot \\ C_{n1} & & & & C_{nn} \end{bmatrix} \tag{A.21}$$

defined in the previous section.

It should be noted that $\mathbf{A}^{-1}\mathbf{A} = \mathbf{A}\mathbf{A}^{-1} = \mathbf{I}$ so that the inverse of a matrix has similar properties to the inverse of a scalar quantity.

A.6 LINEAR ALGEBRAIC EQUATIONS

Much of the motivation behind developing matrix operations is due to their particular usefulness in handling sets of linear algebraic equations. Let us consider the three linear equations

$$\begin{aligned} a_{11}x_1 + a_{12}x_2 + a_{13}x_3 &= b_1 \\ a_{21}x_1 + a_{22}x_2 + a_{23}x_3 &= b_2 \\ a_{31}x_1 + a_{32}x_2 + a_{33}x_3 &= b_3 \end{aligned} \tag{A.22}$$

containing the unknowns x_1, x_2, x_3. This set of equations can be represented by the single matrix equation

$$\mathbf{A}\mathbf{x} = \mathbf{b} \tag{A.23}$$

† If $|\mathbf{A}| = 0$ for any matrix **A**, the matrix is said to be *singular*.

where

$$\mathbf{A} = \begin{bmatrix} a_{11} & a_{12} & a_{13} \\ a_{21} & a_{22} & a_{23} \\ a_{31} & a_{32} & a_{33} \end{bmatrix}; \qquad \mathbf{x} = \begin{bmatrix} x_1 \\ x_2 \\ x_3 \end{bmatrix}; \qquad \mathbf{b} = \begin{bmatrix} b_1 \\ b_2 \\ b_3 \end{bmatrix} \tag{A.24}$$

The solution of these equations can be written very simply by premultiplying Eq. A.23 by $\mathbf{A}^{-1}$ to yield

$$\mathbf{A}^{-1}\mathbf{A}\mathbf{x} = \mathbf{I}\mathbf{x} = \mathbf{x} = \mathbf{A}^{-1}\mathbf{b} \tag{A.25}$$

There are a large number of standard, computer oriented numerical procedures for performing the operation $\mathbf{A}^{-1}\mathbf{b}$ to produce the solution given in Eq. A.25. Indeed, the availability of these computer subroutines makes the use of matrix techniques very attractive from a practical viewpoint.

A.7 EIGENVALUES

Many properties of the square $n \times n$ matrix $\mathbf{A}$ can be related to the eigenvalues of the matrix. The n eigenvalues of $\mathbf{A}$ are defined as the solutions to the characteristic equation

$$|\mathbf{A} - \lambda\mathbf{I}| = 0 \tag{A.26}$$

Equation A.26 may be expressed as a polynomial of the form:

$$\lambda^n + \alpha_1\lambda^{n-1} + \alpha_2\lambda^{n-2} + \cdots + \alpha_{n-1}\lambda + \alpha_0 = 0 \tag{A.27}$$

which has n roots λ.

EXAMPLE. Compute the eigenvalues of

$$\mathbf{A} = \begin{bmatrix} 1 & 2 \\ 3 & -4 \end{bmatrix}$$

$$|\mathbf{A} - \lambda\mathbf{I}| = \left| \begin{bmatrix} 1 & 2 \\ 3 & -4 \end{bmatrix} - \begin{bmatrix} \lambda & 0 \\ 0 & \lambda \end{bmatrix} \right| = \left| \begin{bmatrix} (1-\lambda) & 2 \\ 3 & (-4-\lambda) \end{bmatrix} \right| = 0$$

or

$$-(1-\lambda)(4+\lambda) - 6 = \lambda^2 + 3\lambda - 10 = 0$$

the solution of which is

$$\lambda = -5, 2$$

A.8 QUADRATIC FORMS

A quadratic form, c, is a scalar quantity made up of the matrix **A** in the following way

$$c = \mathbf{x}^T \mathbf{A} \mathbf{x} \tag{A.28}$$

For example if

$$\mathbf{x} = \begin{bmatrix} x_1 \\ x_2 \end{bmatrix}; \qquad \mathbf{A} = \begin{bmatrix} a_{11} & a_{12} \\ a_{21} & a_{22} \end{bmatrix} \tag{A.29}$$

Then Eq. A.28 becomes

$$c = [x_1 x_2] \begin{bmatrix} a_{11} & a_{12} \\ a_{21} & a_{22} \end{bmatrix} \begin{bmatrix} x_1 \\ x_2 \end{bmatrix} \tag{A.30}$$

Quadratic forms are interesting because the sign of the scalar c can be determined often by examining the eigenvalues of **A**. For example, the quadratic form c will always be positive if the matrix **A** has only positive eigenvalues (thus **A** is termed *positive definite*). Similarly, c will always be negative if **A** has only negative eigenvalues (A negative definite). Finally c will be of mixed sign if the eigenvalues of A have mixed signs.

If **A** has both *zero* and positive eigenvalues, then it is termed *positive semidefinite* and c can be nonnegative only. Similarly, if **A** has both *zero* and negative eigenvalues, then it is *negative semidefinite* and c will be nonpositive.

The striking feature of these properties is that the sign of c can be determined regardless of values assumed by **x**. This feature is very important in determining necessary and sufficient conditions for optimality.

A.9 SUMMARY

In this Appendix we have attempted to introduce all of the matrix manipulations used in the text. For a more thorough study of matrix algebra the reader is referred to the bibliography at the end of this Appendix. We shall conclude this section with a brief glossary of the terms used.

A.10 GLOSSARY

Determinant	the absolute value of a matrix
Square matrix	a matrix whose number of columns and rows is the same

Symmetric matrix	a matrix whose off-diagonal elements are mirror-images, for example, for the matrix **A**, $a_{ij} = a_{ji}$.
Singular matrix	a matrix the determinant of which is zero.
Positive (semi) definite matrix	a matrix whose eigenvalues are all positive (or zero).
Negative (semi) definite matrix	a matrix whose eigenvalues are all negative (or zero).
Identity matrix, **I**	the matrix whose elements are all zero except for 1's along the main diagonal.

BIBLIOGRAPHY†

Kreysig, E., *Advanced Engineering Mathematics*, 2nd Ed., Wiley, 1967, Chap. 7.

Amundson, N. R., *Mathematical Methods in Chemical Engineering*, Prentice-Hall, 1966.

Frazer, R. Q., W. J. Duncan, and A. R. Collar, *Elementary Matrices*, Cambridge University Press, 1963.

Bellman, R., *Introduction to Matrix Analysis*, McGraw-Hill, 1960.

Gantmacher, F. R., *The Theory of Matrices*, Vol. I and II, Chelsea Publishing Company, 1960.

† Listed in increasing order of depth.

Appendix B
Supplementary Reading

Chapters 1 and 2

Edelbaum, T. N., "Theory of Maxima and Minima," in *Optimization Techniques*, George Leitmann, Ed., Academic Press, 1962.

King, R. P., "Necessary and Sufficient Conditions for Inequality Constrained Extreme Values," *I and EC Fund.* **5**, 484 (1966).

Saaty, T. L., and J. Bram, *Non-Linear Mathematics*, McGraw-Hill, 1964, p. 93.

Chapters 3 to 5

Abadie, J., *Non-linear Programming*, Wiley, 1967.

Abadie, J., *Integer and Non-linear Programming*, American Elsevier, 1970.

Aris, R., *Discrete Dynamic Programming*, Blaisdell, 1964.

Bellman, R. E., and S. E. Dreyfus, *Applied Dynamic Programming*, Princeton University Press, 1962.

Beveridge, G. S. G., and R. S. Schechter, *Optimization: Theory and Practice*, McGraw-Hill, 1970.

Dantzig, G. B., *Linear Programming and Extensions*, Princeton University Press, 1963.

Fan, L. T., and C. S. Wang, *Discrete Maximum Principle*, Wiley, 1964.

Fiacco, A. V., and G. P. McCormick, *Non-linear Programming*, Wiley, 1968.
Garvin, W. W., *Linear Programming*, McGraw-Hill, 1960.
Hadley, G., *Linear Programming*, Addison-Wesley, 1962.
Hadley, G., *Non-linear and Dynamic Programming*, Addison-Wesley, 1964.
Himmelblau, D. M., *Applied Non-Linear Programming*, McGraw-Hill, 1972.
Kunzi, H. P., W. Krelle, and W. Oettli, *Non-linear Programming*, Blaisdell, 1966.
Lasdon, L. S., *Optimization Theory for Large Systems*, Macmillan, 1970.
Wismer, D. A., *Optimization Methods for Large-Scale Systems*, McGraw-Hill, 1971.

Chapters 6 and 7

Bryson, A. E., and Y.-C. Ho, *Applied Optimal Control*, Blaisdell, 1969.
Denn, M. M., *Optimization by Variational Methods*, McGraw-Hill, 1969.
Fan, L. T., *The Continuous Maximum Principle*, Wiley, 1966.
Jacobson, D. H., and D. Q. Mayne, *Differential Dynamic Programming*, American Elsevier, 1970.
Kopp, R. E., and H. G. Moyer, *Trajectory Optimization Techniques*, *Adv. Control Syst.* **4,** 104 (1966).
Sage, A. P. *Optimum Systems Control*, Prentice-Hall, 1969.

Appendix C A Partial List of General Subroutines for Optimization

Chapter 3 (Unconstrained Optimization)

	CODE
1. Single Variable Search:	
(a) Program GOLD 1 (golden section search)	*
(b) Program COMB 1 (discrete variable search)	*
2. Multivariable Search (gradient available):	
(a) Program FMFP (Fletcher-Powell)	*, ++
(b) Program FMCG (Davidon)	*, ++
(c) Program MEMGRAD—(conjugate gradient with memory)	*, +
3. Multivariable Search (only function available):	
(a) Program BOTM (Powell's conjugate direction)	*, ±
(b) Program SEEK 1 (Hooke and Jeeves search)	*, +
(c) Program SIMPLEX (Simplex procedure)	*, +
(d) Program PATERN (Pattern search)	*, ±

Chapter 4 (Constrained Optimization Techniques)

1. Penalty Function Methods:	
(a) Program SUMT (Fiacco and McCormick)	*, ±

(b) Program SEEK 3 (Hooke and Jeeves with SUMT penalties) *, +
(c) Program SIMPLEX (SIMPLEX with SUMT penalties) *, +
(d) Program CLIMB (Rosenbrock orthogonal search) ±

2. Linear Programming:
   (a) Program SIMPLE (Simplex method) *, +
   (b) Program MPS (modified simplex) ±
   (c) Program NYBLPC (modified simplex) *
   (d) Program MINIT (Primal-dual) *
3. Quadratic Programming:
   (a) Program BEALE (Beale algorithm) *
   (b) Program WOLFE (Wolfe algorithm) *
4. Separable Programming:
   (a) Program SCM3 (expanded linear programming) ±
5. General Nonlinear Programming:
   (a) Program OPTM (Rosen's gradient projection) *
   (b) Program POP II (Iterative linear programming) ±
   (c) Program APPROX (Griffith and Stewart iterative L.P.) *, +, ±
   (d) Program RICCHT (Ricochet gradient method) *, ±
6. Geometric Programming
   (a) Program GEOM (geometric programming algorithm) +
7. Integer Programming:
   (a) Program LIP 1 (discrete linear programming) ±

Chapter 6 (Trajectory Optimization)

1. Modified Gradient Method
   (a) Program IBM 0037 (Lapidus and Wu) ±

Code: * = Available at SUNYAB Computing Center.
± = Available from IBM Share Library.
+ = System OPTIPAC, OPTISEP available from McMaster University Computing Center.
++ = Available in IBM Scientific Subroutine Package.

Appendix D
A Partial Bibliography of Optimization Case Studies on Metallurgical Processes

---

Alvarado, F. L., and R. Mukundan, "An Optimization Problem in Distributed Parameter Systems," *Int. J. Control* **9,** 665 (1969).

Dzhikaev, A. K., and S. A. Malyi, "Heating with Minimal Oxidation," *Auto. Remote. Control,* 1044 (1970).

Frohberg, V. M. G., D. Papamantellos, and E. Hanert, "Determination of the Optimal Conditions of a Metallurgical Process Elucidated by the Example of Chromium Removal from Iron Melts Containing Carbon," *Arch. Eisenhüttenwesen* **38,** 91 (1967).

Gosiewski, A., and A. Wierzbicki, "Dynamic Optimization of a Steel-Making Process in an Electric Arc Furnace," *Automatica* **6,** 767 (1970).

Kwakernaak, H., P. Tyssen, and R. C. W. Strijbos, "Optimal Operation of Blast Furnace Stoves," *Automatica* **6,** 33 (1970).

Kwasnoski, D., and R. W. Bouman, "Application of Linear Programming to Distribution of Iron Making Raw Materials," *Ironmaking Proc.* **1967** 26.

Lerner, A. J., and S. A. Maly, "The Optimum Control of the Thermal and Chemical Interaction During the Massive Billets Heating," *IFAC Symposium on the Control of Distributed Parameter Systems*, Banff, Canada, June 1971.

Mular, A. L., "The Selection of Optimization Methods for Mineral Processes," *Proceedings of Symposium on Automatic Control Systems in Mineral Processing Plants*, Brisbane, Australia, May 1971.

Nenonen, L. K., and B. Pagurek, "Conjugate Gradient Optimization Applied to a Copper Converter Model," *Automatica* **5,** 801 (1969).

Petrov, B. N., G. M. Ulanov, and R. U. Yakooleva, "Some Aspects of Controlling Thermal Plants with Distributed Parameters," *IFAC*, Banff, 1971.

Pike, H. E., and S. J. Citron, "Optimization Studies of a Slab Reheating Furnace," *Automatica* **6,** 41 (1970).

Tsao, C. S., and R. H. Day, "A Process Analysis Model of the U.S. Steel Industry," *Manage. Sci.* **17,** B588 (1971).

Vyrk, A. Kh., "Optimal Heating of Massive Bodies in Continuous Furnaces," *Autom. Remote Control* 1132 (1971).

Woodburn, E. T., R. P. King, E. M. Buchalter, and S. E. Piper, "Integrated Grinding and Floatation Control—Computational Problems in an Optimum Decision Making Model," in *A Decade of Digital Computing in the Mineral Industry*, A. Weiss, Ed., A.I.M.E., 1969.

Woodside, C. M., B. Pagurek, J. Pauksens, and A. N. Ogale, "Singular Arcs Occurring in Optimal Electric Steel Refining," *IEEE Trans. Auto. Control* **AC-15,** 549 (1970).

Work, A. H., "The Optimal Control of Slab Reheating Furnaces," *IFAC*, Banff, 1971.

Yavin, Y., and Y. Rasis, "The Bounded Energy Optimal Control for a Class of Heat Conduction Systems," *Int. J. Control* **11,** 153 (1970).

Appendix E
The Numerical Solution of Ordinary and Partial Differential Equations

In the problems discussed in Chapters 6, 7, and 8 the systems were modeled by ordinary or partial differential equations and the solution of the optimization problem necessarily involved the solution of these differential equations. The majority of the modeling equations were nonlinear, and for this reason numerical methods had to be employed for their integration. In this section we shall present a very brief review of the techniques available for the numerical solution of differential equations. Extensive literature references will be given throughout and the emphasis will be on the principles of the techniques. It is not expected that the reader who is totally inexperienced in the numerical solution of differential equations will be able to master this subject on studying the few pages of this section. It is hoped, however, that we can provide an introduction to this important subject—with suggestions for further reading.

E.1 ORDINARY DIFFERENTIAL EQUATIONS

Initial Value Problems

Let us consider the first order differential equation,

$$\frac{dy}{dt} = f(y, t) \tag{E.1}$$

with

$$y = y_0, \qquad \text{at} \quad t = t_0 \tag{E.2}$$

where the form of the function $f(y, t)$ is such that analytical integration is impossible. Quite a number of techniques may be employed for the numerical integration of Eq. E.1.

"Marching" Techniques

These techniques make use of the fact that if we know the function at the initial point, that is, $y = y_0$ at $t = t_0$, we also know the derivative at $t = t_0$. Then over a short interval, $\Delta t = t_1 - t_0$, we may calculate the corresponding increment in y, that is, $y_1 - y_0$, from the values of the function in this interval and from the derivative at $t = t_0$. Once $y = y_1$ is known at $t = t_1$, we may advance another time step, $\Delta t = t_2 - t_1$, and the solution obtained by "marching" forward in time.

(a) *Euler Method*

The simplest of these techniques, though not usually the most efficient, is the *Euler Method*, which can be written as

$$y_1 = y_0 + f(y_0, t_0)\, \Delta t \tag{E.3}$$

for the first step or, more generally,

$$y_{n+1} = y_n + f(y_n, t_n)\, \Delta t \tag{E.4}$$

The basic premise of the *Euler Method* is that the function $f(y, t)$ may be approximated by a constant $f(y, t_i)$ over the interval $(t_{i+1} - t_i)$ so that the solution $y(t)$ is approached by a piecewise linear approximation (Eq. E.4). The application of the technique is illustrated in Fig. E.1, where it is seen that the approximation involved could lead to serious errors, unless the interval chosen is very small.

The Euler method was discussed for the purpose of illustration; more sophisticated methods of numerical integration are available. For a detailed

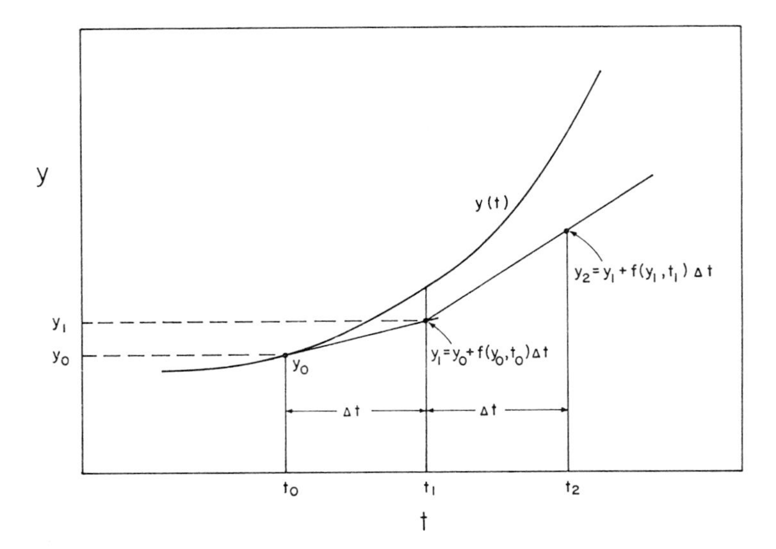

Fig. E.1 Illustration of Euler's method.

description of these the reader is referred to the texts listed at the end of this section. Here, we shall confine ourselves to a brief description of only one of the most commonly used procedures, the method of Runge-Kutta.

(b) *Runge-Kutta Method*

Let us expand the solution to Eq. E.1 in a Taylor series, about the point t, with an interval Δt:

$$y(t + \Delta t) = y(t) + \Delta t y'(t) + \frac{(\Delta t)^2}{2!} y''(t) + \frac{(\Delta t)^3}{3!} y'''(t) + \cdots \tag{E.5}$$

which may be written as

$$\Delta y \equiv y(t + \Delta t) - y(t) = \Delta t y'(t) + \frac{(\Delta t)^2}{2!} y''(t) + \frac{(\Delta t)^3}{3!} y'''(t) + \cdots \tag{E.6}$$

where Δy is the desired increment in y, corresponding to the increment Δt, that we wish to calculate.

Let us recall from the chain rule of differentiation, that since

$$y' = \frac{dy}{dt} = f \tag{E.7}$$

$$y'' = f_t + f_y f \tag{E.8}$$

$$y''' = f_{tt} + 2f_{ty}f + f_{tt}f^2 + f_y(f_t + f_y f) \tag{E.9}$$

where

$$f_t \equiv \frac{\partial f}{\partial t}, \quad f_y \equiv \frac{\partial f}{\partial y} \quad \text{etc.}$$

On combining Eqs. E.6–E.9, we have the following:

$$\Delta y = (\Delta t)f + \frac{(\Delta t)^2}{2}(f_t + f_y f) + \frac{(\Delta t)^3}{6}[f_{tt} + 2f_{yt}f + f_{yy}f^2 + f_y(f_t + f_y f)] + \cdots \quad \text{(E.10)}$$

Clearly one could evaluate f, f_t, f_y, etc., and produce a high-order approximation for Δy from Eq. E.10. The error in the approximation depends on the number of terms retained. However, it is useful to put the scheme in a more convenient form for computation—one that requires only the evaluation of f. One such formula, which has errors on the order of $(\Delta t)^3$, termed Heun's method, takes the following form:

$$\begin{aligned} \Delta y &= \tfrac{1}{4}(k_1 + 3k_3) \\ k_1 &= \Delta t[f(y, t)] \\ k_2 &= \Delta t[f(t + \tfrac{1}{3}\Delta t, y + \tfrac{1}{3}k_1)] \\ k_3 &= \Delta t[f(t + \tfrac{2}{3}\Delta t, y + \tfrac{2}{3}k_2)] \end{aligned} \quad \text{(E.11)}$$

Thus the calculation proceeds as:

$$\begin{aligned} t_1 &= (t_0 + \Delta t) \\ y_1 &= (y_0 + \Delta y) \end{aligned} \quad \text{(E.12)}$$

or

$$\begin{aligned} t_{i+1} &= (t_i + \Delta t) \\ y_{i+1} &= (y_i + \Delta y) \end{aligned} \quad \text{(E.13)}$$

A large number of such Runge-Kutta formulae are available, each having its own stability and convergence criteria; for the details of these, the reader is referred to the texts by Ketter and Prawel, or Lapidus and Seinfeld. The Runge-Kutta technique is of considerable appeal, especially as the availability of subroutines renders the actual computational labor quite minimal. Under certain conditions, problems may arise in the stability of the solution. Although stability problems can often be overcome by selecting a sufficiently small interval Δt, or by using lower order methods, sometimes alternative methods have to be explored.

(c) *Extension to Higher Order Equations*

The marching methods can be readily extended to higher order differential equations by a change of variable, which reduces the problem to a set of simultaneous first-order differential equations. As an illustration, let us consider the problem,

$$\frac{d^2y}{dt^2} + 5y^4\left(\frac{dy}{dt}\right)^2 + y^{3/2} = 0 \tag{E.14}$$

with

$$y = y_0, \qquad t = 0 \tag{E.15}$$

$$\frac{dy}{dt} = C_1, \qquad t = 0 \tag{E.16}$$

Let us define

$$\frac{dy}{dt} = Z, \qquad \text{thus} \quad \frac{d^2y}{dt^2} = \frac{dZ}{dt}$$

then Eqs. (E.14)–(E.16) may be written as:

$$\frac{dZ}{dt} + 5y^4Z^2 + y^{3/2} = 0 \tag{E.17}$$

$$\frac{dy}{dt} = Z \tag{E.18}$$

with

$$y = y_0, \qquad t = 0 \tag{E.19}$$

and

$$Z = C_1, \qquad t = 0 \tag{E.20}$$

Equations E.17 and E.18 may then be solved together for these boundary conditions, using (for example) a multivariable version of the Runge-Kutta method (cf. Lapidus and Seinfeld). We note that this technique can be used for *initial value problems* only, that is, for systems where all the boundary conditions are specified at one particular value of the independent variable, $t = 0$ in the present case.

Problems involving "split boundary conditions" termed, *boundary value problems*, represent a more complex task, and will be discussed subsequently.

Alternatives to "Marching" Methods

Here we shall confine ourselves to a brief mention of the various alternatives that are available for the numerical solution of ordinary differential equations. This section should be regarded as a glossary of notation and a guide to the literature.

(a) *Methods of Weighted Residuals*

In using this technique, we define the domain of interest and then seek the solution in a particular form, such as a power series, or some other series with a finite number of terms. The coefficients of the various terms initially are unknown, but must be so selected as to satisfy the differential equation in some approximate way. One very effective weighted residual method is the collocation technique which requires the differential equation to be satisfied at certain fixed "collocation" points in t.

As an example, let us consider the differential equation:

$$\frac{dy}{dt} = 2 - 5yt^2 \tag{E.21}$$

with

$$y = 2, \quad t = 0 \tag{E.22}$$

and let the domain of interest be $0 \leq t \leq 1$. Then we may wish to seek the solution in the following form:

$$y = a_0 + a_1 t + a_2 t^2 \tag{E.23}$$

Inspection of Eq. E.22 shows immediately that

$$a_0 = 2$$

and the remaining coefficients, a_1 and a_2, may be evaluated by stipulating that the differential equation must be exactly satisfied at two given values of t. On substituting from Eq. E.23 into Eq. E.21, we have

$$a_1 + 2a_2 t - 2 + 5t^2(2 + a_1 t + 2a_2 t) = R \tag{E.24}$$

where R is the *residual*.

Thus a_1 and a_2 may be evaluated by setting R zero at two given values of t, say,

$$R = 0 \quad \text{at} \quad t = 0.4$$

$$R = 0 \quad \text{at} \quad t = 0.6$$

which gives the equations for the two unknowns a_1 and a_2.

A very good discussion of collocation and related methods may be found in the monographs by Crandall and Finlayson.

(b) *Predictor-Corrector Methods*

The predictor-corrector methods involve an iterative procedure:

(a) An estimate (i.e., prediction) is made of the increment in the dependent variable and then the error associated with this prediction is calculated.

(b) The sign and magnitude of the error is then used to make an improved prediction.

Predictor-corrector methods may be the best technique when there are stability problems with marching methods.

One cannot overemphasize the need *for independent checks* of any numerical solutions to differential equations. These independent checks may involve solution by two unrelated methods, or solutions with several different "time steps" to insure that the numerical solution has converged and is stable.

Boundary Value Problems

In boundary value problems, the boundary conditions specify values or relationships at two or more points of the independent variable. As noted earlier, these *boundary value problems* tend to be rather more difficult to solve than the initial value problems discussed in the preceding sections.

In order to illustrate the nature of boundary value problems, let us consider the following differential equation:

$$\frac{d^2y}{dt^2} + f_1(y)\frac{dy}{dt} + f_2(y, t) = 0 \tag{E.25}$$

with

$$y = C_1 \quad \text{at} \quad t = 0, \tag{E.26}$$

and

$$y = C_2 \quad \text{at} \quad t = 1 \tag{E.27}$$

Problems of this type arise in (steady state) heat conduction, diffusion, and in chemical reactor design.

Equations E.25 to E.27 cannot be solved by forward integration, because in order to do this, we would need to know both y and dy/dt at $t = 0$ (viz. the discussion of the Runge-Kutta method, as applied to higher order equations). Several techniques are available for tackling problems of this type.

Shooting Techniques

Let us return to the preceding example to illustrate the nature of the "shooting techniques."

Let us assume a value for the unknown gradient at $t = 0$, and then perform the forward integration of the differential equation, up to $t = 1$. Then we compare the value of y obtained by this method at $t = 1$ with that specified in the boundary condition, viz. ($y = C_2$ at $t = 1$) and repeat the procedure, with a new value of the gradient, until satisfactory agreement is reached. This procedure is illustrated in Fig. E.2. While the shooting method is straightforward in concept, it may involve excessive computation, especially if the solution is quite sensitive to the initial value of dy/dt chosen, which is often the case.

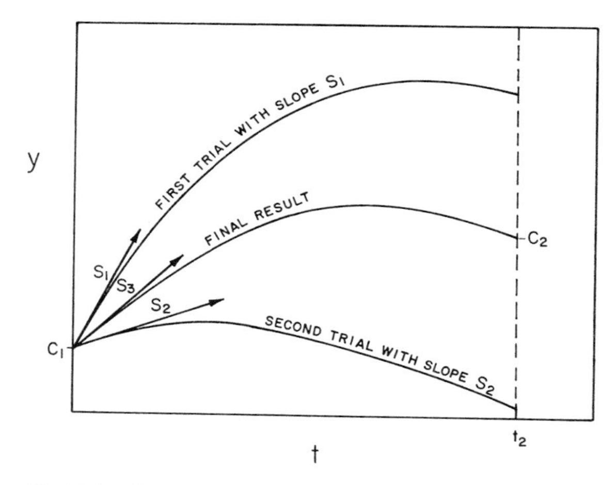

Fig. E.2 Illustration of the "shooting method."

Weighted Residual Methods

These methods are particularly suitable for solving boundary value problems, because one can often choose trial functions (e.g., Eq. E.23) so as to automatically satisfy the boundary conditions.

Other Methods

There are a number of other methods which can be used for boundary value problems. The quasilinearization technique, discussed in Chapter 6, has been successful in solving these problems. In addition, relaxation methods to be discussed below have also been found efficient for the solution of boundary value problems.

References

In the following we shall list a few key references for the solution of ordinary differential equations by numerical methods, together with some comments about their suitability.

General Texts

Ketter, R. L., and S. P. Prawel, *Modern Methods of Engineering Computation*, McGraw-Hill, 1969. A very good introductory book with heavy emphasis on the integration of numerical analysis with actual computational schemes. Especially recommended to those with no prior experience in computation.

Lapidus, L., and J. H. Seinfeld, *Numerical Solution of Ordinary Differential Equations*, Academic Press, 1971. A good general treatment; recommended for more detailed study or in case of more complex problems. A particularly good treatment of "stiff" differential equations.

Hildebrand, F. B., *Introduction to Numerical Analysis*, McGraw-Hill, 1956. A fairly rigorous derivation of algorithms with error bounds, etc.

Scarborough, J. B., *Numerical Mathematical Analysis*, Johns Hopkins Press, 1950. A "cookbook" full of many recipes for solving numerical problems—the easiest book for the novice.

Finlayson, B. A., *The Method of Weighted Residuals and Variational Principles*, Academic Press, 1972. A good discussion of weighted residual, and in particular orthogonal collocation methods.

Crandall, S. H., *Engineering Analysis*, McGraw-Hill, 1956. A good general treatment of numerical methods including weighted residual methods.

Treatment of Boundary Value Problems

Wachpress, E. in Ralston and Wilf, Ed., *Mathematical Methods for Digital Computers*, Wiley, New York, 1960, Chap. 10.

Keller, H. B., *Numerical Methods for Two-Point Boundary Value Problems*, Blaisdell, 1968.

Fox, L., *The Numerical Solution of Two Point Boundary Value Problems*, Oxford University Press, 1957.

E.2 THE NUMERICAL SOLUTION OF PARTIAL DIFFERENTIAL EQUATIONS

The numerical solution of partial differential equations is in general a great deal more difficult task than the numerical integration of ordinary differential equations.

Second order partial differential equations in two independent variables may be written as

$$\alpha \frac{\partial^2 f}{\partial x^2} + \beta \frac{\partial^2 f}{\partial x \, \partial y} + \gamma \frac{\partial^2 f}{\partial y^2} = \psi(x, y, f_x, f_y) \tag{E.28}$$

This equation can be classified hyperbolic, parabolic, or elliptic depending on the sign of the discriminant Λ,

$$\Lambda = \beta^2 - 4\alpha\gamma \tag{E.29}$$

The equation is hyperbolic, when $\Lambda > 0$
The equation is parabolic, when $\Lambda = 0$, and
The equation is elliptic, when $\Lambda < 0$

As typical examples, the one dimensional heat conduction or diffusion equations are *parabolic*, viz:

$$\begin{aligned} \alpha \frac{\partial^2 T}{\partial x^2} &= \frac{\partial T}{\partial t} \\ D \frac{\partial^2 C}{\partial x^2} &= \frac{\partial C}{\partial t} \end{aligned} \tag{E.30}$$

The equations describing the propagation of a temperature or concentraiton front through a packed bed, in the absence of dispersion would be *first order hyperbolic* equations, namely,

$$-u\frac{\partial T}{\partial y}=\frac{\partial T}{\partial t}$$
$$-u\frac{\partial c}{\partial y}=\frac{\partial c}{\partial t} \tag{E.31}$$

Whereas the description of wave motion (e.g., sound waves, hydrodynamic phenomena, etc.) would be a *second order hyperbolic* equation

$$\frac{\partial^2 u}{\partial t^2}=c^2\frac{\partial^2 u}{\partial y^2} \tag{E.32}$$

where u is the wave height, and c is the wave velocity.

Finally, steady-state diffusion or heat conduction in more than one dimension would lead to *elliptic* equations, viz.

$$\frac{\partial^2 T}{\partial x^2}+\frac{\partial^2 T}{\partial y^2}=0$$
$$\frac{\partial^2 c}{\partial x^2}+\frac{\partial^2 c}{\partial y^2}=0 \tag{E.33}$$

Let us now survey, briefly, the methods available for solving these equations.

Parabolic Equations

Let us consider the linear differential equation

$$\frac{\partial^2 f}{\partial x^2}=\frac{\partial f}{\partial t}; \qquad 0\le x\le 1 \tag{E.34}$$

with the boundary conditions

$$f=0, \qquad t=0 \tag{E.35}$$

$$f=0, \qquad x=1, \qquad t>0 \tag{E.36}$$

$$f=1; \qquad x=0; \qquad t>0 \tag{E.37}$$

While this equation can be readily solved analytically, let us illustrate some of the numerical methods available for its solution.

Finite Difference Methods

Let us put Eq. E.34 in a finite difference form by establishing a time-space grid, sketched in Fig. E.3, and by using the calculus of finite differences. By using a forward difference equation, we have that

$$\left(\frac{\partial^2 f}{\partial x^2}\right)_{m,n} = \frac{f_{m+1,n} - 2f_{m,n} + f_{m-1,n}}{(\Delta x)^2} \tag{E.38}$$

and

$$\left(\frac{\partial f}{\partial t}\right)_{m,n} = \frac{f_{m,n+1} - f_{m,n}}{\Delta t} \tag{E.39}$$

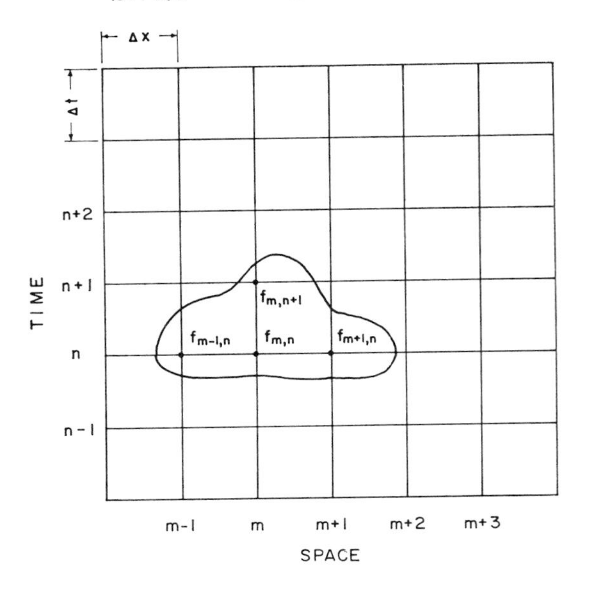

Fig. E.3 A time-space grid.

where Δt and Δx are the time and spatial increments, respectively.

On substituting for $\partial^2 f/\partial x^2$ and $\partial f/\partial t$ from Eqs. E.38 and E.39, and after some algebra, we obtain the following expression:

$$f_{m,n+1} = f_{m,n} + \frac{\Delta t}{(\Delta x)^2}(f_{m+1,n} - 2f_{m,n} + f_{m-1,n}) \tag{E.40}$$

Equation E.40 provides an *explicit relationship* between $f_{m,n+1}$ [the value of f at spatial grid point m and at time step $(n + 1)$] and the values of f at spatial grid points $(m - 1)$, m, and $(m + 1)$ for the preceding time step, n. In general, the right-hand side of Eq. E.40 is known; thus by the successive application of Eq. E.40 we may evaluate f_{n+1} for all values of m and then proceed in a similar manner with the evaluation of f_{n+2}, etc. and march forward in time to produce the solution.

Treatment of the Boundary Conditions

The procedure outlined above holds for the interior points within the grid. In the present example the treatment of the boundary conditions posed no problem, as these specified fixed values of f at the bounding surfaces, so that the calculations were necessarily restricted to the interior points.

In order to illustrate some of the methods of dealing with boundary conditions, let us consider the differential equation (Eq. E.34), but replace the boundary condition given in Eq. E.36 by

$$\frac{\partial f}{\partial x} = 0; \qquad x = 1 \tag{E.41}$$

while retaining the boundary condition contained in Eq. E.37. Let us denote the (spatial) grid point corresponding to the boundary ($x = 1$) by e; then the forward difference form of Eq. E.41 is given as

$$\left.\frac{\partial f}{\partial x}\right|_{x=1} = \frac{f_{e+1,n} - f_{e-1,n}}{2\,\Delta x} = 0 \tag{E.42}$$

However, as seen in Fig. E.4, $(e + 1)$ is a fictitious point, since it is not contained in the actual spatial domain under consideration. The function

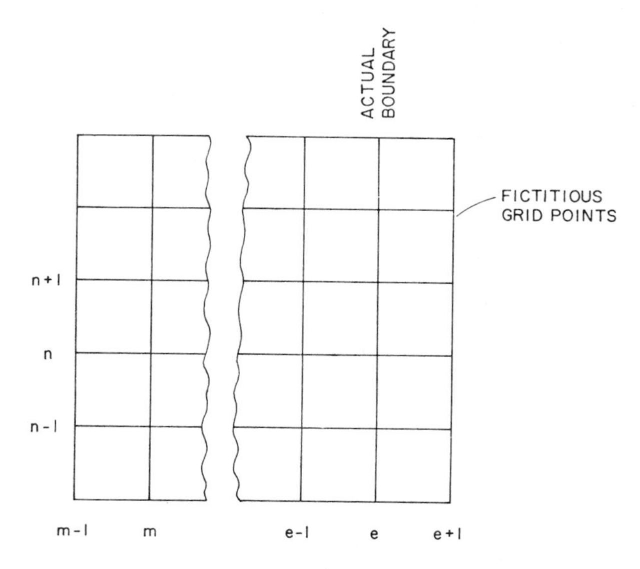

Fig. E.4 The use of a fictitious grid point for dealing with boundary conditions.

at this fictitious point may be eliminated through the use of Eq. E.40, so that we obtain

$$f_{e,n+1} = f_{e,n} + \frac{\Delta t}{\Delta x^2}(2f_{e-1,n} - 2f_{e,n}) \tag{E.43}$$

which is the desired explicit relationship between $f_{e,n+1}$ and the known $f_{e,n}$ and $f_{e-1,n}$. There are other techniques which can be used for treating boundary conditions (cf. the Bibliography).

The explicit technique described above is straightforward to apply in concept and may be adequate for a number of problems, even when non-constant coefficients introduce nonlinearities, viz. temperature dependent specific heat or conductivity in Fourier's equation.

The principle disadvantage of the explicit technique is that problems are often encountered with *stability*. It has been shown for the example problem that it is necessary for stability to select the ratio $\Delta t/\Delta x^2$ so that

$$\frac{\Delta t}{\Delta x^2} < \frac{1}{2}$$

However, in more complex problems this stability condition may either be unknown or more stringent so that the "time steps" would become very small, with a corresponding increase in the computer time requirement.

Implicit Methods

One alternative to the explicit technique is provided by implicit methods in which an *implicit relationship* is provided between the values of adjacent grid points at the subsequent and the prior steps. In these implicit methods the solution is advanced to the subsequent time step through the solution of a set of simultaneous equations.

Of the numerous implicit techniques available, we shall confine our attention to the *Crank-Nicolson method.* Let us express Eq. E.33 in a finite difference form, about the point x_m and $t_{n+1/2}$, i.e., halfway between the time steps n and $(n + 1)$. Thus we have

$$\left(\frac{\partial f}{\partial t}\right)_{m,n+1/2} = \frac{f_{m,n+1} - f_{m,n}}{\Delta t} \tag{E.44}$$

and

$$\left(\frac{\partial^2 f}{\partial x^2}\right)_{m,n+1/2} = \frac{1}{2}\left[\frac{f_{m+1,n} - 2f_{m,n} + f_{m-1,n}}{(\Delta x)^2} + \frac{f_{m+1,n+1} - 2f_{m,n+1} + f_{m-1,n+1}}{(\Delta x)^2}\right] \tag{E.45}$$

where the right hand side of Eq. E.45 is just the arithmetic average of the second difference, evaluated at n and $(n + 1)$ time steps, respectively.

Upon substituting Eqs. E.44 and E.45 into our original differential equation, we obtain the following:

$$f_{m-1,n+1} - 2\left[1 + \frac{(\Delta x)^2}{\Delta t}\right] f_{m,n+1} + f_{m+1,n+1}$$
$$= -f_{m-1,n} + 2\left[1 - \frac{(\Delta x)^2}{\Delta t}\right] f_{m,n} - f_{m+1,n} \quad \text{(E.46)}$$

Equation E.46 provides a relationship between the values of f at $(m-1)$, m and $(m+1)$ at time steps n and $(n+1)$, respectively.

When written over the whole domain, we have a set of linear equations to solve, a task which is conveniently undertaken with the aid of matrix methods. We note that the Crank-Nicolson method is unconditionally stable for all values for the $\Delta t/(\Delta x)^2$ ratio, and thus is usually more economical from the viewpoint of computer time requirements than explicit procedures.

Additional, more sophisticated techniques are described in the references, listed at the end of this section.

Parabolic Equations with More than One Spatial Variable

Many practical problems are represented by parabolic equations with more than one spatial variable, such as

$$\frac{\partial^2 f}{\partial x^2} + \frac{\partial^2 f}{\partial y^2} = \frac{\partial f}{\partial t} \quad \text{(E.47)}$$

which would correspond to transient heat conduction or diffusion in a two-dimensional domain.

Several specialized techniques are available for the integration of equations of this type.

(a) We may use *implicit relaxation techniques* [4] to solve the left hand side of Eq. E.47 for each time step, and then advance the solution to the subsequent time step.

(b) Perhaps the most popular method of tackling the problem is through the use of the *Alternating Direction Implicit Method*. This technique, which is described in detail in the references given, involves the alternate use of two implicit finite difference formulae and provides good stability and computational economy.

Hyperbolic Equations

Let us consider the first order hyperbolic equation of the following form:

$$\frac{\partial f}{\partial t} + a\frac{\partial f}{\partial x} = R \quad \text{(E.48)}$$

which could describe the propagation of a temperature or concentration front through a packed bed for very rapid fluid to particle transfer rates. The finite difference methods discussed previously could be used to solve these problems; however, the *method of characteristics* would, perhaps, be more efficient. This technique makes use of the fact that along *characteristic curves* in the x, t plane defined (for this example) by

$$\frac{dx}{dt} = a \tag{E.49}$$

(sketched in Fig. E.5) the solution is given by the solution to the ordinary differential equation

$$\frac{df}{dt} = R \tag{E.50}$$

Thus by the repetitive solution of Eqs. E.49 and E.50 for different initial conditions, one can generate the numerical solution to Eq. E.48.

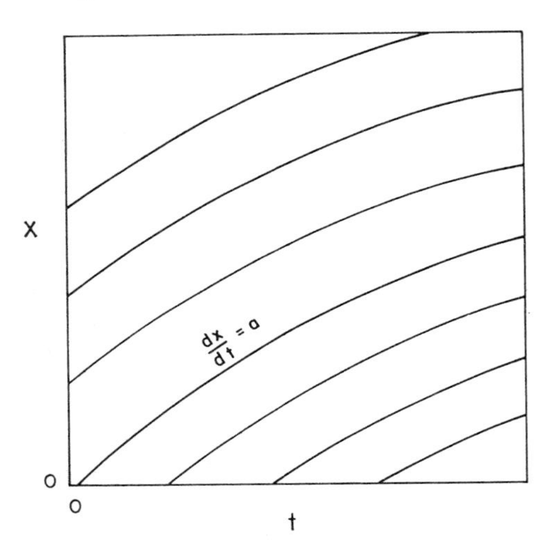

Fig. E.5 Characteristic curves defined by $dx/dt = a$.

Elliptic Equations

Elliptic equations of the type:

$$\frac{\partial^2 f}{\partial x^2} + \frac{\partial^2 f}{\partial y^2} = 0 \tag{E.51}$$

arise in steady-state heat conduction or steady-state diffusion problems; in addition elliptic equations are of importance in various stress calculations and in fluid mechanics. Elliptic equations can be treated by finite difference techniques in a similar way to the parabolic equations.

Concluding Remarks

In concluding this brief review it should be noted that the relatively simple techniques described here may have to be modified when applied to strongly nonlinear problems, such as radiative heat transfer, nonlinear kinetics in reactor design, etc. These nonlinear problems are usually best tackled by incorporating iterative procedures into these methods. Some of these techniques are described in the references.

Finally, a brief mention should also be made of the alternative methods that are available for the integration of partial differential equations.

Collocation and other Weighted Residual Methods

These techniques, mentioned previously, have been found very efficient in solving parabolic systems. They often give much greater accuracy than finite difference schemes with the same number of grid points.

Partial Discretization

This technique converts all but one of the independent variables to a finite difference form (e.g., a two-variable partial differential equation would become a set of coupled ordinary differential equations) so that ordinary differential equation numerical techniques can be used.

The reader is advised to consult the references for further details about these methods.

References

Rosenberg, D. U., *Methods for the Numerical Solution of Partial Differential Equations*, Elsevier, New York, 1969. A very good introductory text with the emphasis on practical applications and digital computation. Several programs are also included.

Clark, M., and K. F. Hanson, *Numerical Methods of Reactor Analysis*, Academic Press, 1964. A good general treatment with an extensive coverage of both theory and practice.

Ames, W. F., *Non-linear Partial Differential Equations in Engineering*, Academic Press, 1964. A general treatment with heavy emphasis on theory.

Mickley, H. S., T. K. Sherwood, and C. E. Reed, *Applied Mathematics in Chemical Engineering*, McGraw-Hill, 1957.

Lapidus, L., *Digital Computation for Chemical Engineers*, McGraw-Hill, 1960.

Author Index

Subject Index

Maximum principle, 220
Melting shop, 6
Metal oxide, reduction of, 277
Method of conjugate directions, 77, 84
Method of Fletcher and Powell, 68, 84
Method of Nelder and Mead, 82, 84
Method of weighted residuals, 268, 350, 353, 361
Minimum, 6
absolute, 26
constrained, 31, 36
global, 24, 26
local, 24, 26
necessary conditions, 29, 33–34, 36–38
relative, 26
sufficient conditions, 29, 33–34, 36–38
unconstrained, 27